T0188936

Computational Fluid Dynamics

Takeo Kajishima · Kunihiko Taira

Computational Fluid Dynamics

Incompressible Turbulent Flows

Springer

Takeo Kajishima
Department of Mechanical Engineering
Osaka University
Osaka
Japan

Kunihiko Taira
Florida State University
Tallahassee, FL
USA

ISBN 978-3-319-83263-0 ISBN 978-3-319-45304-0 (eBook)
DOI 10.1007/978-3-319-45304-0

Printed on acid-free paper

This Springer imprint is published by Springer Nature
The registered company is Springer International Publishing AG
The registered company address is: Gewerbestrasse 11, 6330 Cham, Switzerland

To our families:
Yoko, Ryohei, and Junpei
Yuki, Kai, and Rio

Preface

The majority of flows encountered in nature and engineering applications are turbulent. Although turbulence has been studied as part of classical mechanics for over a century, it is still one of the unsolved problems in physics and remains to be an active area of research. The field of study that uses numerical simulation to examine fluid dynamics is called Computational Fluids Dynamics (CFD). There is great demand in utilizing CFD for analyzing problems ranging from turbulent flows around aircraft and ground vehicles to larger scale problems related to weather forecasting and environmental assessment. Such demands are very likely to grow in the coming years as the engineering community at large pursues improvements in energy efficiency and performance for various fluid-based systems.

Presently, there are several commercial CFD solvers released with turbulence analysis capability. With software creating beautiful visualizations of turbulent flows, it may appear that any type of flows can be numerically predicted. While there may be some truth to such capability, it is still difficult to solve most turbulent flow problems without relying on companion experiments. That raises a question of why we still are not able to perfectly predict the behavior of turbulent flows. The dynamics of turbulent flows obeys the Navier–Stokes equations, upon which CFD solvers are based. However, turbulence exhibits flow structures over a wide range of spatial and temporal scales that all interacts amongst them in a complex nonlinear manner. That means that the spatial grid must be fine enough to resolve the smallest scales in turbulent flows while ensuring that the computational domain is large enough to encompass the largest flow structures. Such grid requirement becomes increasingly costly as we tackle flows at a higher Reynolds number. Despite the significant improvement in the computational capability with recent high-performance computers, we still do not expect computers to be able to handle these large grids for very high Reynolds number flows.

For this particular reason, turbulence is not likely to be completely solved in the near future. Flow physics taking place at scales below the resolvable scales must be represented with appropriate models, referred to as turbulence models. Presently, there is not a universally accepted turbulence model or numerical algorithm that can yield a solution unaffected by discretizations of the flow field. Thus, CFD should continue to be an active field of research with efforts focused towards predicting the essential features of turbulent flows with turbulence models. As such, engineers and scientists using CFD must understand how the governing equations are numerically solved. We must also be equipped with the ability to correctly interpret the numerical solution. With these points in mind, we should construct a necessary and sufficient computer program appropriate to simulate the fluid flow of interest. For commercial software, sufficient details on the solver technique should be provided in the reference manual so that users can determine whether the solver can be appropriately used for the problem at hand.

This book describes the fundamental numerical methods and approaches used to perform numerical simulations of turbulent flows. The materials presented herein are aimed to provide the basis to accurately analyze unsteady turbulent flows. This textbook is intended for upper level undergraduate and graduate students who are interested in learning CFD. This book can also serve as a reference when developing incompressible flow solvers for those already active in CFD research. It is assumed that readers have some knowledge of fluid mechanics and partial differential equations. This textbook does not assume the readers to have advanced knowledge of numerical analysis.

This textbook aims to enable readers to construct his or her own CFD code from scratch. The present textbook covers the numerical methods required for CFD and places emphasis on the incompressible flow solver with detailed discussions on discretization techniques, boundary conditions, and turbulent flow physics. The introduction to CFD and the governing equations are offered in Chap. 1, followed by the coverage of basic numerical methods in Chap. 2. Incompressible flow solvers are derived and discussed in detail in Chaps. 3 and 4. We also provide discussions on the immersed boundary methods in Chap. 5. A brief overview on turbulent flows is given in Chap. 6 with details needed for analyzing turbulent flows using Reynolds-Averaged Navier–Stokes equations (RANS) and Large-Eddy Simulation (LES) provided in Chaps. 7 and 8, respectively. At the end of the book, an appendix is attached to offer details on the generalized coordinate system, Fourier analysis, and modal decomposition methods.

A large portion of the present book is based on the material taught over the years by the first author for the course entitled "Computational Fluid Dynamics and Turbulent Flows" at Osaka University and his textbook entitled, "Numerical Simulations of Turbulent Flows" (1st and 2nd editions in Japanese) that has been available in Japan since 1999. Chapters 1–4 and 6–8 as well as Appendices A and B in the present book are founded heavily on the Japanese version by Kajishima. The present textbook enjoys additions of stability analysis (Sect. 2.5), immersed boundary methods (Chap. 5), and modal decomposition methods (Sect. 6.3.7 and Appendix C) by Taira based on the courses taught at the Florida State University. Furthermore, exercises have been added after each chapter to provide supplemental materials for the readers.

The preparation of this book has benefited greatly from comments, feedback, and encouragements from Takashi Ohta, Shintaro Takeuchi, Takeshi Omori, Yohei Morinishi, Shinnosuke Obi, Hiromochi Kobayashi, Tim Colonius, Clarence

Rowley, Steven Brunton, Shervin Bagheri, Toshiyuki Arima, and Yousuff Hussaini. The stimulating discussions with them on various topics of CFD over the years have been invaluable in putting together the materials herein. We must also thank our research group members and students for providing us with detailed comments on the drafts of this book that helped improve the organization and correctness of the text. Finally, we would like to acknowledge Michael Luby and Brian Halm at Springer for working with us patiently and Nobuyuki Miura and Kaoru Shimada at Yokendo for their support on the earlier Japanese versions.

Osaka, Japan   Takeo Kajishima
Tallahassee, Florida, USA   Kunihiko Taira
July 2016

Contents

Chapter 1
Numerical Simulation of Fluid Flows

1.1 Introduction

Numerical simulations, along with experiments and theoretical analysis, are often used as a tool to support research and development in science and engineering. The use of simulations has been popularized by the development and wide-spread availability of computers. Since numerical computations are advantageous to experiments from the aspects of speed, safety, and cost in many cases, their uses have been widely accepted in the industry. Simulations have also become a valuable tool in fundamental research due to its ability to analyze complex phenomena that may be difficult to study with experimental measurements or theoretical analysis. Reflecting upon these trends, the adjective *computational* is now widely used to describe sub-fields that utilize simulation in various disciplines, such as computational physics and computational chemistry.

The field of study concerned with analyzing various types of fluid flows with numerical simulations and developing suitable simulation algorithms is known as *computational fluid dynamics* (CFD). Applications of CFD can be found in the analysis of the following studies but not limited to

- Flows around aircraft, ships, trains, and automobiles;
- Flows in turbo-machineries;
- Biomedical and biological flows;
- Environmental flows, civil engineering, and architecture;
- Large-scale flows in astrodynamics, weather forecasting, and oceanography.

The flows in these settings usually do not have analytical expressions because of the complex physics arising from boundary geometry, external forcing, and fluid properties. In CFD, flow physics is analyzed and predicted by numerically solving the governing equations and reproducing the flow field with the use of computers.

In general, a fluid can be viewed as a continuum, for which there are established conservation laws for mass, momentum, and energy. For many fluids that are used in engineering applications, there are well-accepted constitutive relations. The equation

© Springer International Publishing AG 2017
T. Kajishima and K. Taira, *Computational Fluid Dynamics*,
DOI 10.1007/978-3-319-45304-0_1

of state for gases or equations for phase transformation or chemical reactions are also included in the system of equations as needed. The objective of flow simulations is to numerically solve such system of equations with appropriate initial and boundary conditions to replicate the actual flow. In the simulation, the flow field is represented by variables such as velocity, pressure, density, and temperature at discrete set of points. The evolution of these variables over time is tracked to represent the flow physics.

There is always the question of whether a simulation correctly reproduces the flow physics, because we are representing the continuum on discrete points. We can only obtain reliable numerical solutions when the simulation methodology is validated against experimental measurements or theoretical solutions. The use of such validated method must be limited to parameters that are within the applicable range. While there has been numerous achievements with computational fluid dynamics in the design fields and fundamental research, the development of accurate simulation methods that are widely applicable and more robust will continue in the future.

1.2   Overview of Fluid Flow Simulations

Numerical simulation of fluid flow follows the steps laid out in Fig. 1.1.

1. Decide to solve the full Navier–Stokes equations, the inviscid approximation, or any other approximation to reproduce the flow physics of interest. Choose a turbulence model and a non-Newtonian constitutive equation if necessary. Based on these choices, the governing partial differential equations to be solved in the simulation are obtained.
2. Discretize the governing equations with the finite-difference, volume or element method and choose the appropriate grid (spatial discretization). The corresponding algebraic equations to be solved are derived. We can then decide on the numerical algorithm to solve these equations and develop a computer program. In some cases, the program can be written to specifically take advantage of the characteristics of the available computer hardware.
3. Numerical simulation of fluid flow can output a large number of numerical values as the solution. Comprehending such solution is difficult with just pure numbers. Hence, graphs and visualizations with computer graphics and animations are used to aid the analysis of the simulation results.

Thus, to perform numerical simulations of fluid flows, it is not sufficient to only have the understanding of fluid mechanics. One must have knowledge of numerical analysis for discretization schemes and numerical algorithms, as well as computer science for programing and visualization. It goes without saying that the combination of these fields is necessary for CFD. To ensure that the results obtained from numerical simulation of fluid flow are reliable, we must also be concerned with verification and validation. We will discuss this in further details in Sect. 1.6.

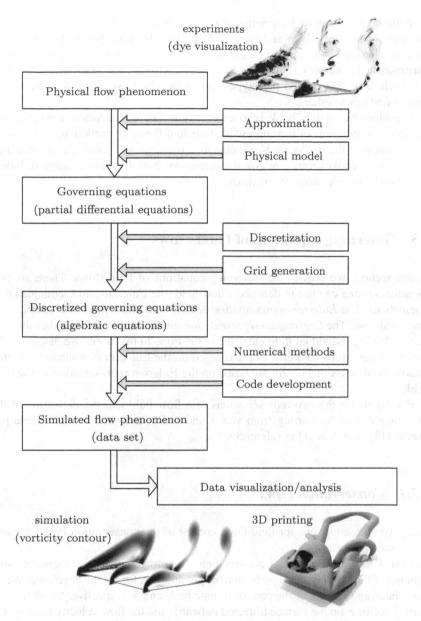

Fig. 1.1 General process of simulating fluid flows. Inserted visualizations are for unsteady flow over a pitching plate (Reprinted with permission from [8]; copyright 2014, AIP Publishing LLC)

With the invention of 3D printers, it is possible to print three-dimensional flow structures that can be held and examined up close for a deeper understanding of the flow physics. An example of a 3D print of the vortex behind a pitching plate is illustrated in the bottom of Fig. 1.1 based on the data from direct numerical simulation [8]. Such novel 3D printed prototypes can expand how we study flow fields and convey findings to others.

The objective of this book is to explain how one can develop a program for numerical simulations of incompressible turbulent flows. In particular, we focus on the treatment of the motion of Newtonian fluid (e.g., air and water) that is described by the Navier–Stokes equations. We also discuss methodologies to simulate turbulent flow with high-order accurate methods.

1.3   Governing Equations of Fluid Flows

In this section, we present the governing equations of fluid flows. There are two formulations one can use to describe a flow field: the Eulerian and Lagrangian representations. The *Eulerian representation* describes the flow field with functions of space and time. The *Lagrangian representation* on the other hand describes the flow field following individual fluid elements in the flow. In this book, we discretize the governing equations on a grid which is based on the Eulerian formulation. For that reason, the discussions herein are based on the Eulerian representation of the flow field.

For details on the two representations of a flow field and the derivation of the governing equations starting from vector and tensor analysis, we list Currie [4], Panton [15], and Aris [1] as references.

1.3.1   Conservation Laws

The governing equations for fluid flows consist of the conservation laws for mass, momentum, and energy.

First, let us consider mass conservation. Denoting the density (mass per unit volume) of the fluid by ρ, we perform a budget analysis for a control volume. We let the volume and surface of the control volume be V and S, respectively, with the unit normal vector n on the surface (directed outward) and the flow velocity u, as shown in Fig. 1.2. The time rate of change of the mass within the control volume consists of the mass flux going in and out of the volume $(\rho u) \cdot n$ through the surface assuming there is no mass source or sink:

$$\iiint_V \frac{\partial \rho}{\partial t} dV = - \iint_S (\rho u) \cdot n dS. \tag{1.1}$$

Fig. 1.2 Control volume for
conservation laws

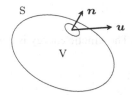

Here, mass influx is negative ($\rho u \cdot n < 0$) and efflux is positive ($\rho u \cdot n > 0$).

Using Gauss' theorem, the above equation can be written only with a volume integral

$$\iiint_V \left[\frac{\partial \rho}{\partial t} + \nabla \cdot (\rho u) \right] dV = 0. \qquad (1.2)$$

Equations (1.1) and (1.2) are integral representations of mass conservation. Since Eq. (1.2) must hold for any arbitrary control volume, the integrand should be zero. Thus, we have

$$\frac{\partial \rho}{\partial t} + \nabla \cdot (\rho u) = 0, \qquad (1.3)$$

which is the differential form of mass conservation.

Following similar control volume analysis, the equations for momentum and energy conservation can be derived. We can represent these mass, momentum, and energy conservation laws all together in a single equation using tensors. The integral representation of the three conservation laws for a control volume V becomes

$$\iiint_V \frac{\partial \Lambda}{\partial t} dV = - \iint_S \Pi \cdot n dS + \iiint_V \Gamma dV \qquad (1.4)$$

$$\iiint_V \left(\frac{\partial \Lambda}{\partial t} + \nabla \cdot \Pi - \Gamma \right) dV = 0 \qquad (1.5)$$

with the corresponding differential form being

$$\frac{\partial \Lambda}{\partial t} + \nabla \cdot \Pi = \Gamma. \qquad (1.6)$$

Here, the vector Λ denotes the conserved quantities (per unit volume)

$$\Lambda = \begin{bmatrix} \rho \\ \rho u \\ \rho E \end{bmatrix}, \qquad (1.7)$$

where E is the total energy per unit mass, which is comprised of the internal energy (per unit mass) e and the kinetic energy k

$$E = e + k. \tag{1.8}$$

The kinetic energy is defined as

$$k = \frac{|u|^2}{2}. \tag{1.9}$$

The term Π is the flux of Λ

$$\Pi = \begin{bmatrix} \rho u \\ \rho u u - T \\ \rho E u - T \cdot u + q \end{bmatrix}. \tag{1.10}$$

The term ρu can be recognized as the momentum per unit volume in the second row of Eq. (1.7) and also as the mass flux vector for the first row of Eq. (1.10). The term $\rho u u - T$ is the momentum flux tensor in the momentum equation which consists of $\rho u u$ that describes the flux of momentum ρu moving at a velocity u and T that represents the momentum exchange due to the stress at the surface of the control volume. In the energy conservation equation, $\rho E u$ is the flux of energy, $T \cdot u$ is the work performed by stress, and q is the heat flux. By taking an inner product of Π and the unit normal vector n, we obtain the physical quantity that passes through the control surface per unit time and area. Since n represents the outward unit normal vector on a control surface S in Eq. (1.4), positive and negative $\Pi \cdot n$ correspond to efflux and influx, respectively, from the control surface.

In what follows, we assume that there is no sink, source, or heat generation within the volume of interest. If a body force f acts on the fluid, there is production of momentum and work performed by the force, making the right-hand side of Eq. (1.6) become

$$\Gamma = \begin{bmatrix} 0 \\ \rho f \\ \rho u \cdot f \end{bmatrix}. \tag{1.11}$$

Note that Eqs. (1.4) and (1.5) represent the change in conserved variables that is attributed to the flux across surface of the control volume and the source within the volume.

1.3.2   Closure of the Governing Equations

In order to solve the governing equations for fluid flow, we need to match the number of unknowns to the number of equations. This is referred to as the closure of the system of equations. For the flow equations, we need to express the stress T and heat flux q in the flux term Π by ρ, u, and E. These relations are called the *constitutive equations*.

For a *Newtonian fluid*, the stress tensor T is expressed as the sum of pressure and viscous stress in the following manner

$$T = -pI + 2\mu\left(D - \frac{1}{3}I\nabla \cdot u\right), \tag{1.12}$$

in which the Stokes relation has been used. Here, I is the identity tensor, p is the static pressure, μ is the dynamic viscosity, and D is the rate-of-strain tensor given by

$$D = \frac{1}{2}\left[(\nabla u)^T + \nabla u\right]. \tag{1.13}$$

For the heat flux q, we can use the Fourier's law

$$q = -k\nabla T, \tag{1.14}$$

where T is the absolute temperature and k is the thermal conductivity. Note that μ and k can be expressed as functions of T for Newtonian fluids (i.e., $\mu(T)$ and $k(T)$).

We have introduced T and p, which can be related to ρ by the *equation of state*. Here, we make an assumption that the fluid is an *ideal gas* in thermodynamic equilibrium. The ideal gas law states that

$$p = \rho RT = (\gamma - 1)\rho e, \tag{1.15}$$

where $\gamma = c_p/c_v$ is the specific heat ratio, c_v is the heat capacity at constant volume, c_p $(= c_v + R)$ is the heat capacity at constant pressure, and R is the gas constant. The internal energy e is

$$e = c_v T \tag{1.16}$$

(or $de = c_v dT$). With the above relations, the system of equations is in closure. Equations (1.4)–(1.6) expressed only in terms of the unknowns ρ, u, and E. Hence, the number of equations and the number of unknowns are equated.

1.3.3   Divergence and Gradient Forms

The mass conservation equation is also referred to as the continuity equation and is

$$\frac{\partial \rho}{\partial t} + \nabla \cdot (\rho u) = 0 \tag{1.17}$$

for flows without sinks or sources. We can consider the time rate of change and transport of some physical quantity ϕ by the velocity field u to be decomposed as

$$\frac{\partial(\rho\phi)}{\partial t} + \nabla \cdot (\rho\boldsymbol{u}\phi) = \rho\left(\frac{\partial\phi}{\partial t} + \boldsymbol{u} \cdot \nabla\phi\right) + \phi\left[\frac{\partial\rho}{\partial t} + \nabla \cdot (\rho\boldsymbol{u})\right], \qquad (1.18)$$

where the second term on the right-hand side becomes zero due to continuity, Eq. (1.17). Accordingly, we have

$$\frac{\partial(\rho\phi)}{\partial t} + \nabla \cdot (\rho\boldsymbol{u}\phi) = \rho\frac{\mathrm{D}\phi}{\mathrm{D}t} = \rho\left(\frac{\partial\phi}{\partial t} + \boldsymbol{u} \cdot \nabla\phi\right), \qquad (1.19)$$

where $\mathrm{D}/\mathrm{D}t$ is defined as

$$\frac{\mathrm{D}}{\mathrm{D}t} \equiv \frac{\partial}{\partial t} + \boldsymbol{u} \cdot \nabla \qquad (1.20)$$

and called the *material derivative* (*substantial derivative*).

The second term on the left-hand side of Eq. (1.19) represents advection by the divergence of $\rho\boldsymbol{u}\phi$. For that reason, this form is called the *divergence form* (*conservative form*). The second term on the right-hand side is expressed as the inner product of the velocity $\boldsymbol{u}$ and the gradient $\nabla\phi$, and is thus referred to as the *gradient form* or *advective form*[1] (*non-conservative form*). The terminology of conservative and non-conservative forms is widely used in the CFD literature.

Equation (1.19) is merely a statement of differentiation rule with the continuity equation incorporated. Therefore, calculations based on either form should be identical. For correctly discretized equations, the identity should hold, which makes the use of the term *non-conservative form* for the right-hand side of Eq. (1.19) somewhat misleading. Depending on the discretization schemes, this relation may not hold. It would appear more appropriate to use the term non-conservative to describe incompatible discretization schemes rather than the form of the right-hand side of Eq. (1.19). The notion of compatible numerical differentiation will be discussed in detail in Chap. 2.

The conservation of momentum provides us with the equation of motion, which is

$$\frac{\partial(\rho\boldsymbol{u})}{\partial t} + \nabla \cdot (\rho\boldsymbol{u}\boldsymbol{u} - \boldsymbol{T}) = \rho\boldsymbol{f} \qquad (1.21)$$

in the divergence form and

$$\rho\frac{\mathrm{D}\boldsymbol{u}}{\mathrm{D}t} = \nabla \cdot \boldsymbol{T} + \rho\boldsymbol{f} \qquad (1.22)$$

in the gradient form. When the constitutive equation (Eq. (1.12)) for a Newtonian fluid is used for $\boldsymbol{T}$, the equation of motion is referred to as the Navier–Stokes equation of motion. Since

$$\frac{\mathrm{D}\boldsymbol{u}}{\mathrm{D}t} = \frac{\partial\boldsymbol{u}}{\partial t} + \boldsymbol{u} \cdot \nabla\boldsymbol{u} \qquad (1.23)$$

---

[1]Advective form is often called convective form. We however use the term advective form to be consistent with the use of the term advection instead of convection (= advection + diffusion) for preciseness.

corresponds to the acceleration of a fluid element, we can notice that Eq. (1.22) describes Newton's second law (mass × acceleration = force) per unit mass.

Taking an inner product of the momentum equation, Eq. (1.22), with the velocity u, we arrive at the conservation of kinetic energy

$$\rho\frac{Dk}{Dt} = u \cdot (\nabla \cdot T) + \rho u \cdot f. \tag{1.24}$$

Subtracting the above equation from the conservation of total energy

$$\rho\frac{DE}{Dt} = \nabla \cdot (T \cdot u) - \nabla \cdot q + \rho u \cdot f, \tag{1.25}$$

we obtain the conservation of internal energy

$$\rho\frac{De}{Dt} = T : (\nabla u) - \nabla \cdot q, \tag{1.26}$$

where $T : S$ denotes the contraction of tensors (i.e., $T_{ij}S_{ji}$). The set of equations consisting of the mass, momentum, and energy equations for a Newtonian fluid is called the *Navier–Stokes equations*.

1.3.4 Indicial Notation

Up to this point, we have used vector notation in the governing equations for fluid flows. We can also utilize what is called *indicial notation* to represent the components for a Cartesian coordinate system, in which we denote the coordinates with $x_1 = x$, $x_2 = y$, and $x_3 = z$ and the corresponding velocity components with $u_1 = u, u_2 = u$, and $u_3 = w$.

We then can express the mass conservation as

$$\frac{\partial \rho}{\partial t} + \frac{\partial(\rho u_j)}{\partial x_j} = 0. \tag{1.27}$$

The momentum conservation equation can be written in the divergence form (Eq. (1.21)) as

$$\frac{\partial(\rho u_i)}{\partial t} + \frac{\partial}{\partial x_j}(\rho u_i u_j - T_{ij}) = \rho f_i \tag{1.28}$$

and in gradient form (Eq. (1.22)) as

$$\rho\left(\frac{\partial u_i}{\partial t} + u_j\frac{\partial u_i}{\partial x_j}\right) = \frac{\partial T_{ij}}{\partial x_j} + \rho f_i. \tag{1.29}$$

The conservation of total energy in indicial notation becomes

$$\frac{\partial(\rho E)}{\partial t} + \frac{\partial}{\partial x_j}(\rho E u_j - T_{ij}u_i + q_j) = \rho u_i f_i \qquad (1.30)$$

for the divergence form and

$$\rho\left(\frac{\partial E}{\partial t} + u_j\frac{\partial E}{\partial x_j}\right) = \frac{\partial}{\partial x_j}(T_{ij}u_i - q_j) + \rho u_i f_i \qquad (1.31)$$

for the gradient form (Eq. (1.25)).

The constitutive relations can also be written in indicial notation. The stress tensor (Eq. (1.12)) is represented as

$$T_{ij} = -\delta_{ij}p + 2\mu\left(D_{ij} - \frac{1}{3}\delta_{ij}\frac{\partial u_k}{\partial x_k}\right), \quad \text{where} \quad D_{ij} = \left(\frac{\partial u_i}{\partial x_j} + \frac{\partial u_j}{\partial x_i}\right) \qquad (1.32)$$

and the Fourier heat conduction law becomes

$$q_j = -k\frac{\partial T}{\partial x_j}. \qquad (1.33)$$

When the same index appears twice in the same term, summation over that index is implied (summation convention). That is, in two dimensions, $a_j b_j = \sum_{j=1}^{2} a_j b_j$, and in three dimensions, $a_j b_j = \sum_{j=1}^{3} a_j b_j$. When summation is not to be performed, it will be noted in the text. The symbol for the index can be different in the summation but results in the same sum (i.e., $a_j b_j = a_k b_k$). The symbol δ_{ij} appears often when using indicial notation and is called the Kronecker delta, defined as

$$\delta_{ij} = \begin{cases} 1 & i = j \\ 0 & i \neq j \end{cases} \qquad (1.34)$$

This is the component-wise representation of the basis tensor I for the Cartesian coordinate system. Note that the trace in three dimensions is $\delta_{kk} = 3$ (contraction of δ_{ij}).

1.3.5 Governing Equations of Incompressible Flow

Here, we summarize the governing equations for incompressible flow of a Newtonian fluid, which is the focus of this book. For incompressible flow, the material derivative of the density does not change

$$\frac{D\rho}{Dt} = \frac{\partial \rho}{\partial t} + \boldsymbol{u} \cdot \nabla \rho = 0. \tag{1.35}$$

Note that incompressibility does not necessarily mean that the density ρ is constant. This continuity equation becomes much more complex for multispecies system with diffusion, even if the flow can be treated as incompressible [9]. With Eq. (1.35), the continuity equation, Eq. (1.17), turns into

$$\nabla \cdot \boldsymbol{u} = 0, \tag{1.36}$$

which is also referred to as the incompressibility constraint or the divergence-free constraint. This relation implies that the volumetric flux budget is enforced instantaneously at each moment in time, since there is no term with time rate of change.

Next, let us consider the momentum equation. For a Newtonian fluid in incompressible flow, the stress tensor is

$$\boldsymbol{T} = -p\boldsymbol{I} + 2\mu\boldsymbol{D}. \tag{1.37}$$

Thus, the momentum equation for incompressible flow becomes

$$\rho \left[\frac{\partial \boldsymbol{u}}{\partial t} + \nabla \cdot (\boldsymbol{u}\boldsymbol{u}) \right] = -\nabla p + \nabla \cdot (2\mu\boldsymbol{D}) + \rho \boldsymbol{f}. \tag{1.38}$$

If we can treat viscosity to be a constant, the above equation can be further simplified to become

$$\frac{\partial \boldsymbol{u}}{\partial t} + \nabla \cdot (\boldsymbol{u}\boldsymbol{u}) = -\frac{\nabla p}{\rho} + \nu \nabla^2 \boldsymbol{u} + \boldsymbol{f}, \tag{1.39}$$

where $\nu = \mu/\rho$ is the kinematic viscosity.

Taking the divergence of Eq. (1.39) with the assumption of ρ and ν being constant and utilize Eq. (1.36), we arrive at an equation for pressure

$$\frac{\nabla^2 p}{\rho} = -\nabla \cdot \nabla \cdot (\boldsymbol{u}\boldsymbol{u}) + \nabla \cdot \boldsymbol{f}, \tag{1.40}$$

where ∇^2 is the Laplacian operator. Here, the flow field is constrained by incompressibility, Eq. (1.36), at all times and the corresponding pressure field is determined from the instantaneous flow field. Equation (1.40) is called the *pressure Poisson equation* and is an elliptic partial differential equation that is solved as a boundary value problem. While pressure for compressible flow is provided thermodynamically by the equation of state, Eq. (1.15), the pressure for incompressible flow is not determined using thermodynamics. This leads to adopting different numerical methods for incompressible and compressible flows. While both incompressible and compressible flows have attributes of wave propagation[2] and viscous diffusion, incompressible flow needs to enforce the divergence-free constraint, Eq. (1.36), which necessitates an elliptic solver. On the other hand, compressible flow does not need an elliptic solver since incompressibility need not be satisfied. Different types of partial differential equations are briefly discussed in Sect. 1.3.6.

The conservation of internal energy for incompressible flow becomes

$$\rho\frac{De}{Dt} = \mu \mathbf{D} : \mathbf{D} - \nabla \cdot \mathbf{q}. \tag{1.41}$$

With the use of Eqs. (1.14) and (1.16), the equation for the temperature field reads

$$\rho c_v \frac{DT}{Dt} = \mu \mathbf{D} : \mathbf{D} + k\nabla^2 T. \tag{1.42}$$

The first term on the right-hand side represents the heat generation due to fluid friction (viscous effect). If the influence of friction on the temperature field is small, the temperature equation simplifies to

$$\rho c_v \frac{DT}{Dt} = k\nabla^2 T. \tag{1.43}$$

Under this assumption, we can consider the kinetic and internal energies separately for incompressible flows. Since the conservation of kinetic energy depends only passively on the conservation of mass and momentum, it is not necessary to explicitly handle the kinetic energy in the fluid flow analysis. The conservation of internal energy can be used as a governing equation for temperature. In the discussions to follow, we do not consider the temperature equation.

Component-wise Representation

As a summary of the above discussion and for reference, let us list the governing equations for incompressible flow with constant density and viscosity for a Cartesian coordinate system using the component-wise representation. The continuity equation (1.36) is

$$\frac{\partial u_i}{\partial x_i} = 0 \tag{1.44}$$

and the momentum equation (1.39) is

$$\frac{\partial u_i}{\partial t} + \frac{\partial(u_i u_j)}{\partial x_j} = \frac{\partial u_i}{\partial t} + u_j\frac{\partial u_i}{\partial x_j} = -\frac{1}{\rho}\frac{\partial p}{\partial x_i} + \nu\frac{\partial^2 u_i}{\partial x_j \partial x_j} + f_i. \tag{1.45}$$

---

[2]For incompressible flow, the Mach number $M = u/a$ (the ratio of characteristic velocity u and sonic speed a) is zero or very small. That means that the acoustic wave propagation is very fast compared to the hydrodynamic wave propagation. In the limit of $M \to 0$, the acoustic propagation is considered to take place instantaneously over the whole domain, which leads to the appearance of ellipticity in the governing equations.

The pressure Poisson equation derived from the above two equations is

$$\frac{1}{\rho}\frac{\partial^2 p}{\partial x_i \partial x_i} = -\frac{\partial^2 (u_i u_j)}{\partial x_i \partial x_j} + \frac{\partial f_i}{\partial x_i} = -\frac{\partial u_i}{\partial x_j}\frac{\partial u_j}{\partial x_i} + \frac{\partial f_i}{\partial x_i}. \tag{1.46}$$

Note that Eqs. (1.45) and (1.46) make use of Eq. (1.44).

1.3.6 Properties of Partial Differential Equations

As we have seen above, the conservation laws that describe the flow are represented by partial differential equations (PDEs). Hence, it is important to understand the characteristics of the governing PDEs to numerically solve for the flow field. It is customary to discuss the classification of PDEs and the theory of characteristic curves, but unless readers go beyond the scope of this book or become involved in the development of advanced numerical algorithms, there is not a critical need to dive deeply into the theory of PDEs.

Depending on how information travels, the second-order PDEs are classified into elliptic, parabolic, and hyperbolic PDEs as shown in Table 1.1. These types are not influenced by the choice of coordinate systems but can be affected by the location in a flow field. For example, let us consider flow over a bluff body. Flow away from the body and outside of the boundary layer (without any vorticity) can be treated as potential flow which is described by an elliptic PDE. However, in the region right next to the surface of the body (boundary layer), viscous diffusion becomes the dominant physics describing the flow which can be captured by a parabolic PDE. Hence, the classification of PDEs is in general performed locally. We will let other textbooks [10, 13] on PDEs describe the classification of PDEs, characteristic curves, and initial/boundary value problems. For the purpose of this book, we will focus on presenting the PDE types of the governing equations for flow in a brief fashion. The classification of PDEs is based on the existence of a unidirectional coordinate [16].

Table 1.1 Canonical second-order partial differential equations in fluid mechanics

Type	Examples	
Elliptic	Poisson equation	$\nabla^2 p = -\rho \nabla \cdot \nabla \cdot (uu) + \nabla \cdot f$
	Laplace equation	$\nabla^2 \phi = 0$
Parabolic	Diffusion equation	$\frac{\partial u}{\partial t} = \nu \nabla^2 u$
	Heat equation	$\frac{\partial T}{\partial t} - \frac{k}{\rho c_v}\nabla^2 T$
Hyperbolic	Wave equation	$\frac{\partial^2 u}{\partial t^2} = U^2 \nabla^2 u$

The temporal axis t is obviously unidirectional (past to future). On the other hand, the spatial coordinate x_i is bidirectional.

For viscous incompressible flows in general, the PDEs for unsteady flow (dependent on t) are parabolic and the PDEs for steady flow (independent of t) are elliptic. In some steady cases where the influence from downstream can be neglected, the flow may be solved by marching from upstream to downstream. This is called parabolic approximation (parabolization). One can also solve for steady flow by leaving the temporal derivative term and reframing the problem as an unsteady one. Once the solution is converged to the steady profile after sufficient time advancement, the time derivative term would vanish and the flow field becomes the solution to the original elliptic PDE. Such approach could be described as a parabolic technique for solving an elliptic PDE.

1.4 Grids for Simulating Fluid Flows

For Eulerian methods, variables such as velocity, pressure, and density are determined at a large number of discrete points to represent the motion of a liquid or a gas as a continuum. A polygon (in two-dimensional and polyhedron in three-dimensional space) made from local collection of discrete points (vertices) is called a cell and the space filled by these cells is referred to as a grid (or mesh). Physical variables of interest are positioned at various locations on the cells chosen to satisfy particular numerical properties.

A few representative Eulerian grids are shown in Fig. 1.3. For a *Cartesian grid*, the governing equations for the flow takes the simplest form and makes spatial discretization effortless. However, it is not suitable for discretizing a flow field around a body of complex geometry, as illustrated in Fig. 1.3a. Even a body with rather simple geometry, such as a sphere or a circular cylinder, would requires a very fine mesh near the body boundary to resolve the flow. One solution to this issue is to employ an *immersed boundary method* that generates a body without regard to the underlying grid. Chapter 5 is devoted to the immersed boundary method. Another remedy to this problem is to use a *curvilinear coordinate grid* that fits around the boundary, as shown in Fig. 1.3b. Such curvilinear grid is called a *boundary-fitted coordinate grid* or a *body-fitted coordinate grid* (BFC). A transformation of such BFC in physical

(a)   **(b)**   **(c)**

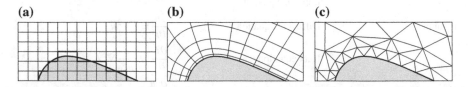

Fig. 1.3 Grids for fluid flow simulations (two-dimensional). **a** Cartesian grid. **b** Curvilinear grid. **c** Unstructured grid

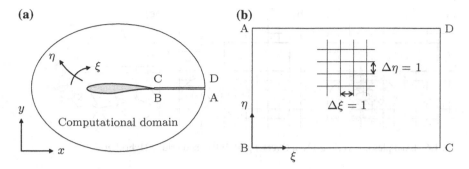

Fig. 1.4 Mapping between the physical space and computation space for a boundary-fitted grid. **a** Physical space. **b** Computational space

domain to another domain (computation domain), becomes convenient, especially if we make the grid size unity in the computational domain (see Fig. 1.4). Grids that has the same topology as a Cartesian grid is referred to as *structured grids*, as shown by Fig. 1.3a, b.

One can also discretize the spatial domain as shown in Fig. 1.3c with cells that are arranged in an irregular manner to accommodate complex boundary geometries. Boundary-fitted grids of this type are called *unstructured grids*. Unstructured grids often utilize triangles, quadrilaterals, and hexagons in two dimensions and tetrahedra and hexahedra in three dimensions.

Structured grids are not as versatile as unstructured grids for discretizing complex geometry. However, structured grids are often preferred for computing highly accurate flow data in fundamental research. Given the same number of grid points, structured grids allow for more orderly access to memory on computers, compared to unstructured grids, and achieve higher computational efficiency. The boundary-fitted coordinate is a generalized curvilinear coordinate which has basis vectors that are not always orthonormal and these vectors can vary in space. Thus, the governing equations for flow become complex with the use of the generalized coordinate system. See Appendix A for details.

Let us present a few types of boundary-fitted grids. Depending on the boundary shape, different types of boundary-fitted grids is chosen. In Fig. 1.5, we show canoni-

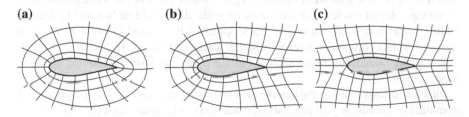

Fig. 1.5 Boundary-fitted grids. **a** O-grid. **b** C-grid. **c** H-grid

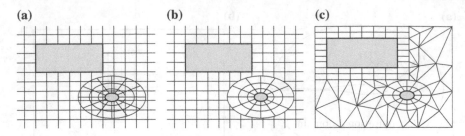

Fig. 1.6 Examples of hybrid grids. **a** Overset grid. **b** Patched grid. **c** Hybrid grid

cal boundary-fitted grids that are used to discretize the computational domain around an airfoil. The O-grid places the grid around a body efficiently with low skewness in general. For bodies with sharp corners (cusp), for example the trailing edge of an airfoil, there can be highly skewed grids. The C-grid is aligned with the flow around the body and is able to generate unskewed grids near the trailing edge. While such arrangement of grids is suitable for viscous flows, there may be unnecessarily large number of grids downstream of the wing. For airfoil cascade (arrangement of a series of blades, such as in turbines) or flow over a body in a channel, multi-block approach is often utilized with H-grids and L-grids (not shown).

It would be ideal to discretize the flow field with only one type of grid. With the development of grid generation techniques, it has become possible to generate a mesh of a single type even for somewhat complex geometries. However, generating high-quality meshes remains a challenge (e.g., mesh with low skewness, mesh with necessary and sufficient resolution). In many cases, one cannot know where the grid should be refined a priori.

For flows with bodies of complex geometry, with multiple bodies, or with bodies that move (relative to each other) or deform, we can consider the use of multiple types of grids, as illustrated in Fig. 1.6. In most cases, information is transmitted during calculation from one grid to the other at the overlap or along the interface of the grids.

Concentrating grids in regions where the flow exhibits changes in its features is effective for attaining accurate solutions. For example, we know a priori that the boundary layer near the wall has large velocity gradients. Thus, it would be beneficial to place a large number of grid points to resolve the flow there. One can also consider adaptively generating additional grids at regions where the flow shows large gradients in the variables of interest. For unstructured grids, this would be handled by simply adding extra vertices in that region. Such approach is referred to as *adaptive mesh refinement* and is often used to capture shock waves and flames. In case of flows with time varying boundary shapes, such as the waves around a ship, a moving adaptive mesh is used.

In what follows, we assume that the grid has been generated for the computational domain. For details on grid generation, readers should consult with [6, 19].

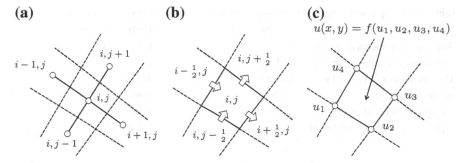

Fig. 1.7 Overview of discretization methods. **a** Finite-difference method. **b** Finite-volume method. **c** Finite-element method

1.5 Discretization Methods

There are three main methods for discretizing the flow field in numerical simulations. We provide brief descriptions of these methods below with a graphical representation in Fig. 1.7.

Finite-Difference Method (FDM)

Finite-difference method [12] is based on the differential form of the governing equation, Eq. (1.6). The numerical solution to the equation is sought on the grid points and information between the grid points is generally not taken into account. The derivative operations such as the gradients of velocity and pressure are approximated by difference quotients. Structured grids are usually chosen for finite-difference methods. The discussions in this book is focused on the finite-difference method.

Finite-Volume Method (FVM)

The integral form of governing equation, Eq. (1.4), is used as a foundation of this method. Instead of placing the variables on grid points, the method solves for the represented value of the variable for the cell volume. The conservation law is expressed as a balance of the influx and efflux through the boundary and the generation and destruction of conserved quantities for each cell. The idea of finite-volume methods is to solve for the unknown variables through such relations. Structured and unstructured grids can be used. For finite-volume methods, LeVeque [11] and Ferziger and Perić [5] are excellent references.

Finite-Element Method (FEM)

This method utilizes the weak formulation of the governing equations, which has a test function multiplied to both sides of the equation [17]. The profile of the variable is provided by the product of the variable and a basis function at the vertex. Substituting these into the governing equation and integrating the equation with a weight function, a relation is obtained for the discrete variables at the cell vertices. The method is applicable to both structured and unstructured grids.

Table 1.2 Comparison of general traits associated with structured and unstructured grids

	Structured grid	Unstructured grid
Suitable computational method	FDM/FVM	FVM/FEM
Handling of complex geometry	Difficult	Flexible
Overall number of grid points	Large	Small
Accuracy (per grid)	High	Low
Computational load (per grid)	Low	High
Approach to enhance resolution	Refine region	Local adaptation

When choosing the type of grid and the computational method, it is important to understand the physical features of the flow of interest and the purpose of the simulation. The general properties of the solvers based on the use of a structured or unstructured grid are shown in Table 1.2. Structured grids, that position cells in an orderly manner, can achieve high-order accuracy and high computational efficiency per cell than unstructured grids in general. Thus, structured grids are suitable for high-fidelity computation in fundamental research, such as the analysis of turbulent flow. With structured grids, the flow field would be limited to simple geometry. On the other hand, unstructured grids can discretize complex geometry and would be more appropriate for practical engineering calculations.

There are also methods that do not solve the PDEs on grids. Methods such as the vortex methods [3] and lattice Boltzmann methods [2] are alternative approaches to simulate fluid flows.

Notation for Approximate Relations

In this book, we will introduce approximations using finite differencing and interpolations. Also to be presented are physical models that approximate turbulent statistics. Instead of representing the approximate relations with ($\sim, \simeq$), we will use equal signs ($=$) throughout the text, as they are used in computer programs.

1.6   Verification and Validation

To ensure that computational solutions are credible, we must carry out certain checks during the course of the computational study. In particular, there are two assessments that should be performed; they are called *verification* and *validation*. Both of these assessments are described in detail in [7, 14, 18].

The first process, verification, examines whether the computational model being solved accurately reproduces the model solution with the expected behavior. Confirming the expected temporal and spatial accuracy and whether the solution converges to a reference solution are parts of verification. The reference solution can be an exact solution to the Navier–Stokes equations, a solution to a model problem (such as the advection-diffusion equation), or numerical solution obtained with very fine temporal and spatial resolutions.

The second process, validation, compares the numerical solution with reliable physical measurements. Validation checks whether the solution replicates the physics of interest well. One can for instance compare the simulated flow field with PIV measurements and computed forces on a body with measurements from force balance or transducers. Validation implicitly assumes that numerical solution has been verified.

It is crucial that both verification and validation are performed to ensure that the numerical solution is credible. When performing these two assessments, it is important to chose reference solutions and experimental measurements that capture the essence of the ultimate problem that will be solved numerically after the computational approach is verified and validated. For example, if a computational model is going to be used for a range of Reynolds numbers and Mach numbers, verification and validation should span across those ranges. The level of accuracy to be sought during the verification and validation processes is problem dependent and should be carefully chosen. There can be cases where a few percent of error in the solution is tolerable, while such error can cause a notable change in the overall physics.

1.7 Remarks

We have provided an overview of numerical flow simulation, discussed the governing equations, and presented different discretization methods. The basic equations for fluid flow are partial differential equations and consist of the conservation laws and the constitutive relations. It is important to maintain the properties of the original differential equations in the discretized system in numerical methods. Further details on discretizations and numerical methods are provided in later chapters.

Besides the finite-difference, finite-volume, and finite-element methods, there are spectral methods, vortex methods, and boundary-element methods. These later methods are somewhat specialized and are not covered here. In this book, we present the fundamentals of numerical simulations of fluid flow with finite-difference methods while integrating concepts from the finite-volume methods for the conservation laws.

At the beginning of this chapter, we have stated that numerical simulation often is viewed as a third set of tools in addition to experiments and theory. First, let us consider the comparison of theoretical research and numerical simulation. Even if we accept the conservation laws as flawless, the constitutive relations contain a number of physical assumptions. There may be cases where the assumption of continuum is not appropriate. Such arguments hold for both theory and simulation. In theoretical studies, we often provide additional assumptions or limit the range of validity for the system of interest, so that analytical approaches can be taken. On the other hand, we discretize the system of equations in simulations for numerical computations to be performed. During the discretization process, error creeps into the computation influencing the results.

Next, let us compare experiments and simulations. We emphasize that there is similarity between experiments and numerical simulations. One can consider most experiments as model experiments (simulations) based on scaled models in similar fluid flow environment. Although, experiments deal with the actual flow, the data can contain some amount of error in the measured signals and the flow can be influenced by the existence of probes or tracers. While the simulation is performed in a virtual world, the data is not affected by the data extraction process. Expressing (or visualizing) the three-dimensional velocity and pressure fields at a desired time is easier with simulations.

Instead of the numerical simulations being a complementary tool to theory and experiment, it would be desirable to develop numerical techniques to establish simulations as an independent tool. Even in industrial uses, we should aim to use simulations as a methodology to develop new machineries or systems, rather than a complimentary approach to reduce the cost of experiments. While we have been able to achieve success in some applications with numerical computation, we must further develop methods to surpass limitations imposed by assumptions on the governing equations and minimize the numerical error that creeps in from the discretization so that simulation results can be accepted as trustworthy as well as experimental measurements. If we perform simulations with a fine mesh to capture the physics at all scales, we should be able to produce results with a high level of confidence. With the continuing growth in computational capabilities, we are able to simulate ever more complex fluid flows.

For the reasons discussed above, it would be crucial for a fluid flow simulation to

1. resolve all essential physical scales in the flow,
2. utilize the most appropriate numerical method, and
3. take full advantage of the available computational resource.

Prior to performing the numerical simulations, determining or estimating the scales to be resolved in the simulations is very important as it is equivalent to the design of experiments. The important physical scales to be analyzed often depend on the physical problem and the engineering application. In this book, we discuss ways to obtain insights into the flow physics through the use of numerical analysis toolsets. It is important to ensure that the numerics impose minimal influence on the essential physical scales and dynamics simulated for the fluid flow problem of interest.

1.8 Exercises

1.1. Derive the set of governing equations, analogous to Eq. (1.6), in differential form when there are mass source (or sink) and heat generation in the domain.
1.2. Find the governing equation for vorticity $\omega \equiv \nabla \times u$ for incompressible flow by taking the curl of the momentum equation. This equation is referred to as the *vorticity transport equation*. Describe what each term in the vorticity transport equation represents.

1.3. Prove that the divergence of the shear stress tensor is zero (i.e., $\text{div}(\tau) = 0$) if the flow is irrotational ($\omega = 0$) and incompressible ($\nabla \cdot u = 0$). Assume constant viscosity.

1.4. Discuss how a nonconstant density field can satisfy $D\rho/Dt = 0$.

1.5. Consider a rigid stationary container of volume V with surface S filled with a quiescent fluid of constant density ρ and constant viscosity ν. At $t = t_0$, the fluid inside the container is perturbed to have nonzero velocity $u(x, t_0) \neq 0$. For an viscous fluid, show that the velocity field always decays to zero by considering

$$\frac{d}{dt} \int_V \frac{\rho}{2} u_i u_i dV$$

to be negative at all times.

1.6. Derive the energy conversation equation in terms of enthalpy $h = e + p/\rho$.

1.7. Show that the solution to the one-dimensional wave equation

$$\frac{\partial^2 u}{\partial t^2} = U^2 \frac{\partial^2 u}{\partial x^2}$$

for a constant U (>0) over $-\infty < x < \infty$ is comprised of two waves traveling to the left and right at speeds $-U$ and U, respectively.

1.8. Consider an inviscid flow (i.e., $\nu = 0$) with no external forcing or heating applied. The equations that describe inviscid flow are known the *Euler's equations*. Here, show that the continuity, momentum, and energy equations in one dimension can be written as

$$\frac{\partial q}{\partial t} + A \frac{\partial q}{\partial x} = 0,$$

where $q = (\rho, u, p)^T$. Assume ideal gas for the state of the flow and find matrix A and its eigenvalues. Comment on your findings.

References

1. Aris, R.: Vectors, Tensors and the Basic Equations of Fluid Mechanics. Dover (1989)
2. Chen, S., Doolen, G.D.: Lattice Boltzmann method for fluid flows. Annu. Rev. Fluid Mech. **30**, 329–364 (1998)
3. Cottet, G.H., Koumoutsakos, P.D.: Vortex Methods: Theory and Practice. Cambridge Univ. Press (2000)
4. Currie, I.G.: Fundamental Mechanics of Fluids. McGraw-Hill (1974)
5. Ferziger, J.H., Perić, M.: Computational Methods for Fluid Dynamics, 3rd edn. Springer, Berlin (2002)
6. Fujii, K.: Numerical Methods for Computational Fluid Dynamics. Univ. Tokyo Press, Tokyo (1994)
7. Guide for the verification and validation of computational fluid dynamics simulations. Tech. Rep. G-077-1998(2002), AIAA (1998)

8. Jantzen, R.T., Taira, K., Granlund, K., Ol, M.V.: Vortex dynamics around pitching plates. Phys. Fluids **26**, 053,696 (2014)
9. Kaviany, M.: Principles of Convective Heat Transfer, 2nd edn. Springer (2001)
10. Kevorkian, J.: Partial Differential Equations: Analytical Solution Techniques, 2nd edn. Springer (2000)
11. LeVeque, R.J.: Finite Volume Methods for Hyperbolic Problems. Cambridge Univ. Press (2002)
12. LeVeque, R.J.: Finite Difference Methods for Ordinary and Partial Differential Equations: Steady-State and Time-Dependent Problems. SIAM (2007)
13. McOwen, R.C.: Partial Differential Equations: Methods and Applications, 2nd edn. Pearson (2002)
14. Oberkampf, W.L., Trucano, T.G.: Verification and validation in computational fluid dynamics. Prog. Aero. Sci. **38**, 209–271 (2002)
15. Panton, R.L.: Incompressible Flow. Wiley-Interscience (1984)
16. Patankar, S.: Numerical Heat Transfer and Fluid Flow. Hemisphere, Washington (1980)
17. Pozrikidis, C.: Introduction to Finite and Spectral Element Methods using MATLAB. Chapman and Hall/CRC (2005)
18. Smith, R.C.: Uncertainty Quantification: Theory, Implementation, and Applications. SIAM (2014)
19. Thompson, J.F., Warsi, Z.U.A., Mastin, C.W.: Numerical Grid Generation: Foundations and Applications. North-Holland (1985)

Chapter 2
Finite-Difference Discretization of the Advection-Diffusion Equation

2.1 Introduction

Finite-difference methods are numerical methods that find solutions to differential equations using approximate spatial and temporal derivatives that are based on discrete values at spatial grid points and discrete time levels. As the grid spacing and time step are made small, the error due to finite differencing becomes small with correct implementation. In this chapter, we present the fundamentals of the finite-difference discretization using the advection-diffusion equation, which is a simple model for the Navier–Stokes equations.

Before we start the presentation of the finite-difference formulation, let us note that in addition to reducing the error in finite differencing, there are two issues in finite-difference methods which may be more important than the actual value of error itself. First is the properties of the error. For schemes with error that behaves diffusively, the solution becomes smoother than the actual physical solution and for schemes with error that acts dispersively, non-physical oscillation develops in the solution leading to its possible blowup. The second important concept is whether the discrete derivative relations are compatible. For example, differentiation rules for continuous functions f and g

$$\frac{\partial (fg)}{\partial x} = f \frac{\partial g}{\partial x} + \frac{\partial f}{\partial x} g \qquad (2.1)$$

$$\frac{\partial^2 f}{\partial x^2} = \frac{\partial}{\partial x} \left(\frac{\partial f}{\partial x} \right) \qquad (2.2)$$

$$\frac{\partial^2 f}{\partial x \partial y} = \frac{\partial}{\partial x} \left(\frac{\partial f}{\partial y} \right) = \frac{\partial}{\partial y} \left(\frac{\partial f}{\partial x} \right) \qquad (2.3)$$

must hold also in a discrete manner. In this book, we refer to those finite-difference schemes that satisfy the derivative properties discretely as *compatible*. We discuss these topics in further detail in this chapter.

© Springer International Publishing AG 2017
T. Kajishima and K. Taira, *Computational Fluid Dynamics*,
DOI 10.1007/978-3-319-45304-0_2

Finite-difference schemes that have minimal errors and satisfy the derivative relations in the discretized settings should be selected. On the other hand, schemes that do not discretely satisfy the derivative relations should be avoided, because they may produce non-physical solutions despite being supposedly highly accurate.

2.2 Advection-Diffusion Equation

For incompressible flow with constant viscosity, the momentum equation in the Navier–Stokes equations, Eq. (1.39), can be written in Cartesian coordinates as

$$
\underbrace{\frac{\partial u_i}{\partial t}}_{\substack{\text{unsteady}\\\text{term}}} + \underbrace{u_j\frac{\partial u_i}{\partial x_j}}_{\substack{\text{advective}\\\text{term}}} = \underbrace{-\frac{1}{\rho}\frac{\partial p}{\partial x_i}}_{\substack{\text{pressure}\\\text{gradient}\\\text{term}}} + \underbrace{\nu\frac{\partial^2 u_i}{\partial x_j\partial x_j}}_{\substack{\text{diffusive}\\\text{term}}} + \underbrace{f_i}_{\substack{\text{external}\\\text{forcing}\\\text{term}}}.
$$
(2.4)

The advective term here is nonlinear and requires special care in numerical calculation. The diffusive term smoothes out the velocity profile and provides a stabilizing effect in numerical computations. The pressure gradient term represents forcing on a fluid element induced by the spatial variation in pressure. The pressure variable is solved by coupling the continuity equation, Eq. (1.40), with the above momentum equation. External forcing includes gravitational, electromagnetic, or fictitious forces (due to non-inertial reference frames). Because of the coupled effects from these terms, the examination of the physical and numerical properties of the governing equation, Eq. (2.4) becomes quite complex. Hence we in general consider a simplified model without the pressure gradient and external forcing terms in one dimension,

$$
\frac{\partial u}{\partial t} + u\frac{\partial u}{\partial x} = \nu\frac{\partial^2 u}{\partial x^2},
$$
(2.5)

which is known as the *Burgers' equation*. This equation is often used to validate numerical methods because the exact solution is available. We can further consider the linear approximation of the above equation with constant advective speed c (≥ 0) and diffusivity a (≥ 0) to obtain the linear advection-diffusion equation[1]

$$
\frac{\partial f}{\partial t} + c\frac{\partial f}{\partial x} = a\frac{\partial^2 f}{\partial x^2}.
$$
(2.6)

When each of the terms is removed from the advection-diffusion equation, the resulting equations have different characteristic properties. When we set $a = 0$, we obtain the advection equation

---

[1]The terms *advection* and *convection* are often used interchangeably. Technically speaking, convection refers to the combined transport by advection and diffusion. Thus, we use advection and convection with careful distinction in this book.

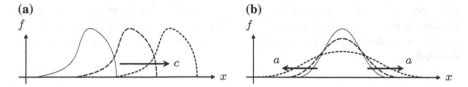

Fig. 2.1 Behavior of the solutions to the advection and diffusion equations. **a** Advection. **b** Diffusion

$$\frac{\partial f}{\partial t} + c\frac{\partial f}{\partial x} = 0, \tag{2.7}$$

which is hyperbolic. When we set $c = 0$, we have the diffusion equation

$$\frac{\partial f}{\partial t} = a\frac{\partial^2 f}{\partial x^2}, \tag{2.8}$$

which is parabolic. Finally, if $\partial f/\partial t = 0$, we find the steady advection-diffusion equation

$$c\frac{\partial f}{\partial x} = a\frac{\partial^2 f}{\partial x^2}, \tag{2.9}$$

which is elliptic. In other words, the advective-diffusion equation as well as the Navier–Stokes equations consist of the combination of these three types of partial differential equations. Properties from one of the models may become more dominant than the others depending on the flow condition.

Solutions for one-dimensional constant coefficient advection and diffusion equations are presented in Fig. 2.1. Since $f(x, t) = f(x - ct, 0)$ satisfies the advection equation, Eq. (2.7), we observe that the solution keeps the initial profile while translating in the x-direction with speed c. For the diffusion equation, Eq. (2.8), the solution decreases ($\partial f/\partial t < 0$) where the solution profile is convex ($\partial^2 f/\partial x^2 < 0$) and increases ($\partial f/\partial t > 0$) where the profile is concave ($\partial^2 f/\partial x^2 > 0$), smoothing out the solution profile over time.

2.3 Finite-Difference Approximation

Let us consider the finite-difference approximation of the spatial derivatives in the advection-diffusion equation, Eq. (2.6). While there are various techniques to derive the finite-difference equations, derivations are commonly based on Taylor series expansions or polynomial approximations.

We note with caution that readers should not use the difference equations that will be presented in Sects. 2.3.1 and 2.3.2 naively. It is important to study the derivations based on Taylor series expansions and the error analyses of the resulting formulas. Similarly, the polynomial approximation to analytically obtain the derivative approximation can provide insights into fundamentals of the numerical methods.

However, the equations presented in Sects. 2.3.1 and 2.3.2 will only be used during the discussion of numerical viscosity and filtering operation for LES in this book. Direct use of these equations in a naive fashion for flow calculations is not recommended.

2.3.1 Taylor Series Expansion

Here, we derive the finite-difference approximations of derivatives using Taylor series expansions and analyze the associated errors. Let us consider the one-dimensional case. Denoting the spatial derivatives of a continuous function $f(x)$ as

$$f'(x) = \frac{\mathrm{d}f(x)}{\mathrm{d}x}, \quad f''(x) = \frac{\mathrm{d}^2 f(x)}{\mathrm{d}x^2}, \quad \cdots, \quad f^{(m)}(x) = \frac{\mathrm{d}^m f(x)}{\mathrm{d}x^m}, \qquad (2.10)$$

the Taylor series expansion about point x evaluated at point $x + y$ is

$$f(x + y) = f(x) + \sum_{m=1}^{\infty} \frac{y^m}{m!} f^{(m)}(x). \qquad (2.11)$$

Consider setting $x = x_j$ and $y = x_k - x_j$, where x_j represents the jth grid point in space. Denoting the functional values at discrete points as $f_k = f(x_k)$, the Taylor series expansion about x_j can be written as

$$f_k = f_j + \sum_{m=1}^{\infty} \frac{(x_k - x_j)^m}{m!} f_j^{(m)}. \qquad (2.12)$$

Using functional values of $f_k = f(x_k)$ at a number of x_k, we will derive approximations for the m-derivative $f_j^{(m)} = f^{(m)}(x_j)$ at point x_k. The geometric arrangement of grid points (x_k) required to construct such finite-difference approximation at a point is called a *stencil*. In this section, we present the derivation of the finite-difference formulas and the discuss the associated order of accuracy.

Big $\mathcal{O}$ Notation

Before we derive the finite-difference formulas, let us briefly introduce a mathematical notation that will be used throughout the book. Below, we will be using what is known as the big $\mathcal{O}$ (Landau notation) [5] to describe the asymptotic behavior of terms and functions. In most of the discussions, this notation will be used to describe the error terms in numerical computation and to study order of magnitude estimates later in examining turbulence.

For example, the big $\mathcal{O}$ is used in the following manner:

$$\sin(x) = x - \tfrac{1}{3!}x^3 + \tfrac{1}{5!}x^5 - \cdots \quad \text{for all } x$$
$$= x - \tfrac{1}{3!}x^3 + \mathcal{O}(x^5) \quad \text{as } x \to 0 \qquad (2.13)$$

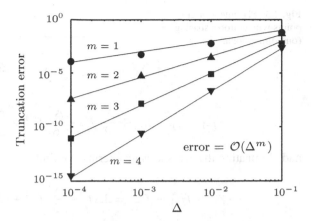

Fig. 2.2 Typical convergence of truncation errors. The slopes of the convergence curves on the log–log plot correspond to the orders of accuracy m

The first line is the Taylor series expansion of $\sin(x)$ about $x = 0$. The second line tells us that one can approximate $\sin(x)$ by $x - \frac{1}{3!}x^3$ in the neighborhood of $x = 0$ and the associated error will decrease faster than a constant times $|x^5|$ as $x \to 0$. With the two-term approximation, we can say $\sin(x) \approx x - \frac{1}{3!}x^3$ with an fifth-order *truncation error* term $\mathcal{O}(x^5)$. Note that it is also possible to use more or less number of terms in the approximation to increase or decrease the accuracy of the computation.

Let us also provide a visual description for the convergence behavior of the truncation error. Consider a truncation error that can be expressed as $\mathcal{O}(\Delta^m)$ for a given numerical approximation with Δ being a small quantity $\ll 1$. If we plot the error over Δ with the use of logarithmic scales for both the x and y-axes, the slope of the error curve (convergence curve) becomes m, as illustrated in Fig. 2.2. The exponent of the leading error term or the slope of the convergence curve m is called the *order of accuracy* of the given numerical method. This term will be used throughout the textbook to assess the accuracies of spatial and temporal numerical discretization schemes.

Central-Difference for Uniform Grid

Taylor series expansion of a function about a point $x_j = j\Delta$ for a uniform mesh, as shown in Fig. 2.3, becomes

$$f_{j+k} = f_j + \sum_{m=1}^{\infty} \frac{(k\Delta)^m}{m!} f_j^{(m)}, \tag{2.14}$$

where $k = \pm 1, \pm 2, \cdots, \pm N$. Using the series expansion of the given function, we can derive numerical approximations for the derivatives.

As an example, let us consider three-point finite difference schemes that are based on f_{j-1}, f_j, f_{j+1}. Taking the difference of following two Taylor series:

$$f_{j+1} = f_j + \Delta f_j' + \frac{\Delta^2}{2} f_j'' + \frac{\Delta^3}{6} f_j^{(3)} + \frac{\Delta^4}{24} f_j^{(4)} + \cdots \tag{2.15}$$

Fig. 2.3 Placement of grid points with uniform spacing (one-dimensional)

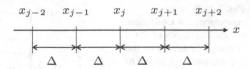

$$f_{j-1} = f_j - \Delta f'_j + \frac{\Delta^2}{2} f''_j - \frac{\Delta^3}{6} f_j^{(3)} + \frac{\Delta^4}{24} f_j^{(4)} - \cdots \qquad (2.16)$$

and eliminating the even-derivative terms, we find

$$f_{j+1} - f_{j-1} = 2\Delta f'_j + \frac{\Delta^3}{3} f_j^{(3)} + \mathcal{O}(\Delta^5). \qquad (2.17)$$

If we had taken the sum of the two Taylor series, we would have found that

$$f_{j+1} + f_{j-1} = 2f_j + \Delta^2 f''_j + \frac{\Delta^4}{12} f_j^{(4)} + \mathcal{O}(\Delta^6). \qquad (2.18)$$

From Eqs. (2.17) and (2.18), we can find the three-point finite-difference formula for f'_j and f''_j

$$f'_j = \frac{-f_{j-1} + f_{j+1}}{2\Delta} - \frac{\Delta^2}{6} f_j^{(3)} + \mathcal{O}(\Delta^4) \qquad (2.19)$$

$$f''_j = \frac{f_{j-1} - 2f_j + f_{j+1}}{\Delta^2} - \frac{\Delta^2}{12} f_j^{(4)} + \mathcal{O}(\Delta^4). \qquad (2.20)$$

With a three-point stencil, we can obtain up to the second-order derivative with finite differencing. The second and third terms on the right-hand side of Eqs. (2.19) and (2.20) represent the truncation errors. As we consider small Δ with bounded high-order derivatives, the truncation error becomes proportional to Δ^2. That means that when the mesh size is reduced by 1/2, the error should become 1/4 in size. When the truncation error is proportional to Δ^n, the finite-difference scheme is said to have nth order of accuracy. The stencil for Eqs. (2.19) and (2.20) is arranged about x_j with symmetry in the positive and negative directions, as well as with symmetry in the corresponding coefficients (their magnitudes). Such schemes are called *central finite-difference schemes*. In particular, Eqs. (2.19) and (2.20) are both second-order accurate central-difference schemes.

If we use a five-point stencil, we can find the finite-difference formulas up to the fourth derivatives

$$f'_j = \frac{f_{j-2} - 8f_{j-1} + 8f_{j+1} - f_{j+2}}{12\Delta} + \frac{\Delta^4}{30} f_j^{(5)} + \mathcal{O}(\Delta^6) \qquad (2.21)$$

$$f''_j = \frac{-f_{j-2} + 16f_{j-1} - 30f_j + 16f_{j+1} - f_{j+2}}{12\Delta^2} + \frac{\Delta^4}{90} f_j^{(6)} + \mathcal{O}(\Delta^6) \qquad (2.22)$$

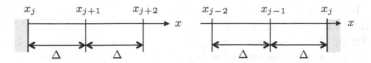

Fig. 2.4 One-sided difference stencils for the *left* and *right edges* of the computational domain

$$f_j^{(3)} = \frac{-f_{j-2} + 2f_{j-1} - 2_{j+1} + f_{j+2}}{2\Delta^3} - \frac{\Delta^2}{4} f_j^{(5)} + \mathcal{O}(\Delta^4) \tag{2.23}$$

$$f_j^{(4)} = \frac{f_{j-2} - 4f_{j-1} + 6f_j - 4_{j+1} + f_{j+2}}{\Delta^4} - \frac{\Delta^2}{6} f_j^{(6)} + \mathcal{O}(\Delta^4). \tag{2.24}$$

The five-point central-difference formulation results in fourth-order accuracy for the first and second derivatives and second-order accuracy for the third and fourth derivatives. Although we do not encounter high-order derivatives in the Navier–Stokes equations, the even derivates such as $f^{(4)}$ with a five-point stencil or $f^{(6)}$ with a seven-point stencil sometimes are used for introducing artificial viscosity or approximating filtering functions, which will be discussed later.

One-Sided Difference for Uniform Grid

When we require the finite-difference approximation for derivatives at the ends of a computational domain, we have to work with one-sided stencils,[2] as shown in Fig. 2.4 when x_j is located at the computational boundary. Given a one-sided n-point stencil, we can find the finite-difference formulas with $(n - 1)$-order accuracy for the first derivative and $(n - 2)$-order accuracy for the second derivative.

Let us present the derivation for the first derivative. We can approximate f_j' with the following two-point formulas:

$$f_j' = \frac{-f_{j-1} + f_j}{\Delta} + \frac{\Delta}{2} f_j'' + \mathcal{O}(\Delta^2) \tag{2.25}$$

$$f_j' = \frac{-f_j + f_{j+1}}{\Delta} - \frac{\Delta}{2} f_j'' + \mathcal{O}(\Delta^2) \tag{2.26}$$

based on Eqs. (2.16) and (2.15), respectively. Since the truncation errors are proportional to Δ, these schemes are first-order accurate. If we incorporate another spatial point into the stencil and eliminate the f_j'' term, we end up with the second-order accurate three-point stencil finite-difference schemes of

$$f_j' = \frac{f_{j-2} - 4f_{j-1} + 3f_j}{2\Delta} + \frac{\Delta^2}{3} f_j^{(3)} + \mathcal{O}(\Delta^3) \tag{2.27}$$

$$f_j' = \frac{-3f_j + 4f_{j+1} - f_{j+2}}{2\Delta} - \frac{\Delta^2}{3} f_j^{(3)} + \mathcal{O}(\Delta^3). \tag{2.28}$$

---

[2]Similar idea is used for evaluating temporal derivatives if data is available only from past time.

Next, let us consider cases where the finite-difference error can be problematic. Take a function $f = x^n$ ($n = 2, 3, 4, \cdots$) whose derivative is $f' = nx^{n-1} = 0$ at $x = 0$. Setting $\Delta = 1$ for ease of analysis, we get $f_j = j^n$. For $n = 2$, Eq. (2.26) returns $f_0' = 1$ which is incorrect, but Eq. (2.28) gives $f_0' = 0$ which is the correct solution. For $n = 3$, Eq. (2.28) provides $f_0' = -2$ while in reality $f' \geq 0$ (=0 only at $x = 0$). The numerical solution turns negative, which is opposite in sign for the gradient.

Furthermore, as we will observe in the next section, if grid-stretching is applied such that the adjacent grid is stretched three-times or larger in size near the wall boundary, one-sided three-point finite-difference schemes cannot correctly compute the coordinate transform coefficient [1]. For turbulent flow over a flat plate, turbulent energy $k \propto y^4$ and the Reynolds stress $\overline{u'v'} \propto y^3$. Therefore, it is important to use enough points in the stencil to track changes in the gradient of functions in the vicinity of the wall.

For the second derivatives, the one-sided three-point schemes are

$$f_j'' = \frac{f_{j-2} - 2f_{j-1} + f_j}{\Delta^2} + \Delta f_j^{(3)} + \mathcal{O}(\Delta^2) \tag{2.29}$$

$$f_j'' = \frac{f_j - 2f_{j+1} + f_{j+2}}{\Delta^2} - \Delta f_j^{(3)} + \mathcal{O}(\Delta^2). \tag{2.30}$$

Observe that these formulas are the same as Eq. (2.20) but just shifted by one grid. That means that $(f_{j-1} - 2f_j + f_{j+1})/\Delta^2$ is an approximation to the second derivative at any of the three points, x_{j-1}, x_j, and x_{j+1}. However, the properties of the truncation error is different for the different points at which the derivative is evaluated. We note in passing that to achieve second-order or higher accuracy for the second derivative with a one-sided stencil, we need at least four points for the finite-difference approximation.

Note that the errors for one-sided finite-difference schemes have the same order of accuracy as the central-difference scheme for the first derivative, while the order of accuracy is reduced by one for the second derivative.

Finite-Difference for Nonuniform Grid

There are two approaches to develop finite-difference schemes for nonuniform grids. The first approach is to employ Taylor series expansion, Eq. (2.12), in physical space, as shown in Fig. 2.5a. The second approach is to introduce a mapping

$$\frac{\partial f}{\partial x} = \frac{\partial \xi}{\partial x} \frac{\partial f}{\partial \xi} \tag{2.31}$$

such that the grid on the transformed variable ξ is spaced uniformly, as illustrated in Fig. 2.5b. In such case, we need to determine the coordinate transform $\partial \xi / \partial x$ and then construct the usual finite-difference scheme on the uniform grid for $\partial f / \partial \xi$. It is convenient to set the grid spacing in ξ to be 1, so that the coefficient $(\partial \xi / \partial x)$ represents the inverse of physical grid spacing.

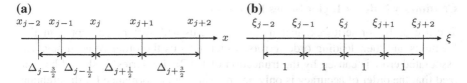

Fig. 2.5 Spatial discretization of a nonuniform grid (one-dimensional). **a** Nonuniform grid spacing in physical space. **b** Uniform grid spacing in transformed computational space

Alternatively, we can derive finite-difference schemes by directly differencing a function over a nonuniform grid in physical space. For a three-point difference method, the Taylor series expansions are

$$f_{j-1} = f_j - \Delta_{j-\frac{1}{2}} f_j' + \frac{\Delta_{j-\frac{1}{2}}^2}{2} f_j'' - \frac{\Delta_{j-\frac{1}{2}}^3}{6} f_j^{(3)} + \cdots \tag{2.32}$$

$$f_{j+1} = f_j + \Delta_{j+\frac{1}{2}} f_j' + \frac{\Delta_{j+\frac{1}{2}}^2}{2} f_j'' + \frac{\Delta_{j+\frac{1}{2}}^3}{6} f_j^{(3)} + \cdots \tag{2.33}$$

These expansions can be used to eliminate f_j'' to derive the finite-difference expression for the first derivative with second-order accuracy

$$f_j' = -\frac{\Delta_{j+\frac{1}{2}} f_{j-1}}{\Delta_{j-\frac{1}{2}}(\Delta_{j-\frac{1}{2}} + \Delta_{j+\frac{1}{2}})} - \frac{(\Delta_{j-\frac{1}{2}} - \Delta_{j+\frac{1}{2}}) f_j}{\Delta_{j-\frac{1}{2}} \Delta_{j+\frac{1}{2}}}$$
$$+ \frac{\Delta_{j-\frac{1}{2}} f_{j+1}}{\Delta_{j+\frac{1}{2}}(\Delta_{j-\frac{1}{2}} + \Delta_{j+\frac{1}{2}})} - \frac{\Delta_{j-\frac{1}{2}} \Delta_{j+\frac{1}{2}}}{6} f_j^{(3)} + \mathcal{O}(\Delta^3). \tag{2.34}$$

If we instead eliminate f_j', the expression for the second derivative can be found

$$f_j'' = -\frac{2 f_{j-1}}{\Delta_{j-\frac{1}{2}}(\Delta_{j-\frac{1}{2}} + \Delta_{j+\frac{1}{2}})} - \frac{f_j}{\Delta_{j-\frac{1}{2}} \Delta_{j+\frac{1}{2}}}$$
$$+ \frac{2 f_{j+1}}{\Delta_{j+\frac{1}{2}}(\Delta_{j-\frac{1}{2}} + \Delta_{j+\frac{1}{2}})} + \frac{\Delta_{j-\frac{1}{2}} - \Delta_{j+\frac{1}{2}}}{3} f_j^{(3)} + \mathcal{O}(\Delta^2). \tag{2.35}$$

Although this approximation (Eq. (2.35)) is first-order accurate, the dominant error term is proportional to $(\Delta_{j-\frac{1}{2}} - \Delta_{j+\frac{1}{2}})$. If the adjacent meshes are of similar size, the resulting first-order error is small and the scheme essentially retains second-order accuracy. For nonuniform grids, the weights derived for the finite-differences become asymmetric. However, if the stencil used is symmetric and if no numerical viscosity is added, some refer to these schemes as central differencing in a broader sense. Based on this argument, some may call Eqs. (2.34) and (2.35) as three-point central-difference schemes with second-order accuracy, even though x_j is not exactly located at the midpoint between $x_{j-\frac{1}{2}}$ and $x_{j+\frac{1}{2}}$. As expected, we recover Eqs. (2.19) and (2.20) by setting $\Delta_{j-\frac{1}{2}} = \Delta_{j+\frac{1}{2}} = \Delta$ in Eqs. (2.34) and (2.35).

Cautionary Note on Taylor Series Expansion

The use of Taylor series expansion can provide us with difference approximation formulas and their leading order errors, which tells us the order of accuracy. The associated error is caused by the truncation of the Taylor series. It should be realized that the order of accuracy is only one measure of performance for differencing schemes. The Taylor series expansion can be very effective for smooth functions. As we have discussed above, the function needs to be smooth over the differencing stencil for the error analysis to be meaningful. It should be noted that Taylor series approximation is not almighty as it can have issues for certain types of functions even if they are smooth.

2.3.2 Polynomial Approximation

In addition to the derivation based on Taylor series expansion, finite-difference methods can also be obtained by analytically differentiating a polynomial approximation $\tilde{f}(x)$ of a function $f(x)$

$$\tilde{f}(x) = a_0 + a_1 x + a_2 x^2 + a_3 x^3 + \cdots \qquad (2.36)$$

over a stencil. There are two ways we can approximate a function with a polynomial, as illustrated in Fig. 2.6. The first approach is *curve fitting* in which case the polynomial coefficients $\{a_0, a_1, a_2, \dots\}$ are determined to minimize the overall difference (residual) between the polynomial and the functional values at discrete spatial points as depicted in Fig. 2.6a. The least squares method is one of the common methods. The second approach is to have the polynomial pass through all discrete points, which is called *interpolation* and is illustrated in Fig. 2.6b. There are advantages and disadvantages to use one or the other. Locally speaking, interpolation satisfies $\tilde{f}(x_j) = f_j$ for all x_j. For global approximation of a function and its derivatives, curve fitting can perform better.

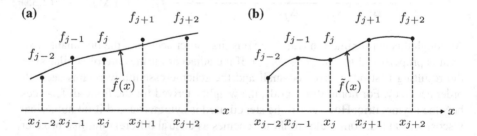

Fig. 2.6 Polynomial approximation using **a** curve fitting and **b** interpolation

Fig. 2.7 Parabolic
approximation using three
points over a uniform grid

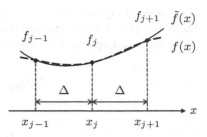

The finite-difference schemes derived from Taylor series expansion are equivalent
to methods based on analytical differentiation of the interpolating polynomials. As an
example, let us look at the three-point central-difference formula for a uniform grid,
as shown in Fig. 2.7. Since there are three degrees of freedom over this stencil, we
can use a quadratic polynomial $\tilde{f}(x) = a_0 + a_1 x + a_2 x^2$ for interpolation. Choosing
this quadratic function to pass through the three points of (x_{j-1}, f_{j-1}), (x_j, f_j), and
(x_{j+1}, f_{j+1}) to determine the coefficients a_0, a_1, and a_2, we find that

$$\tilde{f}(x) = f_j + \frac{-f_{j-1} + f_{j+1}}{2\Delta}(x - x_j) + \frac{f_{j-1} - 2f_j + f_{j-1}}{2\Delta^2}(x - x_j)^2. \quad (2.37)$$

Using the derivatives of $\tilde{f}(x)$ as approximations to the derivatives of $f(x)$, we now
have

$$f'(x) = \frac{-f_{j-1} + f_{j+1}}{2\Delta} + \frac{f_{j-1} - 2f_j + f_{j+1}}{\Delta^2}(x - x_j) \quad (2.38)$$

$$f''(x) = \frac{f_{j-1} - 2f_j + f_{j+1}}{\Delta^2}. \quad (2.39)$$

Setting $x = x_j$, we reproduce Eqs. (2.19) and (2.20), and substituting $x = x_{j\pm 1}$, we
obtain the formulas for the end points, Eqs. (2.27)–(2.30). Hence we observe that
the derivatives of the interpolating polynomial provide the same finite-difference
formulas that are derived using Taylor series expansions.

In general, we can consider an interpolating polynomial of degree N and use its
derivative to derive the finite-difference formula. The interpolating polynomial can
be written as

$$\tilde{f}(x) = \sum_{k=1}^{N} \phi_k(x) f_k \quad (2.40)$$

with

$$\phi_k(x) = \frac{\pi(x)}{(x - x_k)\pi'(x_k)},$$

$$\text{where } \pi(x) = \prod_{m=1}^{N}(x - x_m), \quad \pi'(x) = \sum_{n=1}^{N}\prod_{\substack{m=1 \\ m \neq n}}^{N}(x - x_m). \quad (2.41)$$

This interpolating formulation in Eq. (2.40) is called the Lagrange interpolation. The finite-difference approximation of derivatives can be found by analytically differentiating the Lagrange interpolation function

$$\tilde{f}'(x) = \sum_{k=1}^{N} \phi_k'(x) f_k,$$ (2.42)

where

$$\phi_k'(x) = \frac{(x - x_k)\pi'(x) - \pi(x)}{(x - x_k)^2 \pi'(x_k)}$$ (2.43)

is the polynomial coefficient. Evaluating these coefficients at discrete spatial points x_j and noticing that $\pi(x_j) = 0$, the coefficients are expressed as

$$\phi_k'(x_j) = \frac{\pi'(x_j)}{(x_j - x_k)\pi'(x_k)}.$$ (2.44)

This procedure can be made into a subroutine so that coefficients for any stencil can be determined automatically for a finite-difference scheme with an arbitrary order of accuracy.

We mention briefly that interpolation can suffer from spatial oscillation, known as the *Runge phenomenon*. For high-order Lagrange interpolation, spatial oscillation can be observed, such as the one in the example illustrated in Fig. 2.8. The example here considers a function $f(x) = 1/(x^2 + 1)$ with uniform grid spacing for $x \in [-5, 5]$. We show here the tenth-order interpolating function for 11 discrete grid points. Although the interpolating function goes through all grid points (x_j, f_j), the interpolating function exhibits significant spatial oscillations between the grid points. Higher order interpolating functions should not be used without care because oscillations can develop. We can avoid this problem by using lower-order interpolation, least squares curve fitting, piecewise interpolation, or changing the interpolation points.

The approximation described in this subsection should not be naively used for computational fluid dynamics. Nonetheless, it is important to understand that

Fig. 2.8 Runge phenomenon observed for the tenth-order Lagrange interpolation of $f(x) = 1/(x^2 + 1)$ with uniform grid

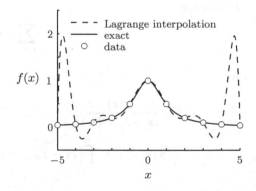

finite-difference approximations can be derived by taking the analytical derivative of approximating polynomials for considering the compatibility of finite difference with analytical differentiation (as it will be mentioned in Sect. 2.3.4).

2.3.3 Central Difference at Midpoints

In this section, we derive finite-difference schemes that are appropriate for the numerical computation of fluid flow (or conservation equations in general). Let us consider a finite-difference formula for a stencil using points of $j \pm \frac{1}{2}$, $j \pm \frac{3}{2}$, $\cdots$ centered about x_j, as shown in Fig. 2.9. Since the functional values at midpoints are unknown, we use the second-order accurate interpolation and the Taylor series expansion in the derivation of central-difference at midpoints.

Using the values from two adjacent points of $j \pm \frac{1}{2}$, the second-order accurate interpolation and finite-difference can be formulated as

$$f_j = \frac{f_{j-\frac{1}{2}} + f_{j+\frac{1}{2}}}{2} - \frac{\Delta^2}{8} f_j'' + \mathcal{O}(\Delta^4) \tag{2.45}$$

$$f_j' = \frac{-f_{j-\frac{1}{2}} + f_{j+\frac{1}{2}}}{2} - \frac{\Delta^2}{24} f_j^{(3)} + \mathcal{O}(\Delta^4). \tag{2.46}$$

If we widen the stencil to include two additional points of $j \pm \frac{3}{2}$, we obtain the fourth-order accurate interpolation and first-derivative approximation, as well as the second-order accurate second and third-derivative approximations

$$f_j = \frac{-f_{j-\frac{3}{2}} + 9f_{j-\frac{1}{2}} + 9f_{j+\frac{1}{2}} - f_{j+\frac{3}{2}}}{16} + \frac{3\Delta^4}{128} f_j^{(4)} + \mathcal{O}(\Delta^6) \tag{2.47}$$

$$f_j' = \frac{f_{j-\frac{3}{2}} - 27f_{j-\frac{1}{2}} + 27f_{j+\frac{1}{2}} - f_{j+\frac{3}{2}}}{24\Delta} + \frac{3\Delta^4}{640} f_j^{(5)} + \mathcal{O}(\Delta^6) \tag{2.48}$$

$$f_j'' = \frac{f_{j-\frac{3}{2}} - f_{j-\frac{1}{2}} - f_{j+\frac{1}{2}} + f_{j+\frac{3}{2}}}{2\Delta^2} + \frac{5\Delta^2}{24} f_j^{(4)} + \mathcal{O}(\Delta^4) \tag{2.49}$$

$$f_j^{(3)} = \frac{-f_{j-\frac{3}{2}} + 3f_{j-\frac{1}{2}} - 3f_{j+\frac{1}{2}} + f_{j+\frac{3}{2}}}{\Delta^3} - \frac{\Delta^2}{8} f_j^{(5)} + \mathcal{O}(\Delta^4). \tag{2.50}$$

While the third derivative is not generally used in fluid flow simulations since it does not appear in the Navier–Stokes equations, Eq. (2.50) can be used in an

Fig. 2.9 Central difference at midpoint on uniform grid (one-dimensional)

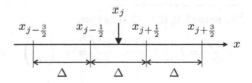

upwinding formulation, which will be discussed later. The use of a second-derivative approximation above in Eq. (2.49) should be avoided.

For the second derivative, we should take the first-derivative finite difference of the first-derivative difference at points $j \pm \frac{1}{2}$ and $j \pm \frac{3}{2}$. By taking the finite difference twice in this manner, we obtain for the second-order accurate formulation

$$f_j'' = \frac{-f_{j-\frac{1}{2}}' + f_{j+\frac{1}{2}}'}{\Delta} = \frac{f_{j-1} - 2f_j + f_{j+1}}{\Delta^2}, \tag{2.51}$$

which matches the second-order second-derivative difference formula, Eq. (2.20). For the fourth-order approximation, we obtain

$$\begin{aligned} f_j'' &= \frac{f_{j-\frac{3}{2}}' - 27f_{j-\frac{1}{2}}' + 27f_{j+\frac{1}{2}}' - f_{j+\frac{3}{2}}'}{24\Delta} \\ &= \frac{f_{j-3} - 54f_{j-2} + 783f_{j-1} - 1460f_j + 783f_{j+1} - 54f_{j+2} + f_{j+3}}{(24\Delta)^2}, \end{aligned} \tag{2.52}$$

which is based on a wider seven-point stencil and is different from Eqs. (2.22) or (2.49). Although the use of a wider stencil may appear cumbersome, this scheme satisfies Eq. (2.2) or $(f')' = f''$ in a discrete sense and constitutes a compatible differencing scheme. Additional details will be offered next in Sect. 2.3.4.

For nonuniform grids, there are two techniques for deriving the finite-difference schemes as mentioned in Sect. 2.3.1. In the present discussion, we choose to map the nonuniform grid onto a uniform computation grid and construct a central-difference approximation. Further details on the implementation of nonuniform grid treatment are offered in Chaps. 3 and 4.

As one may have noticed, the sum of coefficients for interpolation is one, and the sum of coefficients for finite difference is zero. If round-off errors from interpolation and finite differencing creep into numerical calculations and cause problems, we can take advantage of the properties of the aforementioned sums. To avoid problems related to such round-off errors, we can set one of the coefficients to be $1 -$ (sum of other interpolation coefficients) or $0 -$ (sum of other differencing coefficients).

2.3.4   Compatibility of Finite Differencing

We mentioned that differentiation rules such as Eqs. (2.1)–(2.3) should be satisfied both in the continuous and discrete settings.[3] For the finite-difference

---

[3]We restate Eqs. (2.1)–(2.3) for clarity

$$\frac{\partial(fg)}{\partial x} = f\frac{\partial g}{\partial x} + \frac{\partial f}{\partial x}g, \quad \frac{\partial^2 f}{\partial x^2} = \frac{\partial}{\partial x}\left(\frac{\partial f}{\partial x}\right), \quad \text{and} \quad \frac{\partial^2 f}{\partial x \partial y} = \frac{\partial}{\partial x}\left(\frac{\partial f}{\partial y}\right) = \frac{\partial}{\partial y}\left(\frac{\partial f}{\partial x}\right).$$

schemes derived above, let us examine whether the analytical derivative relations are satisfied discretely.

Using the second-order differencing from Eq. (2.19) for the differentiation of a product of two functions f and g shown in Eq. (2.1), we find that

$$\frac{-(fg)_{j-1} + (fg)_{j+1}}{2\Delta} \neq f_j \frac{-g_{j-1} + g_{j+1}}{2\Delta} + \frac{-f_{j-1} + f_{j+1}}{2\Delta} g_j, \quad (2.53)$$

which tells us that the product rule does not hold discretely for the chosen differencing scheme. For the same differencing scheme from Eq. (2.19), let us also examine if Eq. (2.2) holds discretely when we apply the first-derivative finite differencing twice. We observe that

$$\frac{1}{2\Delta}\left(-\frac{-f_{j-2} + f_j}{2\Delta} + \frac{-f_j + f_{j+2}}{2\Delta}\right) = \frac{f_{j-2} - 2f_j + f_{j+2}}{4\Delta^2}$$

$$\neq \frac{f_{j-1} - 2f_j + f_{j+1}}{\Delta^2}, \quad (2.54)$$

which is not equivalent to Eq. (2.20) that directly derived the finite-difference scheme for the second derivative. Thus, it can be said that the first-derivative finite-difference scheme about x_j is not compatible if the stencil is based on $j \pm 1, j \pm 2, \cdots$.

Revisiting this issue with the interpolation and difference operations from Eqs. (2.45) and (2.46), respectively, the finite-difference approximation of the product becomes

$$\left[\frac{\partial(fg)}{\partial x}\right]_j = \frac{1}{2}\left\{\left[\frac{\partial(fg)}{\partial x}\right]_{j-\frac{1}{2}} + \left[\frac{\partial(fg)}{\partial x}\right]_{j+\frac{1}{2}}\right\}$$

$$= \frac{1}{2}\left(\frac{-f_{j-1}g_{j-1} + f_j g_j}{\Delta} + \frac{-f_j g_j + f_{j+1}g_{j+1}}{\Delta}\right), \quad (2.55)$$

which agrees with

$$\left[f\frac{\partial g}{\partial x} + \frac{\partial f}{\partial x}g\right]_j = \frac{1}{2}\left(\left[f\frac{\partial g}{\partial x} + \frac{\partial f}{\partial x}g\right]_{j-\frac{1}{2}} + \left[f\frac{\partial g}{\partial x} + \frac{\partial f}{\partial x}g\right]_{j+\frac{1}{2}}\right)$$

$$= \frac{1}{2}\left[\left(\frac{f_{j-1} + f_j}{2}\frac{-g_{j-1} + g_j}{\Delta} + \frac{-f_{j-1} + f_j}{\Delta}\frac{g_{j-1} + g_j}{2}\right) \quad (2.56)\right.$$

$$\left. + \left(\frac{f_j + f_{j+1}}{2}\frac{-g_j + g_{j+1}}{\Delta} + \frac{-f_j + f_{j+1}}{\Delta}\frac{g_j + g_{j+1}}{2}\right)\right].$$

The above discretization scheme exhibits compatibility for differentiation in discrete sense. We also note that Eq. (2.55) is equivalent to the left-hand side of Eq. (2.53) as shown below

$$\left[\frac{\partial(fg)}{\partial x}\right]_j = \frac{1}{\Delta}\left(-[fg]_{j-\frac{1}{2}} + [fg]_{j+\frac{1}{2}}\right)$$

$$= \frac{1}{\Delta}\left(-\frac{f_{j-1}g_{j-1} + f_j g_j}{2} + \frac{f_j g_j + f_{j+1}g_{j+1}}{2}\right). \tag{2.57}$$

However, it should be observed that, for Eqs. (2.55) and (2.56), we are not directly performing finite difference about x_j but rather interpolating the difference approximations at $x_{j\pm\frac{1}{2}}$ to determine the derivative at x_j. As discussed in Sect. 2.3.3, finite-difference formulas can be derived using polynomial approximations. From that point of view, finite-difference schemes can be thought of as being the analytical derivatives of the polynomial approximation. Therefore, if we employ the same stencil for the same point with polynomial approximation, the differentiation rules should hold discretely. As a general rule, difference schemes that satisfy $(f')' = f''$, such as Eqs. (2.51) and (2.52), should be used in practice. The satisfaction of compatibility for different spatial directions (e.g., Eq. (2.3)) will be examined in Sect. 4.1.1.

2.3.5 Spatial Resolution

The spatial resolution of finite-difference methods is determined by the grid spacing and the schemes themselves. Any physical phenomena with structures having length scales smaller than the grid size cannot be captured. For scales larger than the grid spacing, spatial resolution varies for different wavelengths. Waves with larger wavelengths can be represented smoothly compared to waves with shorter wavelengths on the same grid, as shown in Fig. 2.10.

The number of waves between 0 to 2π is called the wave number and has a dimension of 1/length. For a given grid size Δ, the smallest scale of fluctuation (smallest wavelength) that can be resolved is 2Δ as shown in Fig. 2.10a. This means that for a grid spacing with Δ, the maximum wave number that can be analyzed would be $k_c = \pi/\Delta$, which is referred to as the cutoff wave number. Below, we examine the consequence of performing finite-difference operation in wave space through Fourier analysis and study how accurately finite difference can represent the derivative operation. For details on Fourier transform, readers should refer to Appendix B.

Fourier Analysis of Finite Difference

Suppose we have a smooth periodic function $f(x)$ with period 2π in one dimension. Expressing the function $f(x)$ with Fourier series

$$f(x) = \sum_{k=0}^{\infty} A_k \exp(ikx), \tag{2.58}$$

Fig. 2.10 Wavelengths that
can be represented for a grid
spacing of Δ. **a** Wavelength
2Δ. **b** Wavelength 4Δ.
c Wavelength 8Δ

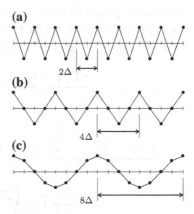

the derivative of the function can be expressed as

$$f'(x) = \sum_{k=1}^{\infty} ikA_k \exp(ikx), \qquad (2.59)$$

where $i = \sqrt{-1}$. We can hence represent differentiation as

$$\mathcal{F}(f') = ik\mathcal{F}(f), \qquad (2.60)$$

where $\mathcal{F}$ denotes the Fourier transform. This indicates that differentiation in wave space is equivalent to the multiplication of the Fourier transformed function and the wave number.

Now, let us examine finite-difference operations in wave space and compare them with the exact expression given by Eq. (2.59). For finite differencing over a uniform grid with spacing $\Delta = 2\pi/N$, we utilize the relations below based on Euler's formula $e^{\pm i\theta} = \cos\theta \pm i\sin\theta$

$$-f_{j-m} + f_{j+m} = \sum_{k=1}^{\infty} 2i \sin(mk\Delta) A_k \exp(ik\Delta j) \qquad (2.61)$$

$$f_{j-m} + f_{j+m} = \sum_{k=0}^{\infty} 2 \cos(mk\Delta) A_k \exp(ik\Delta j). \qquad (2.62)$$

For analyzing the first-derivative finite-difference schemes in wave space, we can use Eq. (2.61) and find that the two- and four-point difference formulas, Eqs. (2.46) and (2.48), respectively, can be expressed as

$$\frac{-f_{j-\frac{1}{2}} + f_{j+\frac{1}{2}}}{\Delta} = \sum_{k=1}^{\infty} \frac{2i}{\Delta} \sin \frac{k\Delta}{2} A_k \exp(ik\Delta j) \tag{2.63}$$

$$\frac{f_{j-\frac{3}{2}} - 27 f_{j-\frac{1}{2}} + 27 f_{j+\frac{1}{2}} - f_{j+\frac{3}{2}}}{\Delta}$$

$$= \sum_{k=1}^{\infty} \frac{i}{12\Delta} \left(27 \sin \frac{k\Delta}{2} - \sin \frac{3k\Delta}{2} \right) A_k \exp(ik\Delta j). \tag{2.64}$$

Comparing Eqs. (2.63) and (2.64) with Eq. (2.59), we notice that instead of multiplying the wave number k for analytical differentiation in wave space, these finite-difference methods multiply

$$K_{(2)} = \frac{2}{\Delta} \sin \frac{k\Delta}{2} \tag{2.65}$$

$$K_{(4)} = \frac{1}{12\Delta} \left(27 \sin \frac{k\Delta}{2} - \sin \frac{3k\Delta}{2} \right) \tag{2.66}$$

to the Fourier coefficients in wave space. For central-difference schemes with sixth or higher order of accuracy, same derivations can be carried out. The variable $K_{(m)}$ is called the *modified wave number* of the m-th order accurate finite-difference scheme.

Following similar procedures, we can further analyze the finite-difference schemes for second derivatives. Analytical differentiation using Fourier series is

$$f''(x) = \sum_{k=1}^{\infty} -k^2 A_k \exp(ikx), \tag{2.67}$$

which means that the second derivative can be computed by multiplying $-k^2$ to each wave number component. In other words, $\mathcal{F}(f'') = -k^2 \mathcal{F}(f)$. For the difference schemes in Eqs. (2.51), (2.52), and (2.62) can be utilized to find the modified squared wave number that should correspond to k^2. For the two schemes, the modified squared wave numbers are

$$K_{(2)}^2 = \frac{2(1 - \cos k\Delta)}{\Delta^2} \tag{2.68}$$

$$K_{(4)}^2 = \frac{2(730 - 783 \cos k\Delta + 54 \cos 2k\Delta - \cos 3k\Delta)}{24^2 \Delta^2}, \tag{2.69}$$

respectively.

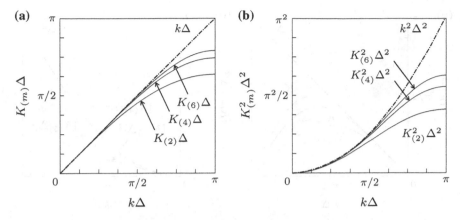

Fig. 2.11 Modified wave numbers for central-difference schemes about x_j using $j \pm 1/2$, $j \pm 3/2$, $\cdots$ ($N = 64$, $\Delta = 2\pi/N$, $k_c = 32$). **a** $K_{(m)}\Delta$. **b** $K_{(m)}^2 \Delta^2 (=[K_{(m)}\Delta]^2)$

We compare these modified wave numbers with the wave numbers from the exact analysis in Fig. 2.11. For finite-difference schemes, high wave number components of the derivatives appear as if they have been dissipated (filtered out) compared to the analytically computed derivatives. Such filtering effects to reduce the effective resolution that we can achieve. If the order of accuracy of the central-difference scheme is increased, we can attain enhanced spatial resolution for high-frequency components using the same grid size. Accuracy improvement becomes relatively smaller as we increase the order of accuracy from fourth order to sixth order in comparison to what we achieve from second order to fourth order.

The compatibility of finite-difference schemes discussed in Sect. 2.3.3 can also be examined with Fourier analysis. It should be noted that discretely satisfying $f'' = (f')'$ translates to satisfying $K_{(m)}^2 = [K_{(m)}]^2$ in discrete wave space. Such relation is satisfied for example with the Eqs. (2.51) and (2.52).

For comparison, let us also perform the Fourier analysis of the first-derivative finite-difference schemes about point x_j that use stencils of $j \pm 1$, $j \pm 2$, $\cdots$ presented in Sect. 2.3.1. For the finite-difference schemes, Eqs. (2.19) and (2.21), the modified wave numbers are

$$K_{(2)} = \frac{\sin k\Delta}{\Delta} \tag{2.70}$$

$$K_{(4)} = \frac{8\sin k\Delta - \sin 2k\Delta}{6\Delta}, \tag{2.71}$$

respectively, and are plotted in Fig. 2.12a. The modified wave numbers for the second derivatives with Eqs. (2.20) and (2.22) are

$$K_{(2)}^2 = \frac{2(1 - \cos k\Delta)}{\Delta^2} \tag{2.72}$$

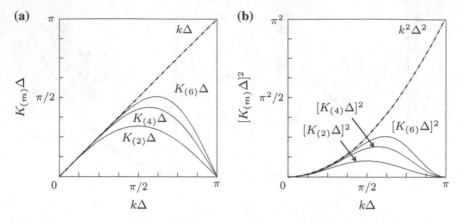

Fig. 2.12 Modified wave numbers for central-difference schemes about x_j using $j \pm 1$, $j \pm 2$, $\cdots$ ($N = 64$, $\Delta = 2\pi/N$, $k_c = 32$). **a** $K_{(m)}\Delta$. **b** $[K_{(m)}\Delta]^2 (\neq K_{(m)}^2 \Delta^2)$

$$K_{(4)}^2 = \frac{15 - 16\cos k\Delta + \cos 2k\Delta}{6\Delta^2}, \qquad (2.73)$$

respectively, which are not equal to the square of the modified wave numbers shown in Fig. 2.12b. The finite-difference schemes derived in Sect. 2.3.1 result in modified wave numbers of $K_{(m)}^2 \neq [K_{(m)}]^2$, implying that $f'' = (f')'$ does not hold discretely.

Based on the above analysis, we can refine the mesh or increase the order of accuracy of the finite-difference schemes to increase the spatial resolution. Widening the finite-difference stencil can increase the amount of memory allocation required on a computer, and increasing the order of accuracy increases the amount of computation at each point. Selecting which approach to follow for the pursuit of better numerical solution is dependent on the problem at hand and on the available computational resource. While high-order accurate schemes in theory can resolve smaller scales, it should be taken with care that such methods are often prone to numerical instabilities.

2.3.6 Behavior of Discretization Error

There are a variety of finite-difference approximations that one can derive as we have seen above. The determination of which finite-difference scheme to select over another is driven by the behavior of their errors and the order of accuracy of the schemes. Here, we examine the error behavior for a few finite-difference methods.

As an example, let us consider the advection equation

$$\frac{\partial f}{\partial t} + c\frac{\partial f}{\partial x} = 0 \qquad (2.74)$$

for which the solution translates in the x-direction with velocity c, as illustrated in Fig. 2.1.

Now, let us take a look at the spatial discretization of the advective term. For ease of analysis, we consider a uniform mesh. Using the central-difference approximation

$$c \left. \frac{\partial f}{\partial x} \right|_j = c \frac{-\bar{f}_{j-\frac{1}{2}} + \bar{f}_{j+\frac{1}{2}}}{\Delta} - c \frac{\Delta^2}{24} f_j^{(3)} + c \mathcal{O}(\Delta^3), \tag{2.75}$$

where we will make different choices for evaluating $\bar{f}$ in discussions to follow.

If we compute $\bar{f}_{j-\frac{1}{2}}$ and $\bar{f}_{j+\frac{1}{2}}$ using interpolation with symmetric stencils about $j \pm \frac{1}{2}$:

$$\bar{f}_{j-\frac{1}{2}} = \frac{f_{j-1} + f_j}{2} - \frac{\Delta^2}{8} f''_{j-\frac{1}{2}} + \mathcal{O}(\Delta^4),$$

$$\bar{f}_{j+\frac{1}{2}} = \frac{f_j + f_{j+1}}{2} - \frac{\Delta^2}{8} f''_{j+\frac{1}{2}} + \mathcal{O}(\Delta^4), \tag{2.76}$$

we obtain a second-order accurate central-difference scheme for the advective term

$$c \left. \frac{\partial f}{\partial x} \right|_j = c \frac{-f_{j-1} + f_{j+1}}{2\Delta} - c \frac{\Delta^2}{6} f_j^{(3)} + c \mathcal{O}(\Delta^3). \tag{2.77}$$

If we instead consider using upstream values for evaluating $\bar{f}_{j\pm\frac{1}{2}}$,

$$\bar{f}_{j-\frac{1}{2}} = f_{j-1} + \frac{\Delta}{2} f'_{j-\frac{1}{2}} + \mathcal{O}(\Delta^2),$$

$$\bar{f}_{j+\frac{1}{2}} = f_j + \frac{\Delta}{2} f'_{j+\frac{1}{2}} + \mathcal{O}(\Delta^2), \tag{2.78}$$

the advective term is expressed as a first-order accurate one-sided discretization

$$c \left. \frac{\partial f}{\partial x} \right|_j = c \frac{-f_{j-1} + f_j}{\Delta} + c \frac{\Delta}{2} f''_j + c \mathcal{O}(\Delta^2). \tag{2.79}$$

It is also possible to use downstream values to derive a first-order accurate one-sided formula of

$$c \left. \frac{\partial f}{\partial x} \right|_j = c \frac{-f_j + f_{j+1}}{\Delta} - c \frac{\Delta}{2} f''_j + c \mathcal{O}(\Delta^2). \tag{2.80}$$

For $c > 0$, Eq. (2.79) uses information only from points upstream of x_j. Such scheme is referred to as the *upstream* or *upwind finite-difference scheme*. On the other hand Eq. (2.80) uses information only from points downstream of x_j and is referred to as the *downwind finite-difference scheme*. Downwind schemes are not generally used in simulations of fluid flows.

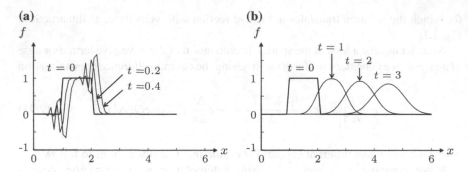

Fig. 2.13 Examples of error appearance due to differencing of the advection equation. **a** Dispersive error from central difference. **b** Diffusive error from upwind difference

For the central-difference formula, Eq. (2.77), the dominant truncation error contains an odd derivative (i.e., third derivative). In such a case, the error generally behaves in a dispersive manner. For the upwind difference formula, Eq. (2.79), the dominant truncation error contains an even derivative (i.e., second derivative), which exhibits a diffusive behavior. When using information from downstream, as it is the case for Eq. (2.80), the error behaves diffusively but with negative diffusivity. In this case, the gradients in the solution becomes steeper as time advances, which eventually causes the solution to blow up from numerical instability. While the exact solution to the advection equation, Eq. (2.74), maintains the solution profile under translation, central differencing introduces spatial oscillations leading to numerical instability and upwind differencing diffuses the solution over time. These error behaviors are illustrated in Fig. 2.13.

Let us further examine the effect of using the upwind finite-difference method

$$\left.\frac{\partial f}{\partial t}\right|_j + c\frac{-f_{j-1} + f_j}{\Delta} = 0 \tag{2.81}$$

by substituting the Taylor series expansion, Eq. (2.15), into f_{j-1}. We accordingly find that Eq. (2.81) becomes

$$\frac{\partial f}{\partial t} + c\frac{\partial f}{\partial x} - \frac{c\Delta}{2}\frac{\partial^2 f}{\partial x^2} + c\mathcal{O}(\Delta^2) = 0. \tag{2.82}$$

Note that we started with a pure advection equation, Eq. (2.74), and did not include any diffusive effects. By employing the upwind difference, Eq. (2.82) now contains a diffusive term $\frac{c\Delta}{2}\frac{\partial^2 f}{\partial x^2}$ with diffusivity of $c\Delta/2$. In contrast to physical diffusivity caused by viscous diffusion, such numerical effect due to truncation error is referred to as *numerical diffusion*. When this effect appears in the momentum equation, numerical diffusivity is called *numerical viscosity*. With the use of the upwinding difference method, the solution profile becomes smoothed out over time due to numerical diffusion, as shown in Fig. 2.13b (see also Sects. 3.5.3 and 4.7.2 for use of upwind difference schemes).

We can also consider the case with viscous diffusion present

$$\frac{\partial f}{\partial t} + c\frac{\partial f}{\partial x} - a\frac{\partial^2 f}{\partial x^2} = 0. \tag{2.83}$$

Discretizing the second term on the left-hand side with first-order accurate upwind difference (and the third term with a second-order or higher scheme), what essentially is solved becomes

$$\frac{\partial f}{\partial t} + c\frac{\partial f}{\partial x} - \left(a + \frac{c\Delta}{2}\right)\frac{\partial^2 f}{\partial x^2} + c\mathcal{O}(\Delta^2) = 0. \tag{2.84}$$

For cases where the advection speed c and the grid size Δ are large in their magnitudes, it is possible to have $c\Delta/2 \gg a$ in which case the numerical diffusion overwhelms physical diffusion. If the contribution from numerical diffusion is large in the momentum equation, this can cause the solution to become insensitive to viscous diffusion.

In the above discussion, we have focused on the influence of the second and third derivatives from the truncation error on the numerical solution of the advection equation. Let us further generalize the analysis and consider the influence of even and odd derivatives on the behavior of errors. For the ease of analysis, we consider the solution to be periodic over $x \in [0, 2\pi]$. First, the solution to original advection equation can be expressed in a separable form as

$$f(x, t) = \sum_{k=0}^{\infty} \widetilde{f}_k(t) \exp(ikx), \tag{2.85}$$

where we denote the wave number as k. Substituting this expression in the advection equation $\frac{\partial f}{\partial t} + c\frac{\partial f}{\partial x} = 0$, we find

$$\frac{\partial \widetilde{f}_k}{\partial t} = -ikc\widetilde{f}_k, \tag{2.86}$$

for each wave number. Solving the above differential equation gives the exact solution of

$$f(x, t) = \sum_{k=0}^{\infty} \widetilde{f}_k(0) \exp[ik(x - ct)]. \tag{2.87}$$

Here, $\widetilde{f}_k(0)$ can be determined from the initial condition.

Next, let us examine the influence of truncation error for the advection equation. In reality, the discretized advection equation results in

$$\frac{\partial f}{\partial t} + c\frac{\partial f}{\partial x} = \alpha_2\frac{\partial^2 f}{\partial x^2} + \alpha_3\frac{\partial^3 f}{\partial x^3} + \alpha_4\frac{\partial^4 f}{\partial x^4} + \cdots, \tag{2.88}$$

where $\alpha_2, \alpha_3, \alpha_4, \ldots$ are dependent on the chosen finite-difference schemes and the grid spacing Δ. Substituting the separable form of f from Eq. (2.85) into Eq. (2.88), for each wave number, we find

$$\frac{\partial \widetilde{f_k}}{\partial t} = \left[-ik(c + k^2\alpha_3 - k^4\alpha_5 + \cdots) + (-k^2\alpha_2 + k^4\alpha_4 - \cdots) \right] \widetilde{f_k}. \quad (2.89)$$

We find that the solution to the above equation is

$$f(x,t) = \sum_{k=0}^{\infty} \widetilde{f_k}(0) \exp\left[ik\left\{x - (c + k^2\alpha_3 - k^4\alpha_5 + \cdots)t\right\}\right]$$
$$\times \exp\left[(-k^2\alpha_2 + k^4\alpha_4 - \cdots)t\right]. \quad (2.90)$$

Comparing Eqs. (2.87) and (2.90), we notice that extra terms appear in the numerical solution. We observe that the advection speed c is modified to be $(c + k^2\alpha_3 - k^4\alpha_5 + \cdots)$ with truncation error. The error is generated when we have nonzero α_j for odd j. Hence, we see that the phase speed is altered by the inclusion of the odd-derivative terms in the truncation error. The profile of a propagating wave can be influenced by this wave number-dependent phase-speed error, leading to the wave becoming dispersive. Furthermore, $(-k^2\alpha_2 + k^4\alpha_4 - \cdots)$ is added as a component to the solution which alters the growth or decay of the solution. For $\alpha_2 > 0$, we introduce numerical diffusion from the second derivative and likewise for $\alpha_4 < 0$ from the fourth derivative. Numerical diffusion can be added to the governing equation by having even-derivative terms.

When we conduct fundamental research using numerical simulation, the aforementioned numerical errors should be kept as low as possible. Industrial or commercial software often incorporates upwind difference schemes intentionally to allow for simulations of a wide variety of flows in a numerically stable manner. When using these softwares, we must understand that numerical viscosity may influence the outcome of the solution. Simulations of fluid flow cannot be trusted if the discretization error displays unphysical behavior. The question that we should pose is: How do we verify our numerical solver? What needs to be addressed is how the numerical error possibly affects the flow physics. It is difficult to predict how the solution may be influenced by the error a priori, but we can focus on the fact that error is driven by how fine the grid spacing is, and examine how the error responds to the change in grid resolution. To state that a method has been verified (converged), the result must be shown not to change when the grid size is altered. We must at least be able to display the trend of error when the grid is refined.

The error discussed here is based on the truncation error from spatial discretization and does not consider the error from time integration (stepping) of the differential equations. Time stepping methods and the associated temporal discretization errors are discussed below.

2.4 Time Stepping Methods

The advection-diffusion equation is a time evolution equation. The time rate of change of the function f can be determined once we evaluate the advective and diffusive terms. Time derivative of f can be numerically integrated in time to find the solution. In this section, we present the time stepping (integration) methods for an example of the advection-diffusion equation written as

$$\frac{\partial f}{\partial t} = g(t, f). \tag{2.91}$$

Here, we represent the spatial derivative terms by g (advective and diffusive terms), the time step by Δt, and the discrete time levels by $t_n = n\Delta t$. The number of time steps is denoted by n and we use a superscript on f to indicate the time level (i.e., $f^n = f(t_n)$). In what follows, we assume that the flow field up to time t_n is known. In other words, $f^n, f^{n-1}, f^{n-2}, \cdots$ and $g^n, g^{n-1}, g^{n-2}, \cdots$ are taken to be known a priori for determining f^{n+1}. Details on time stepping schemes can also be found in Lambert [2].

2.4.1 Single-Step Methods

First, let us discuss single-step time stepping methods that use only data from t_n to compute the solution at t_{n+1}. Generalizing the trapezoidal rule for integrating Eq. (2.91) in time, we have

$$\frac{f^{n+1} - f^n}{\Delta t} = (1 - \alpha)g^n + \alpha g^{n+1}, \tag{2.92}$$

where $0 \leq \alpha \leq 1$. For $\alpha = 0$, the next solution f^{n+1} is determined from the known f^n and g^n with $f^{n+1} = f^n + \Delta g^n$. Such method that computes the solution at the next time level using information from known time levels is called an *explicit method*. For $0 < \alpha \leq 1$, finite-difference discretization of the differential equation leads to an equation to be solved for f^{n+1}. Such method that computes the solution at the next time level using information from both known and future time levels is called an *implicit method*.

When we set $\alpha = 0$ in Eq. (2.92), we obtain the *explicit Euler's method*

$$f^{n+1} = f^n + \Delta t g^n. \tag{2.93}$$

Using $\alpha = 1$ in Eq. (2.92), we obtain the *implicit Euler's method*

$$f^{n+1} = f^n + \Delta t g^{n+1}. \tag{2.94}$$

Both of these Euler's methods, Eqs. (2.93) and (2.94), are first-order accurate in time. Setting $\alpha = 1/2$, we derive the *Crank–Nicolson method*

$$f^{n+1} = f^n + \Delta t \frac{g^n + g^{n+1}}{2}, \qquad (2.95)$$

which is second-order accurate in time. The parameter α is generally set to 0, 1/2, or 1. It is rare to find other values of α being used.

Consider an example of implicitly solving the diffusion equation, Eq. (2.8), with second-order accurate central differencing. We can write

$$\frac{f_j^{n+1}}{\Delta t} - \alpha a \frac{f_{j-1}^{n+1} - 2f_j^{n+1} + f_{j+1}^{n+1}}{\Delta^2} = \frac{f_j^n}{\Delta t} + (1 - \alpha)a \frac{f_{j-1}^n - 2f_j^n + f_{j+1}^n}{\Delta^2}, \quad (2.96)$$

which results in an algebraic equation with a tridiagonal coefficient matrix on the left-hand side for f^{n+1}. Implicit methods require solving an equation such as the one above, which can be linear or nonlinear depending on the form of g. Solving such system of equations adds computational time per time step. As discussed later, explicit methods can be prone to numerical instability and often needs the time step to be set small, which in turn increases the number time steps for computations. Whether an explicit or implicit scheme is better suited is dependent on the governing equation and how the mesh is created. For terms found in equations of fluid dynamics, linear terms are suited for implicit schemes. On the other hand, the nonlinear advective term, which is often a source of instability, is not suited for implicit treatment due to the added cost of solving the resulting nonlinear equation.

Predictor-Corrector Methods

There are methods that possess the ease of use of explicit methods and the stability property similar to that of implicit methods. These methods are called *predictor-corrector methods* and achieve the benefits by introducing intermediate steps within a single time advancement to carry out prediction and correction of the time-integrated solution.

For example, we can use the explicit Euler's method as a prediction and use the predicted solution $\tilde{f}$ as a correction to compute $\tilde{g} = g(t_{n+1}, \tilde{f})$ for the right-hand side of Eq. (2.94)

$$\left.\begin{array}{l} \tilde{f} = f^n + \Delta t g^n \\ f^{n+1} = f^n + \Delta t \tilde{g} \end{array}\right\} \qquad (2.97)$$

which is known as Matsuno's method in the field of weather forecasting. One can also construct a predictor–corrector method that is similar to the Crank–Nicolson method

$$\left.\begin{array}{l} \tilde{f} = f^n + \Delta t g^n \\ f^{n+1} = f^n + \Delta t \dfrac{g^n + \tilde{g}}{2} \end{array}\right\} \qquad (2.98)$$

which is called Heun's method. Since the form of Eq. (2.98) requires memory allocation for g^n, one can use the equivalent form of

$$f^{n+1} = \frac{1}{2}(f^n + \tilde{f} + \Delta t \tilde{g}) \qquad (2.99)$$

that uses less memory allocation.

Runge–Kutta Methods

We can further extend upon the above concept to introduce multiple predictions per time step in an explicit fashion, which is the basis for a class of time stepping schemes, called the *Runge–Kutta methods*. For example, the two-step Runge–Kutta method is

$$\left. \begin{aligned} f^{(1)} &= f^n + \frac{\Delta t}{2} g^n \\ f^{n+1} = f^{(2)} &= f^n + \Delta t g^{(1)} \end{aligned} \right\} \qquad (2.100)$$

where m in $f^{(m)}$ denotes the level of the intermediate step. This scheme in Eq. (2.100) has second-order temporal accuracy.

The second-order time integration methods introduced up until now use some form of approximation to the average value between time level t^n and t^{n+1} of the right-hand side of the evolution equation. We can summarize them as

- Crank–Nicolson method $\qquad f^{n+1} = f^n + \Delta t \dfrac{g^n + g^{n+1}}{2} \qquad (2.101)$

- Heun's method $\qquad f^{n+1} = f^n + \Delta t \dfrac{g^n + \tilde{g}}{2} \qquad (2.102)$

- 2nd-order Runge–Kutta method $\quad f^{n+1} = f^n + \Delta t g^{n+1/2} \qquad (2.103)$

where $g^{n+1/2} = g(t_n + \frac{\Delta t}{2}, f_n + \frac{\Delta t}{2} g^n)$.

Another widely used time stepping method is the classic *four-step Runge–Kutta method*

$$\left. \begin{aligned} f^{(1)} &= f^n + \frac{\Delta t}{2} g^{(1)} \\ f^{(2)} &= f^n + \frac{\Delta t}{2} g^{(2)} \\ f^{(3)} &= f^n + \Delta t g^{(2)} \\ f^{n+1} = f^{(4)} &= f^n + \Delta t \frac{g^n + 2g^{(1)} + 2g^{(2)} + g^{(3)}}{6} \end{aligned} \right\} \qquad (2.104)$$

which is fourth-order accurate in time. This four-step method uses the Euler prediction, Euler correction, leapfrog prediction, and Milne correction within a single integration step.

The above formulation necessitates f^n, g^n, as well as $g^{(1)}$, $g^{(2)}$, and $g^{(3)}$ to be stored during computation. We note that there are also low-storage Runge–Kutta methods have been proposed by Williamson [6] and have been used widely to achieve high-order temporal accuracy with reduced memory consumption.

2.4.2 Multi-Step Methods

There are time integration methods that use the current state g^n along with the past data g^{n-1}, g^{n-2}, $\cdots$ in an explicit formulation. These methods are referred to as multi-step methods and the *Adams–Bashforth methods* are widely used.

Consider expanding f^{n+1} about f^n with Taylor series and substitute g into $\partial f / \partial t$ to find

$$
\begin{aligned}
f^{n+1} &= f^n + \Delta t \left. \frac{\partial f}{\partial t} \right|^n + \frac{\Delta^2}{2} \left. \frac{\partial^2 f}{\partial t^2} \right|^n + \frac{\Delta^3}{6} \left. \frac{\partial^3 f}{\partial t^3} \right|^n + \cdots \\
&= f^n + \Delta t g^n + \frac{\Delta^2}{2} \left. \frac{\partial g}{\partial t} \right|^n + \frac{\Delta^3}{6} \left. \frac{\partial^2 g}{\partial t^2} \right|^n + \cdots
\end{aligned}
\tag{2.105}
$$

Truncating the series at the second term yields the first-order Adams–Bashforth method which is the explicit Euler's method, Eq. (2.93). Retaining the expansion up to the third term and inserting

$$
\left. \frac{\partial g}{\partial t} \right|^n = \frac{g^n - g^{n-1}}{\Delta t}
\tag{2.106}
$$

provides the second-order Adams–Bashforth method:

$$
f^{n+1} = f^n + \Delta t \frac{3g^n - g^{n-1}}{2}.
\tag{2.107}
$$

If we keep the expansion up to the fourth-order term in the Taylor series and substitute

$$
\left. \frac{\partial g}{\partial t} \right|^n = \frac{3g^n - 4g^{n-1} + g^{n-2}}{2\Delta t}, \quad \left. \frac{\partial^2 g}{\partial t^2} \right|^n = \frac{g^n - 2g^{n-1} + g^{n-2}}{\Delta t^2},
\tag{2.108}
$$

we arrive at the third-order Adams–Bashforth method:

$$
f^{n+1} = f^n + \Delta t \frac{23g^n - 16g^{n-1} + 5g^{n-2}}{12}.
\tag{2.109}
$$

Similar derivations can be followed to find higher-order Adams–Basthforth methods.

In order to initiate time integration with Adams–Bashforth methods with second or higher temporal accuracy, one must have the initial condition f_0 and g_0 as well as past information g^{-1}, g^{-2}, $\cdots$, which may not be available. Thus, an alternate time integration scheme needs to be implemented for the first few steps until all information is gathered for the Adams–Bashforth method to start performing the computation. If the computation is particularly sensitive to the initial condition or transients, it may be desirable to have another high-order accurate method selected initially, instead of using a low-order accurate Adams–Bashforth method.

2.5 Stability Analysis

Once the advection-diffusion equation is discretized in space and time, we can determine the solution over time through numerical integration. However, there can be situations where numerical instability contaminates the solution during the calculation. In some cases, oscillations due to numerical instability can increase the amplitude of the numerical solution, eventually making its value larger than the maximum realizable number on a computer, which is known as overflow. In order to avoid such issues, we examine the stability properties of a few numerical solvers in this section.

The analysis here is motivated by the need to examine the stability of numerical solvers for the advection-diffusion equation

$$\frac{\partial f}{\partial t} + c\frac{\partial f}{\partial x} = a\frac{\partial^2 f}{\partial x^2}. \tag{2.110}$$

By assuming the solution to be of the form

$$f(x, t) = \sum_{k=0}^{\infty} \widetilde{f}_k(t)\exp(ikx), \tag{2.111}$$

we find for each wave number k,

$$\frac{d\widetilde{f}_k}{dt} = (-k^2 a - ikc)\widetilde{f}_k. \tag{2.112}$$

We can then express the above equation with a mathematical abstraction of

$$\frac{df}{dt} = \lambda f \tag{2.113}$$

as a linear model problem by setting $\lambda = (-k^2 a - ikc)$. Note that we have removed the subscript k and tilde to generalize the problem formulation. For the advection-diffusion equation, we have

$$\operatorname{Re}(\lambda) = -k^2 a \quad \text{and} \quad \operatorname{Im}(\lambda) = -kc, \tag{2.114}$$

which means that the real part of λ represents the effect of diffusion and the imaginary part constitutes the effect of advection.

In this section, we first analyze the stability characteristics of time stepping methods. We then examine the stability of finite-difference methods (combined effect of temporal and spatial discretization) using the *von Neumann stability analysis*. Strictly speaking, the analysis is only valid for linear problems and assumes that Fourier analysis is applicable. Since the actual governing equations for fluid mechanics are nonlinear, the analysis cannot be directly applied. Nonetheless, the results from the von Neumann analysis provide us with great insight into the numerical stability of finite-difference methods and can be used as a guideline for choosing an appropriate size of time step Δt.

2.5.1 Stability of Time Stepping Methods

For characterizing the stability of the time stepping methods, we consider a linear model problem of

$$\frac{df}{dt} = \lambda f \quad \text{with} \quad f(t_0) = 1, \tag{2.115}$$

where we let λ be complex and f to be a function of time only. This model problem can capture the fundamental behavior of most differential equations, including the advection-diffusion equation, Eq. (2.110), which was briefly described in terms of this model problem in Sect. 2.5. The solution to this model problem is

$$f(t) = e^{\lambda t} = e^{\text{Re}(\lambda)t}[\cos(\text{Im}(\lambda)t) + i\sin(\text{Im}(\lambda)t)]. \tag{2.116}$$

This expression tells us that the solution is stable (i.e., $\lim_{t \to 0} |f(t)| = 0$) for $\text{Re}(\lambda) < 0$ and the solution is unstable for $\text{Re}(\lambda) > 0$ (solution is neutrally stable for $\text{Re}(\lambda) = 0$). The imaginary component of λ denotes the frequency at which the solution oscillates. Thus, we can observe that $\text{Re}(\lambda)$ and $\text{Im}(\lambda)$ capture the diffusive and advective behaviors of the flow, respectively.

First, let us consider the stability of the explicit Euler's method. For this model problem, we have

$$f^{n+1} = (1 + \lambda\Delta t)f^n = (1 + \lambda\Delta t)^n f^0. \tag{2.117}$$

For this method to be stable, we need

$$|1 + \lambda\Delta t| < 1 \quad \text{(Explicit Euler's method)}. \tag{2.118}$$

Hence, for any value of $\lambda\Delta t$ that lies within a unit circle centered at -1 on the complex plane, the method is stable.

We can perform a similar analysis for the implicit Euler's method to find that

$$f^{n+1} = f^n + \Delta t \lambda f^{n+1} \quad \Rightarrow \quad f^{n+1} = (1 - \lambda \Delta t)^{-1} f^n, \tag{2.119}$$

which requires

$$\left| (1 - \lambda \Delta t)^{-1} \right| < 1 \quad \Rightarrow \quad |\lambda \Delta t - 1| > 1 \quad \text{(Implicit Euler's method)} \tag{2.120}$$

which means that for $\lambda \Delta t$ that is outside of a unit circle centered at 1 on the complex plane, the method is stable.

Let us also consider the Crank–Nicolson method where we have

$$f^{n+1} = f^n + \Delta t \frac{1}{2} \lambda (f^n + f^{n+1}) \quad \Rightarrow \quad f^{n+1} = \frac{1 + \frac{1}{2} \lambda \Delta t}{1 - \frac{1}{2} \lambda \Delta t} f^n, \tag{2.121}$$

which necessitates

$$\left| \frac{1 + \frac{1}{2} \lambda \Delta t}{1 - \frac{1}{2} \lambda \Delta t} \right| < 1 \quad \Rightarrow \quad \text{Re}(\lambda \Delta t) < 0 \quad \text{(Crank–Nicolson method)} \tag{2.122}$$

for stability. This means that the Crank–Nicolson method is stable if $\lambda \Delta t$ is anywhere on the left-half plane of the complex plane.

Note that the analysis can be performed for multistep and predictor–corrector methods also. For example, the second-order Runge–Kutta method gives

$$f^{n+1} = \left[1 + \lambda \Delta t + \frac{(\lambda \Delta t)^2}{2} \right] f^n \tag{2.123}$$

and correspondingly

$$\left| 1 + \lambda \Delta t + \frac{(\lambda \Delta t)^2}{2} \right| < 1 \quad \text{(2nd-order Runge–Kutta method)} \tag{2.124}$$

is needed for stability.

We graphically summarize the stability region in Fig. 2.14 for the explicit and implicit Euler's methods, the Crank–Nicolson method, and the leapfrog method (discussed later in Sect. 2.5.7). The stability contours for predictor-corrector schemes and multistep methods, namely the Runge–Kutta methods and the Adams–Bashforth methods are also shown.

There are several observations we can make about the time stepping methods. First, the Crank–Nicolson method has the entire left-hand side of the complex plane of $\lambda \Delta t$ covered for stability. The leapfrog method is only stable for purely oscillatory (advective) problems. It is stable only along the imaginary axis and would not be able to handle diffusive effects. We can also note that the Runge–Kutta methods can

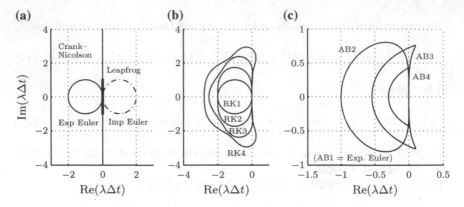

Fig. 2.14 Stability contours of time stepping methods. **a** Stability regions for the explicit Euler's method is inside the circle (*solid*); the implicit Euler is outside the *circle* (*dash-dotted*); the Crank–Nicolson method is the entire *left-hand side* of the complex plane; the leapfrog method is stable along the *thick line*. **b** First to fourth-order Runge–Kutta (RK) methods are stable inside the stability contours. **c** Second to fourth-order Adams–Bashforth (AB) methods are stable inside the contours

enlarge the region of stability by increasing the order of accuracy. This is of course achieved with the added cost of computation required by the use of intermediate steps. For the Adams–Bashforth methods, the scales on the axes used in Fig. 2.14c are different for those in Fig. 2.14a, b. As the temporal accuracy of the Adams–Bashforth method is increased, the region of stability becomes smaller.

If advection is a important feature in the flow, we should choose a time stepping method that covers the imaginary axis. Schemes such as the third-order Adam–Bashforth and Runge–Kutta methods, and their higher order versions cover the imaginary axis as shown in Fig. 2.14b, c. On the other hand, if diffusion is a significant feature in the flow, we should select a scheme that enlarges the stability region in the left-hand side of the complex plane.

2.5.2 von Neumann Analysis

Let us now consider $f(x, t)$ to be a function of space and time and expand it with a Fourier series in the spatial direction (assuming that the variables x and t can be separated)

$$f(x, t) = \sum_{k=0}^{\infty} A_k(t) \exp(ikx), \qquad (2.125)$$

where k is the wave number. The coefficients $A_k(t)$ are taken to be functions of time. For a scheme to be numerically stable, we require the amplitude $A_k(t)$ to be bounded over time. For linear problems, we can examine the stability of the amplitude for each wave number k, instead of considering the sum of all wave number components.

For a discrete point $x_j = \Delta j$ and time $t_n = n\Delta t$, we let

$$f_j^n = A^n \exp(ik\Delta j). \tag{2.126}$$

Using identities such as Eqs. (2.61) and (2.62) in analyzing finite-difference schemes, we can find the amplitude $A = A^{n+1}/A^n$. Stability of numerical methods are classified as

$$|A| \begin{cases} < 1 & \text{stable} \\ = 1 & \text{neutrally stable} \\ > 1 & \text{unstable} \end{cases} \tag{2.127}$$

If a numerical method is stable for all wave numbers, then that method is *unconditionally stable*. Below we present what is referred to as the *von Neumann stability analysis* of various discretization schemes.

2.5.3   Stability of the Discrete Advection Equation

Here, we consider the forward-in-time central-in-space (FTCS) method for the advection equation, Eq. (2.7), with a constant advection speed of c

$$\frac{f_j^{n+1} - f_j^n}{\Delta t} = -c\frac{-f_{j-1}^n + f_{j+1}^n}{2\Delta}. \tag{2.128}$$

This method discretizes the evolution equation in time with explicit Euler's method (forward in time) and in space with central difference (central in space). Fourier transform of the above equation leads to the evolution equation for the amplitude of the solution

$$\frac{A^{n+1} - A^n}{\Delta t} = -ic\frac{\sin(k\Delta)}{\Delta}A^n. \tag{2.129}$$

Here, we have used Eq. (2.61) for the right-hand side. The amplitude is found to be a complex number of

$$A = 1 - i\frac{c\Delta t}{\Delta}\sin(k\Delta). \tag{2.130}$$

We can observe that unless $\Delta t = 0$, clearly $|A| > 1$. Therefore, the FTCS method for the advection equation is unconditionally unstable.

It is possible to find a stable discretization of the advection equation with an explicit method. Replacing f_j^n in the left-hand side of Eq. (2.128) with a spatial average $\bar{f}_j^n = (f_{j-1}^n + f_{j+1}^n)/2$, we can consider

$$\frac{f_j^{n+1} - \bar{f}_j^n}{\Delta t} = -c\frac{-f_{j-1}^n + f_{j+1}^n}{2\Delta}, \tag{2.131}$$

which is called the *Lax–Friedrichs method*. Using Eqs. (2.61) and (2.62), we find

$$\frac{A^{n+1} - A^n \cos(k\Delta)}{\Delta t} = -ic \frac{\sin(k\Delta)}{\Delta} A^n, \tag{2.132}$$

which yields

$$|A| = \left\{ 1 + \left[\left(\frac{c\Delta t}{\Delta} \right)^2 - 1 \right] \sin^2(k\Delta) \right\}^{1/2}. \tag{2.133}$$

We now notice that

$$C \equiv \frac{c\Delta t}{\Delta} \leq 1 \tag{2.134}$$

must be satisfied for $|A| \leq 1$ so that the scheme is stable for all wave numbers. This condition is known as the *Courant–Friedrichs–Lewy (CFL) condition*, and tells us that the information traveling at speed c over a time of Δt must not translate farther than Δ. The ratio of the distance traveled by the information on a spatial grid is defined as $C \equiv c\Delta t/\Delta$, the *Courant number (CFL number)*, and is a very important non-dimensional number in evaluating the numerical stability of solvers for the advection equation.

If we consider the implicit Euler's method

$$\frac{f_j^{n+1} - f_j^n}{\Delta t} = -c \frac{-f_{j-1}^{n+1} + f_{j+1}^{n+1}}{2\Delta} \tag{2.135}$$

then the amplitude for wave number k satisfies

$$\frac{A^{n+1} - A^n}{\Delta t} = -ic \frac{\sin(k\Delta)}{\Delta} A^{n+1} \tag{2.136}$$

resulting in

$$A = \left[1 + i \frac{c\Delta t}{\Delta} \sin(k\Delta) \right]^{-1}. \tag{2.137}$$

Hence, it can be seen that Euler's implicit method always satisfies $|A| < 1$ for all wave numbers and is unconditionally stable. Real fluid flows however have advective velocity that is varying in space and time making the governing equation nonlinear. This makes implementation of implicit methods difficult or computationally expensive, due to the necessity to solve a nonlinear equation at every time step. For instance, the nonlinear Burgers' equation, Eq. (2.5), would be discretized with the implicit Euler's equation as

$$\frac{u_j^{n+1} - u_j^n}{\Delta t} = -u_j^{n+1} \frac{-u_{j-1}^{n+1} + u_{j+1}^{n+1}}{2\Delta}, \tag{2.138}$$

which requires a nonlinear solver. Such formulation can become computationally expensive for large-scale problems. On the other hand, we can consider a semi-implicit method, such as

$$\frac{u_j^{n+1} - u_j^n}{\Delta t} = -u_j^n \frac{-u_{j-1}^{n+1} + u_{j+1}^{n+1}}{2\Delta}. \tag{2.139}$$

From a practical point of view, it is often appropriate to select an explicit method for the advective term that has stability and high order of accuracy.

2.5.4 Stability of the Discrete Diffusion Equation

Let us consider the FTCS method for the diffusion equation, Eq. (2.8), with diffusivity $a > 0$,

$$\frac{f_j^{n+1} - f_j^n}{\Delta t} = a \frac{f_{j-1}^n - 2f_j^n + f_{j+1}^n}{\Delta^2}. \tag{2.140}$$

If we analyze the time evolution of the amplitude with Fourier transform, we observe that

$$\frac{A^{n+1} - A^n}{\Delta t} = 2a \frac{\cos(k\Delta) - 1}{\Delta^2} A^n \tag{2.141}$$

and the amplitude A is found to be

$$A = 1 - \frac{4a\Delta t}{\Delta^2} \sin^2 \frac{k\Delta}{2}. \tag{2.142}$$

For this scheme to be stable ($|A| \leq 1$), we have

$$\frac{4a\Delta t}{\Delta^2} \sin^2 \frac{k\Delta}{2} \leq 2. \tag{2.143}$$

This inequality holds for all wave numbers if

$$\Delta t \leq \frac{\Delta^2}{2a} \tag{2.144}$$

because $0 \leq \sin^2(k\Delta/2) \leq 1$. We can further draw an analogy to the CFL condition by considering a/Δ as the characteristic diffusion speed and express

$$\frac{(a/\Delta)\Delta t}{\Delta} \leq \frac{1}{2}. \tag{2.145}$$

The above analysis means that the diffusion equation can be integrated stably with FTCS by choosing the time step according to Eq. (2.144). However, it should be noted that the time step Δt needs to be selected such that it is proportional to Δ^2. That is, if we choose the spatial discretization to be reduced by half, the time step must be reduced by a quarter. This discretization scheme is thus at a disadvantage for finer grids because of the larger number of steps required for time advancement.

We can switch the time integration scheme to be implicit and show that the diffusion equation can be discretized in a stable manner for all wave numbers. Utilizing the implicit Euler's method, the diffusion equation can be discretized as

$$\frac{f_j^{n+1} - f_j^n}{\Delta t} = a \frac{f_{j-1}^{n+1} - 2f_j^{n+1} + f_{j+1}^{n+1}}{\Delta^2} \tag{2.146}$$

with the corresponding equation for the amplitude being

$$\frac{A^{n+1} - A^n}{\Delta t} = 2a \frac{\cos(k\Delta) - 1}{\Delta^2} A^{n+1}. \tag{2.147}$$

In this case, the amplitude becomes

$$A = \left(1 + \frac{4a\Delta t}{\Delta^2} \sin^2 \frac{k\Delta}{2}\right)^{-1}, \tag{2.148}$$

which satisfies $|A| \leq 1$ for all wave numbers.

If we choose the Crank–Nicolson method, the diffusion equation becomes

$$\frac{f_j^{n+1} - f_j^n}{\Delta t} = \frac{a}{2} \frac{f_{j-1}^n - 2f_j^n + f_{j+1}^n}{\Delta^2} + \frac{a}{2} \frac{f_{j-1}^{n+1} - 2f_j^{n+1} + f_{j+1}^{n+1}}{\Delta^2} \tag{2.149}$$

with

$$A = \left(1 - \frac{2a\Delta t}{\Delta^2} \sin^2 \frac{k\Delta}{2}\right)\left(1 + \frac{2a\Delta t}{\Delta^2} \sin^2 \frac{k\Delta}{2}\right)^{-1}, \tag{2.150}$$

which also satisfies $|A| \leq 1$ for all wave numbers. Hence, the implicit Euler's and Crank–Nicolson methods are unconditionally stable for the diffusion equation.

For the analysis of viscous flows, the grids are generally refined near the wall to resolve the thin boundary layers. In those regions, the diffusive term becomes the dominant term in the governing equation. For fine grids, computations based on explicit methods become inefficient due to the restrictive time stepping constraint ($\Delta t \propto \Delta^2$). In comparison to the nonlinear advective term, the diffusive term is linear which makes the implementation of implicit method straightforward. For these reasons, it is beneficial to treat the diffusive term implicitly. Even for cases where the viscosity is nonconstant, we can implicitly treat the diffusive terms about the baseline viscosity value and explicitly treat diffusion term related to the variation in viscosity from the baseline value.

2.5.5 Stability of the Discrete Advection-Diffusion Equation

We have discussed that explicit treatment is suitable for the advective term and implicit treatment is desirable for the diffusive term. Let us now consider the advection-diffusion equation, Eq. (2.6), where both of these terms appear. Here, we integrate the advective and diffusive terms using different time stepping schemes. Such approach is called *splitting* and is used to apply the appropriate method for each term in the equation.[4]

Let us consider three splitting methods [4]. First, we consider a comparison of cases where the advective term is integrated in time with the explicit Euler but the diffusive term is integrated in time with the explicit Euler's method (FTCS) or with the implicit Euler's method. Next, we consider second-order accurate time integration with the Crank–Nicolson method for the diffusive term and the second-order Adams–Bashforth method for the advective term. Comparisons are made and the stability characteristics are discussed.

First, consider utilizing the explicit Euler's method for both the advective and diffusive terms (FTCS). We find that the discretization results in

$$\frac{f_j^{n+1} - f_j^n}{\Delta t} = -c \frac{-f_{j-1}^n + f_{j+1}^n}{2\Delta} + a \frac{f_{j-1}^n - 2f_j^n + f_{j+1}^n}{\Delta^2}. \tag{2.151}$$

Representing the solution to be of the form $f_j^n = A^n \exp(ik\Delta j)$, we can rewrite the above equation as

$$\frac{A^{n+1} e^{ik\Delta j} - A^n e^{ik\Delta j}}{\Delta t} = -c \frac{-A^n e^{ik\Delta(j-1)} + A^n e^{ik\Delta(j+1)}}{2\Delta}$$
$$+ a \frac{A^n e^{ik\Delta(j-1)} - 2A^n e^{ik\Delta j} + A^n e^{ik\Delta(j+1)}}{\Delta^2}. \tag{2.152}$$

Solving the equation for the amplitude A, we find

$$A = 1 - iC \sin k\Delta - 2\frac{C}{R}(1 - \cos k\Delta), \tag{2.153}$$

where C is the Courant number that we saw in the discussion of the advection equation (Sect. 2.5.4) and R is the non-dimensional number that compares the viscous and inertial effects over a grid

$$R \equiv \frac{c\Delta}{a}, \tag{2.154}$$

---

[4]Splitting methods are not limited to time domains. One can treat upwind and downwind fluxes in a different fashion, which results in *flux splitting methods*.

which is called the *cell Reynolds number*. The magnitude of A can then be expressed as

$$|A|^2 = \left[1 - 2\frac{C}{R}(1 - \cos k\Delta)\right]^2 + C^2 \sin^2 k\Delta. \tag{2.155}$$

When $k\Delta \approx 0$ and π, the stability of scheme becomes critical ($|A| \approx 1$). Expanding Eq. (2.155) using Taylor series about these critical points in terms of $k\Delta$, we observe that

$$|A|^2 \rightarrow 1 + C^2(k\Delta)^2 - 2\frac{C}{R}(k\Delta)^2 + \mathcal{O}((k\Delta)^4) \quad \text{for} \quad k\Delta \rightarrow 0 \tag{2.156}$$

$$|A|^2 \rightarrow 1 - 8\frac{C}{R} + 16\left(\frac{C}{R}\right)^2 + \mathcal{O}((k\Delta - \pi)^2) \quad \text{for} \quad k\Delta \rightarrow \pi \tag{2.157}$$

which we need them to satisfy $|A|^2 \leq 1$ for stability. Therefore, we find that

$$C \leq \frac{2}{R} \quad \text{and} \quad C \leq \frac{R}{2} \tag{2.158}$$

must be satisfied for the stable numerically integration based on the explicit Euler's method for both advective and diffusive terms. The shaded region in Fig. 2.15a illustrates the combination of the Courant number and cell Reynolds number for which the method is stable. With $C \leq R/2$, the stability region is cut off significantly for $R < 2$ at higher values of Courant number.

Next, let us consider integrating the diffusive term implicitly as we have recommended in Sect. 2.5.4. We choose the implicit Euler's method to demonstrate the change in stability characteristics. The finite-difference formulation becomes

$$\frac{f_j^{n+1} - f_j^n}{\Delta t} = -c\frac{-f_{j-1}^n + f_{j+1}^n}{2\Delta} + a\frac{f_{j-1}^{n+1} - 2f_j^{n+1} + f_{j+1}^{n+1}}{\Delta^2}. \tag{2.159}$$

Utilizing the Fourier representation in the above equation, we can find

$$A = \frac{1 - iC \sin k\Delta}{1 + 2\frac{C}{R}(1 - \cos k\Delta)}. \tag{2.160}$$

For a stable calculation, we require that

$$|A| = \frac{\sqrt{1 + C^2 \sin^2 k\Delta}}{1 + 4\frac{C}{R}\sin^2\frac{k\Delta}{2}} \leq 1. \tag{2.161}$$

This inequality is critical near $k\Delta \approx 0$ (or 2π) and upon expanding the above relations with respect to $k\Delta$ near 0, we find

$$1 + C^2(k\Delta)^2 \leq 1 + 2\frac{C}{R}(k\Delta)^2 + \mathcal{O}((k\Delta)^4). \tag{2.162}$$

Hence we observe that for numerical stability, we require

$$C \leq \frac{2}{R} \tag{2.163}$$

for solving the advection-diffusion equation in a stable manner using the explicit Euler's method and the implicit Euler's method for the advective and diffusive terms, respectively. The stability region is illustrated in Fig. 2.15b. Comparing this scheme with the previous scheme that used explicit Euler's method for both the advective and diffusive terms, the restriction of $C \leq R/2$ is absent. Thus, we achieve a larger region of stability for $R < 2$.

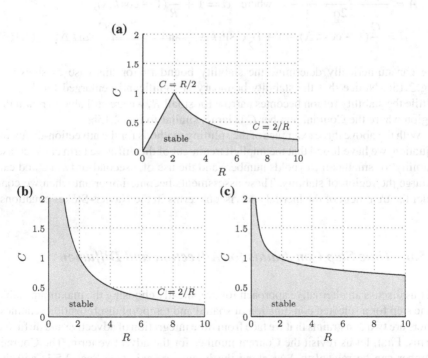

Fig. 2.15 Stability of splitting methods using **a** the explicit Euler's method for advective and diffusive terms, **b** the explicit Euler's method for advective term and the implicit Euler's method for diffusive term, and **c** the second-order Adams–Bashforth method for advective term and the Crank–Nicolson method for diffusive term. *Shaded regions* represent where the methods are stable

In the above analysis, we utilized the first-order accurate explicit and implicit methods to assess the stability properties. We further consider the use of second-order accurate methods that are commonly used to integrate the incompressible Navier–Stokes equations. Below, let us examine the use of the second-order Adams–Bashforth method and the Crank–Nicolson method for advective and diffusive terms, respectively. Based on these schemes, the finite-difference formulation can be written as

$$
\frac{f_j^{n+1} - f_j^n}{\Delta t} = -c \left(\frac{3}{2} \frac{-f_{j-1}^n + f_{j+1}^n}{2\Delta} - \frac{1}{2} \frac{-f_{j-1}^{n-1} + f_{j+1}^{n-1}}{2\Delta} \right)
$$
$$
+ a \left(\frac{1}{2} \frac{f_{j-1}^{n+1} - 2f_j^{n+1} + f_{j+1}^{n+1}}{\Delta^2} + \frac{1}{2} \frac{f_{j-1}^n - 2f_j^n + f_{j+1}^n}{\Delta^2} \right). \tag{2.164}
$$

Again, solving for the amplitude using the Fourier analysis, we find that

$$
A = \frac{-\beta \pm \sqrt{\beta^2 - 4\alpha\gamma}}{2\alpha}, \quad \text{where} \quad \alpha = 1 + \frac{C}{R}(1 - \cos k\Delta),
$$
$$
\beta = \frac{C}{R}(1 - \cos k\Delta) - 1 + i\frac{3}{2}C \sin k\Delta, \quad \text{and} \quad \gamma = -i\frac{C}{2} \sin k\Delta. \tag{2.165}
$$

We can numerically determine the stability boundary for this case as shown in Fig. 2.15c. Notice that the stability boundary is significantly enlarged for $R \gtrsim 2$. While the stability region becomes narrow for small R, we are still able to retain the region where the Courant number C is large, similar to Fig. 2.15b.

With the above three examples of the splitting methods for the advection-diffusion equation, we have found that the implicit treatment of the diffusive term can increase stability for small cell Reynolds numbers and the use of a second-order method can enlarge the region of stability. These assessments become important when we consider the treatment of the inertial and viscous terms in the Navier–Stokes equations.

2.5.6 Time Step Constraints for Advection and Diffusion

Let us discuss an alternative approach for graphically choosing the maximum stable time step for a selected combination of spatial and temporal discretization schemes. There are two constraints that we face from time integration of advective and diffusive terms. First, let us revisit the Courant number for the advective term. The Courant number can be related to $\lambda \Delta t$ along the imaginary axis (see Sect. 2.5.1) and the maximum modified wave number:

$$
C = \frac{c\Delta t}{\Delta} < \frac{\max_{\text{Re}(\lambda)=0} \text{Im}(\lambda \Delta t)}{\max(K\Delta)} = \frac{|\lambda_A|\Delta t}{\max(K\Delta)}, \tag{2.166}
$$

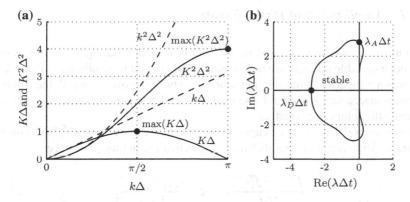

Fig. 2.16 Graphical representation of advective and diffusion time step constraints for the **a** spatial and **b** temporal discretization schemes

where λ_A is determined at the intersect of the stability boundary of the time stepping method and the imaginary axis as depicted in Fig. 2.16.

For the diffusive term, we noticed that the time step is restricted by the diffusive time scale of

$$C/R = \frac{a\Delta t}{\Delta^2} < \frac{- \min_{\mathrm{Im}(\lambda)=0} \mathrm{Re}(\lambda \Delta t)}{\max(K^2\Delta^2)} = \frac{|\lambda_D|\Delta t}{\max(K^2\Delta^2)}, \tag{2.167}$$

by noticing that the decay of the solution is represented by $\lambda \Delta t$ along the real axis. Here, the value of λ_D can be found from the intersect of the stability boundary and the real axis, as shown in Fig. 2.16.

If we choose to discretize the advection-diffusion equation with second-order central-difference schemes of

$$f'_j = \frac{-f_{j-1} + f_{j+1}}{2\Delta} \tag{2.168}$$

$$f''_j = \frac{f_{j-1} - 2f_j + f_{j+1}}{\Delta^2} \tag{2.169}$$

with the fourth-order Runge–Kutta method for time stepping, we can find graphically from Fig. 2.16 that

$$|\lambda_A|\Delta t = 2.83, \quad |\lambda_D|\Delta t = 2.79, \quad \max(K\Delta) = 1, \quad \max(K^2\Delta^2) = 4, \tag{2.170}$$

providing us with

$$C = \frac{c\Delta t}{\Delta} < 2.83 \quad \text{and} \quad C/R = \frac{a\Delta t}{\Delta^2} < 0.698 \tag{2.171}$$

as restrictions on the choice of time step to perform the simulation in a stable manner.

2.5.7   Amplitude and Phase Errors

Up until now, we have focused on the amplitude error during time integration. Although unstable schemes are not useful, it does not automatically mean that stable schemes are superior. If the solution scheme is stable but over-damped, the solution becomes inaccurate. We should also analyze whether neutrally stable schemes can compute the solutions correctly.

Let us consider the next example, known as the *leapfrog method*,[5] based on the second-order accurate central-difference discretization in time and space

$$\frac{f_j^{n+1} - f_j^{n-1}}{2\Delta t} = -c\frac{-f_{j-1}^n + f_{j+1}^n}{2\Delta}. \tag{2.172}$$

Utilizing the von Neumann analysis and realizing that $A = A^{n+1}/A^n = A^n/A^{n-1}$, we obtain a second-order equation

$$A^2 + 2iCA\sin k\Delta - 1 = 0. \tag{2.173}$$

The solution to this equation is

$$A = -iC\sin k\Delta \pm \sqrt{1 - C^2\sin^2 k\Delta}. \tag{2.174}$$

Now let us write $A = |A|e^{i\varphi}$, where φ represents the phase difference. From Eq. (2.174), we observe that the leapfrog method for $C \leq 1$ leads to

$$|A| = 1, \quad \varphi = -\tan^{-1}\frac{C\sin k\Delta}{(1 - C^2\sin^2 k\Delta)^{1/2}}. \tag{2.175}$$

Since the exact solution to the advection equation for $f_j^n = A^n\exp(ik\Delta j)$ is $f_j^{n+1} = A^n\exp[ik(\Delta j - c\Delta t)]$, we know that $|A|_{\text{exact}} = 1$ and $\varphi_{\text{exact}} = -Ck\Delta$. The ratio between Eq. (2.175) and the exact solutions, $|A|/|A|_{\text{exact}}$ and $\varphi/\varphi_{\text{exact}}$ correspond to the relative amplitude error and the relative phase error, respectively [1, 4].

For the leapfrog scheme with $C \leq 1$, the method is neutrally stable with no amplitude error for all wave numbers k. However, for $C < 1$, $\varphi/\varphi_{\text{exact}}$ becomes small in the neighborhood of $k\Delta = \pi$ (near the cutoff frequency $k_c = \pi/\Delta$), resulting in delay of high-frequency components being advected. Such numerical schemes having different wave speeds for different wavelengths are called *dispersive* and the associated error is referred to as *dispersive error*. The central-difference leapfrog method suffers from oscillations due to the phase error and leads to eventual blowup of the solution, even though the method is analytically speaking neutrally stable

---

[5]The stability of the leapfrog method is depicted in Fig. 2.14.

Fig. 2.17 Numerical solution to the advection equation using the leapfrog method (Courant number of 0.5)

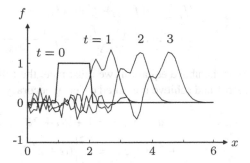

(i.e., $|A| = 1$). As shown in Fig. 2.17, we can observe oscillations for short wavelengths do not get transported as fast as they should in the x-direction. This tells us that the phase delay is appearing in high-frequency components.

Even for numerical schemes with growth rates of less than 1, dispersive methods introduce numerical oscillations into the solution. In general, oscillations due to dispersive errors appear behind the wave packet if the scheme is phase-delayed and in front of the wave packet if the scheme is phase-advanced.

2.6 Higher-Order Finite Difference

In this section, we present an alternative approach for deriving higher-order finite-difference schemes. The approach below considers an implicit formulation of derivative approximation to achieve spectral-like accuracy using a compact stencil. Such schemes have been proven useful to study turbulence and aeroacoustics. High-order accuracy is particularly important since pressure disturbance associated with acoustics waves are orders of magnitude smaller than the hydrodynamic pressure fluctuations. Therefore, solvers need to be designed with high-order accuracy to correctly predict the overall flow physics. The method in this section is not discussed further in other chapters of this book, as we focus on conservation properties of the discrete equations. Here, we briefly present the main idea behind deriving higher-order accurate methods as they continue to be a subject of active research.

We have thus far derived finite-difference formulas in an explicit manner such that the discrete derivative can be computed by summing discrete functional values with appropriate weights. It is also possible to implicitly solve for the derivate values by constructing a sparse matrix equation.[6] Following the work of Lele [3], let us present an example of implicit finite-difference method on a five-point stencil

---

[6]Implicitly solving for the derivative may increase the computational cost. However, this translates to the differencing stencil being all coupled, which shares similarity with spectral methods.

$$\beta f'_{j-2} + \alpha f'_{j-1} + f'_j + \alpha f'_{j+1} + \beta f'_{j+2}$$

$$= c \frac{f_{j+3} - f_{j-3}}{6\Delta} + b \frac{f_{j+2} - f_{j-2}}{4\Delta} + a \frac{f_{j+1} - f_{j-1}}{2\Delta}. \tag{2.176}$$

For the above equation, we must meet the following condition to cancel truncation error and achieve the noted order of accuracy:

$$a + b + c = 1 + 2\alpha + 2\beta \qquad \text{(second order)} \tag{2.177}$$

$$a + 2^2 b + 3^2 c = 2\frac{3!}{2!}(\alpha + 2^2\beta) \qquad \text{(fourth order)} \tag{2.178}$$

$$a + 2^4 b + 3^4 c = 2\frac{5!}{4!}(\alpha + 2^4\beta) \qquad \text{(sixth order)} \tag{2.179}$$

$$a + 2^6 b + 3^6 c = 2\frac{7!}{6!}(\alpha + 2^6\beta) \qquad \text{(eighth order)} \tag{2.180}$$

$$a + 2^8 b + 3^8 c = 2\frac{9!}{8!}(\alpha + 2^8\beta) \qquad \text{(tenth order)} \tag{2.181}$$

Finite-difference schemes based on this formulation is called the *compact finite-difference schemes* and is a generalization of the *Padé approximation*. The corresponding modified wave number for Eq. (2.176) becomes

$$K\Delta = \frac{a\sin(k\Delta) + \frac{b}{2}\sin(2k\Delta) + \frac{c}{3}\sin(3k\Delta)}{1 + 2\alpha\cos(k\Delta) + 2\beta\cos(2k\Delta)}. \tag{2.182}$$

For illustration of the accuracy of the compact scheme, let us consider a tridiagonal system on the left-hand side of Eq. (2.176) with $\beta = 0$ and $c = 0$. We can then find a family of fourth-order accurate finite-difference schemes with

$$a = \frac{2}{3}(\alpha + 2), \quad b = \frac{1}{3}(4\alpha - 1), \quad c = 0, \quad \beta = 0 \tag{2.183}$$

Note that for $\alpha = 0$, we recover the fourth-order central-difference scheme, Eq. (2.21), and for $\alpha = 1/4$, we find the classical Padé approximation. If we choose $\alpha = 1/3$, the truncation errors cancel and provide us with the sixth-order compact finite-difference scheme, where

$$a = \frac{14}{9}, \quad b = \frac{1}{9}, \quad c = 0, \quad \alpha = \frac{1}{3}, \quad \beta = 0. \tag{2.184}$$

This scheme requires us to solve a tridiagonal system but provides us with high order of accuracy. For illustration of the scheme, we provide a comparison of the modified wave number in Fig. 2.18 for higher order schemes and the classical Taylor series-based formulations. It can be observed that the compact finite-difference schemes can closely position the modified wave number to the exact wave number even for high wave number.

Fig. 2.18 Comparison of
modified wave numbers for
the compact finite-difference
methods with sixth, eighth,
and tenth-order accuracy
(*solid*) and Taylor
series-based
central-difference schemes
of second, fourth, and
sixth-order accuracy
(*dash-dot*)

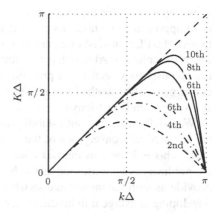

In general, the above approach can be taken to derive higher derivative approximations and higher-order accurate finite-difference methods. For non-periodic boundaries, special care must be taken in the construction of the compact finite-difference methods. Additional details can be found in Lele [3].

2.7 Consistency of Finite-Difference Methods

Finite-difference methods are called *consistent* when the discretized differential equation converges to the original continuous differential equation in the limit of $\Delta \to 0$ and $\Delta t \to 0$. In other words, the truncation error from the finite-difference method should converge to zero for the method to be consistent.

To illustrate how an *inconsistent* finite-difference method can produce incorrect results, we consider a method called the *DuFort–Frankel method* for the diffusion equation

$$\frac{f_j^{n+1} - f_j^{n-1}}{2\Delta t} = a \frac{f_{j-1}^n - 2\frac{f_j^{n+1} + f_j^{n-1}}{2} + f_{j+1}^n}{\Delta^2}. \tag{2.185}$$

The spatial discretization is similar to the second-order central-difference scheme but replaces the middle value with time average value. For, the time derivative, the past and next time step values are employed. While this scheme is unconditionally stable, the Taylor series analysis reveals that in the limit of Δt and Δ approaching 0 the DuFort–Frankel method leads to

$$\frac{\partial f}{\partial t} = u \frac{\partial^2 f}{\partial x^2} - a \frac{\Delta t^2}{\Delta^2} \frac{\partial^2 f}{\partial t^2}. \tag{2.186}$$

This implies that for the method to produce the correct solution, the time step Δt must approach 0 at a faster rate than Δ approaching 0. Otherwise, we would have a hyperbolic PDE instead of a parabolic PDE, which completely changes the behavior of the solution. Although the Dufort–Frankel method is unconditionally stable, the inconsistency seen above prohibits us from achieving convergence. To achieve convergence, this method requires us to satisfy $\Delta t/\Delta \to 0$ in addition to Δt and $\Delta \to 0$, which is restrictive.

We briefly note that both stability and consistency of finite-difference schemes are required for convergence of the numerical solution to exact solution of a linear PDE. This is known as the *Lax equivalence theorem*. The theorem does not hold for nonlinear equations, such as the Navier–Stokes equations. Nonetheless, it does provide us with the insight that satisfying stability and consistency is important for developing convergent finite-difference methods to solve fluid flow problems.

2.8 Remarks

Every finite-difference method has some level numerical error from the discretization process. While one may wonder whether we can obtain a reliable solution from flow simulations, in reality there are stable and accurate computational schemes that allow us to confidently simulate flow physics. It is important that we understand how the results from different spatial and temporal discretization schemes behave. Such insights enable us to develop numerical solvers that can arrive at the incompressible flow solution in a stable and accurate manner, as we will see in the next chapter.

Before we end this chapter, we should discuss how we can validate simulation results. Numerical validation can be performed relatively easily for those simulations where (1) experiments are possible but only limited data can be obtained and (2) the time and cost required by experiments are so large that simulations are preferred to experiments. For the first case, numerical results can provide deeper insight into the flow physics. In these cases, close comparison of numerical and experimental data can provide confidence in using the simulation.

If we do not limit the use of flow simulations to reproduce experiments or theoretical results and use simulations to gain new knowledge of flow physics, it is not enough to simply check the correctness of our solution against experiments or theory. In these cases, the validity of a simulation must be confirmed with extra care because numerical simulation can be the only means to obtain information about the flow. The simplest yet realistic way to check the solution is to examine how the numerical solution may change for different choices of spatial discretization, as mentioned in Sect. 2.3.6. Specifically, we can change the number of grid points in the domain, how the grids are distributed (nonuniformly), or the size of the computational domain. In order to analyze the results, one would need to use insights from fluid mechanics and mathematics to examine how the solution could be affected. What we discussed in this chapter focused only on whether the governing equation has been discretized correctly. Whether the equation that we solve reproduces the flow physics correctly requires insights from fluid mechanics and is a different problem, which we will discuss in the following chapters.

2.9 Exercises

2.1 Burgers' equation captures nonlinear advection and diffusion

$$\frac{\partial u}{\partial t} + u \frac{\partial u}{\partial x} = a \frac{\partial^2 u}{\partial x^2},$$

where a is a positive constant. Show that using a transform (Cole–Hopf transform) of

$$u = -2a \frac{1}{\phi} \frac{\partial \phi}{\partial x},$$

the nonlinear Burgers' equation can be transformed to a linear PDE.

2.2 For real x and integers m, n with $m > n$, prove the following identities:

1. $\mathcal{O}(x^m) \pm \mathcal{O}(x^n) = \mathcal{O}(x^m)$
2. $\mathcal{O}(x^m) \cdot \mathcal{O}(x^n) = \mathcal{O}(x^{m+n})$
3. $\mathcal{O}(x^m)/\mathcal{O}(x^n) = \mathcal{O}(x^{m-n})$

2.3 Given the values of a function $f_j = f(x_j)$ at equally space points $x_j = j\Delta x$, consider the finite-difference scheme for the first derivative below

$$f'(x_j) = \frac{1}{\Delta x}(a f_{j-2} + b f_{j-1} + c f_j).$$

1. Perform Taylor series expansions of f_{j-2} and f_{j-1} about x_j and find the relationships that the coefficients a, b, and c need to satisfy for the above scheme. Show the first four equations for the coefficients starting from relationship for the lowest order terms.
2. Determine the coefficients a, b, and c such that the above scheme achieves the highest possible order of accuracy.
3. Find the coefficients a, b, and c such that the third derivative f''' does not influence the error.

2.4 Given the values of a function $f_j^n = f(x_j, t_n)$ at equally space points $x_j = j\Delta x$ and time levels $t_n = n\Delta t$, consider the linear advection equation with constant advection velocity u

$$\frac{\partial f}{\partial t} + u \frac{\partial f}{\partial x} = 0$$

and derive an explicit time stepping method of the following form:

$$f_j^{n+1} = a f_{j-2}^n + b f_{j-1}^n + c f_j^n.$$

1. Perform Taylor series expansions of f_j^{n+1}, f_{j-2}^n, and f_{j-1}^n about (x_j, t_n) and find the relationships that the coefficients a, b, and c need to satisfy for the above time stepping scheme. Show the first four equations for the coefficients starting from relationship for the lowest order terms.
2. Determine the coefficients a, b, and c such that the above scheme achieves the highest possible order of accuracy.
3. Find the coefficients a, b, and c such that the second (spatial) derivative f'' does not influence the error.

2.5 Consider a function $f_j = f(x_j)$ given at equally spaced points $x_j = j \Delta x$.

1. Determine the coefficients $(a_0, a_1, a_2,$ and $a_3)$ and the order of accuracy for

$$\frac{d^2 f}{dx^2}(x_j) = a_0 f_j + a_1 f_{j+1} + a_2 f_{j+2} + a_3 f_{j+3}.$$

2. For a periodic function $f(x) = \exp(ikx)$ with wave number k, the second derivative is given by $\frac{d^2 f}{dx^2} = -k^2 \exp(ikx)$. Find the expression K_2 that the above finite-difference scheme yields as an approximation (i.e., $-K_2 \exp(ikx)$). Moreover, show that K_2 approaches k^2 as $\Delta x \to 0$.

2.6 Let us consider a function F evaluated at uniform grid points $x_j = j\Delta$ and write $F_j = F(x_j)$. We denote the derivatives of F as $F^{(n)} = d^n F/dx^n$.

1. Express $F_{j\pm1}$ using the Taylor series expansion up to the fourth derivative about $x = x_j$.
2. Denoting $F^{(1)} = U$ and $F^{(2)} = V$, find the finite-difference approximations of $F^{(3)}$ and $F^{(4)}$ as the second derivatives of $F^{(1)}$ and $F^{(2)}$, respectively, using U_j and V_j.
3. Find the fourth-order finite-difference formulas for $F^{(1)}$ and $F^{(2)}$ using U_j and V_j.

2.7 Consider the ordinary differential equation given by Eq. (2.115) $\frac{df}{dt} = \lambda f$ with $f(t_0) = 1$.

1. Show that the Crack–Nicolson method is stable for $Re(\lambda \Delta t) < 0$.
2. Find the stability criteria of first, second, third, and fourth-order Runge–Kutta methods in terms of $\lambda \Delta t$. Plot the stability contours.

2.8 Consider discretizing the two-dimensional linear advection equation

$$\frac{\partial f}{\partial t} + u \frac{\partial f}{\partial x} + v \frac{\partial f}{\partial y} = 0,$$

for constant $u > 0$ and $v > 0$ using the first-order upwind finite differencing in space and forward Euler integration in time to yield

$$\frac{f_{i,j}^{n+1} - f_{i,j}^{n}}{\Delta t} = -u \frac{-f_{i-1,j}^{n} + f_{i,j}^{n}}{\Delta x} - v \frac{-f_{i,j-1}^{n} + f_{i,j}^{n}}{\Delta y}.$$

Based on the von Neumann analysis, find the condition needed to achieve numerical stability using Courant numbers, $C_x \equiv u\Delta t/\Delta x$ and $C_y \equiv v\Delta t/\Delta y$. Assume uniform grid for spatial discretization.

2.9 Given the linear advection-diffusion equation

$$\frac{\partial f}{\partial t} + u \frac{\partial f}{\partial x} = a \frac{\partial^2 f}{\partial x^2},$$

let us consider the forward-in-time central-in-space (FTCS) method with uniform spacing to yield

$$f_j^{n+1} = f_j^n - \frac{C}{2}(-f_{j-1}^n + f_{j+1}^n) + D(f_{j-1}^n - 2f_j^n + f_{j+1}^n),$$

where $C \equiv u\Delta t/\Delta$ and $D \equiv a\Delta t/\Delta^2$. Assuming solution of the form

$$f_j^n = A^n \exp(ik\Delta j),$$

answer the following questions:

1. Find the expression for the amplitude coefficient A and show that it represents an ellipse on the complex plane. Determine its axes.
2. Using the result from part 1, derive the condition for the above FTCS scheme to be stable.
3. What can the cell Reynolds number, $R \equiv C/D = c\Delta/a$, reveal in terms of the analysis here.

2.10 For the advection-diffusion equation, consider the use of

1. The explicit Euler's method for the advection term and the implicit Euler's method for the diffusion term (see: Eq. (2.159)).
2. The second-order Adam–Bashforth method for the advection term and the Crank–Nicolson method for the diffusion term (see: Eq. (2.164)).

Find the expressions for the amplitude A for the above two cases. Hint: final answers are given in Sect. 2.5.5.

References

1. Fujii, K.: Numerical Methods for Computational Fluid Dynamics. Univ. Tokyo Press, Tokyo (1994)
2. Lambert, J.D.: Computational Methods in Ordinary Differential Equations. Wiley, London (1973)
3. Lele, S.K.: Compact finite difference schemes with spectral-like resolution. J. Comput. Phys. **103**, 16–42 (1992)
4. Lomax, H., Pulliam, T.H., Zingg, D.W.: Fundamentals of Computational Fluid Dynamics. Springer, New York (2001)
5. Wade, W.R.: Introduction to Analysis, 4th edn. Pearson (2009)
6. Williamson, J.H.: Low-storage Runge-Kutta schemes. J. Comput. Phys. **35**(1), 48–56 (1980)

Chapter 3
Numerical Simulation of Incompressible Flows

3.1 Introduction

For compressible and incompressible flows, there is a difference in how the numerical solution techniques are formulated, based on whether or not the mass conservation equation includes a time-derivative term. Fluid motion is described by the conservation equations for mass, momentum, and energy. For incompressible flow, the conservation equation for kinetic energy can be derived from the momentum conservation equation. Hence, we only need to be concerned with the mass and momentum conservation equations. Furthermore, if the temperature field is not a variable of interest, we do not need to consider the internal energy in the formulation. We do note that the treatment of momentum conservation should be consistent with the conservation of kinetic energy in a discrete manner, as it influences the achievement of reliable solution and numerical stability.

In this chapter, we present numerical solution techniques for incompressible flow on Cartesian grids. First, we explain how the velocity and pressure fields can be coupled in the incompressible flow solvers. Next, detailed discussions are offered on how to perform finite-difference approximations of each term in the governing equations.

3.2 Time Stepping for Incompressible Flow Solvers

Recall that the mass and momentum conservation equations for compressible flow are

$$\frac{\partial \rho}{\partial t} = -\nabla \cdot (\rho \boldsymbol{u}) \tag{3.1}$$

$$\frac{\partial (\rho \boldsymbol{u})}{\partial t} = -\nabla \cdot (\rho \boldsymbol{u}\boldsymbol{u} - \boldsymbol{T}). \tag{3.2}$$

© Springer International Publishing AG 2017
T. Kajishima and K. Taira, *Computational Fluid Dynamics*,
DOI 10.1007/978-3-319-45304-0_3

Here, we assume that there is no source or external force. For compressible flow, the right-hand sides of the above equations can be computed for a known flow field to determine the time rate of change of density and momentum. For example, we can use the first-order explicit Euler's method for Eqs. (3.1) and (3.2) to obtain

$$\rho^{n+1} = \rho^n - \Delta t \nabla \cdot (\rho \boldsymbol{u})^n \tag{3.3}$$

$$(\rho \boldsymbol{u})^{n+1} = (\rho \boldsymbol{u})^n - \Delta t \nabla \cdot (\rho \boldsymbol{u} \boldsymbol{u} - \boldsymbol{T})^n, \tag{3.4}$$

where Δt is the time step with $t_n = n \Delta t$. These equations enable us to determine ρ^{n+1} and $\boldsymbol{u}^{n+1}$ at the next time step. Similarly, the internal energy e^{n+1} can be determined from the energy equation. Pressure p^{n+1} and temperature T^{n+1} can be found from the equation of state. Once the initial condition is prescribed, we can repeatedly time step the equations to simulate the flow field. The above discussion is given only as a proof of concept, because in practice the explicit Euler's method lacks in accuracy and stability. We later show how to utilize numerical techniques with improved accuracy and stability for actual simulations of fluid flows.

For incompressible flow, the mass and momentum conservation equations, Eqs. (3.1) and (3.2), become

$$\nabla \cdot \boldsymbol{u} = 0 \tag{3.5}$$

$$\frac{\partial \boldsymbol{u}}{\partial t} = -\nabla \cdot (\boldsymbol{u} \boldsymbol{u}) + \frac{1}{\rho} \nabla \cdot \boldsymbol{T}. \tag{3.6}$$

As discussed in Sect. 1.3.5, the energy equation decouples from the mass and momentum equations. Hence the energy equation does not need to be explicitly considered. The continuity equation, Eq. (3.5), does not include the time-derivative term and kinematically constraints the flow to be divergence-free (solenoidal). Observe that there is no time derivative for pressure appearing in the governing equations for incompressible flow. The flow field evolves as described by the momentum equation, Eq. (3.6), while satisfying the incompressibility constraint imposed by Eq. (3.5). The pressure needs to be determined such that the computed flow is consistent with the two conservation equations.

Let us consider a simple time stepping example of using the explicit Euler's method for Eq. (3.6). In discussions below, we assume density ρ to be constant for simplicity. We then have

$$\boldsymbol{u}^{n+1} = \boldsymbol{u}^n + \Delta t (\boldsymbol{A}^n - \nabla P^n + \boldsymbol{B}^n), \tag{3.7}$$

where $P = p/\rho$. The nonlinear advective and viscous terms are denoted by

$$\boldsymbol{A} = -\nabla \cdot (\boldsymbol{u} \boldsymbol{u}), \quad \boldsymbol{B} = \nabla \cdot \left\{ \nu \left[\nabla \boldsymbol{u} + (\nabla \boldsymbol{u})^T \right] \right\}, \tag{3.8}$$

respectively. For constant viscosity, the viscous term reduces to $\boldsymbol{B} = \nu \nabla^2 \boldsymbol{u}$. Even if the known velocity field $\boldsymbol{u}^n$ satisfies the continuity equation $\nabla \cdot \boldsymbol{u}^n = 0$, the predicted u^{n+1} based on Eq. (3.7) would contain discretization and round-off errors, leading to error in enforcing incompressibility. If these errors accumulate over time, the computation would result in a blow up. In order to prevent such a failure, the pressure field should be determined such that $\nabla \cdot u^{n+1} = 0$ is satisfied.

Now, consider replacing P^n with P^{n+1} in Eq. (3.7),

$$u^{n+1} = u^n + \Delta t (A^n - \nabla P^{n+1} + B^n), \tag{3.9}$$

where P can be regarded as a scalar potential[1] that enforces the incompressibility constraint $\nabla \cdot u^{n+1} = 0$. To determine the true pressure field p, we must solve the pressure Poisson equation using the velocity field that satisfies the continuity equation as well as appropriate pressure boundary conditions, which will be discussed in detail in Sect. 3.8.2. The variable P can be thought of the sum of an intermediate pressure at time $t \in (t_n, t_{n+1})$ and a scalar potential needed to enforce incompressibility. Hence, we do not dwell on discussing whether P should have a superscript of n or $n + 1$. From the point of view of P being the variable needed to satisfy $\nabla \cdot u^{n+1} = 0$ at the new time step, we denote P^{n+1} for notational purpose.

Taking the divergence of Eq. (3.9) and utilizing $\nabla \cdot u^{n+1} = 0$, we obtain the pressure Poisson equation

$$\nabla^2 P^{n+1} = \frac{\nabla \cdot u^n}{\Delta t} + \nabla \cdot \left(A^n + B^n \right). \tag{3.10}$$

Notice that we have kept the first term on the right-hand side. This indicates that even if we have an error in mass conservation ($\nabla \cdot u^n \neq 0$) at the current time step t_n, the solution to Eq. (3.10) enforces the flow field to satisfy $\nabla \cdot u^{n+1} = 0$ at the next time level t_{n+1}. It is unavoidable to have round-off errors and residuals from the iterative solver (discussed later) when solving Eq. (3.10). The divergence of the velocity field $\nabla \cdot u^{n+1}$ will inherent these errors. As long as we solve Eq. (3.10) with sufficient accuracy, the error in mass conservation does not grow as the solution is advanced over time.

The pressure field for incompressible flow is governed by Eq. (3.10), which is an elliptic partial differential equation (a boundary-value problem). This means that the local changes in pressure affect the flow field in a global manner. Such behavior of the pressure field tells us that the sonic speed is "infinite" for incompressible flow.

What we have described above is the essence of solving the Navier–Stokes equations, Eqs. (3.5) and (3.6), by numerically time stepping the incompressible flow field. Thus, we now have a general algorithm to numerically solve for the unsteady flow. If the flow ceases to change over time, we then have a steady flow field. In order to make the solution technique reliable and accurate, we must address a couple of points; namely the compatibility of the spatial discretization schemes and the

---

[1]The variable P can be regarded as a Lagrange multiplier that is needed to enforce the incompressibility constraint [5, 24].

stability of the time integration. Below, we discuss a few techniques that take these points into consideration.

Treatment of Low-Mach-Number Flows

While there is no perfectly incompressible flow, compressibility effects can be negligible when the Mach number M ($\equiv u/c$, the ratio of characteristic flow speed u and acoustic speed c) of the flow is low. Empirically speaking, flows with $M < 0.3$ can be approximated to be incompressible. The utilization of compressible flow solvers for low-Mach-number flows requires the use of very small time steps to resolve the acoustic waves, which results in the overall scheme to be *stiff*.[2] This is one of the reasons for which incompressible flow solvers have been developed. The treatment of low-Mach-number flows can be challenging when compressibility effects cannot be neglected (e.g., aeroacoustics). In such cases, one can perform preconditioning of the compressible flow solvers or extend incompressible flow solvers to include weak compressibility effects. Because the main focus of this book is on incompressible flow, we do not cover weakly compressible flow in this book.

3.3   Incompressible Flow Solvers

In this section, we introduce incompressible flow solvers that originated from the Marker and Cell (MAC) method. The original MAC method is an incompressible flow solution technique proposed by Harlow and Welch from 1965 [10]. The highlights of this method include

- use of the staggered grid to avoid spurious oscillation in the pressure field and the accumulation of mass conservation error; and
- use of marker particles to be able to solve for free-surface flows. A dam break example that uses this solution method is shown in Fig. 3.1.

While the original MAC method is a fairly complex algorithm, progress has been made to simplify the original formulation with the staggered grid and develop more advanced techniques to analyze free-surface flows. Although the MAC method originally referred to methods that focused on the use of marker particles, it has become somewhat common to refer to a method that employs staggered grids as a MAC method. Below, we describe methods that are founded on the basic idea behind using the staggered grid. We will not focus on the treatment of free-surface flow using the MAC method.

---

[2]Numerically solving stiff differential equations requires excessively small time steps due to the disparity in the magnitude of the characteristic eigenvalues [17, 19]. In the case of low-Mach-number flow, the time step necessary to satisfy the CFL condition with the acoustic speed c is much smaller than that with the advective speed u, leading to the requirement of very small time steps to be used for time integration.

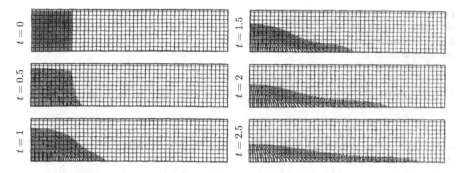

Fig. 3.1 The dam break example solved with the MAC scheme by Harlow and Welch [10]. (Reprinted with permission from [10]. Copyright 1965, AIP Publishing LLC.)

3.3.1 Fractional-Step (Projection) Method

Let us introduce the fractional-step method that couples the continuity equation and the pressure field in the solution algorithm for incompressible flows. In the momentum equation, Eq. (3.9), the pressure gradient term is unknown, which makes time-stepping non-trivial for u^{n+1}. Hence, we decompose the advancement in two steps.

$$u^F = u^n + \Delta t (A^n + B^n) \tag{3.11}$$

$$u^{n+1} = u^F - \Delta t \nabla P^{n+1} \tag{3.12}$$

Between the two steps, P^{n+1} needs to be determined. In order to satisfy the continuity equation with u^{n+1}, we substitute Eq. (3.12) into the continuity equation to obtain the pressure Poisson equation

$$\nabla^2 P^{n+1} = \frac{1}{\Delta t} \nabla \cdot u^F. \tag{3.13}$$

Using the solution P^{n+1} from this Poisson equation, the inclusion of the pressure gradient term in Eq. (3.12) ensures mass conservation to be satisfied. Since the time stepping takes place in *fractional steps*, this method is referred to as the *fractional-step method*.

It should be kept in mind that $\nabla \cdot u^F$ in Eq. (3.13) contains $\nabla \cdot u^n$ because the first term on the right-hand side of Eq. (3.11) contains u^n. In general, Eq. (3.13) is solved with an iterative scheme (e.g., SOR method, multi-grid method, and conjugate gradient method [7, 28]) with the convergence tolerance being a few orders of magnitude larger than machine epsilon (or a desired tolerance). Since most of the computational time is spent on iteratively solving the pressure Poisson equation for incompressible flow, one would like to perform as few iterations as possible to proceed to the next time step. Unless the error in $\nabla \cdot u$ does not grow over time, MAC methods allow for

Fig. 3.2 Illustration of the
projection method enforcing
the incompressibility
constraint on u^{n+1}

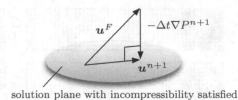

solution plane with incompressibility satisfied

a relatively quick time advancement by removing the continuity error at each time step.

Broadly speaking, the name fractional-step method can refer to more than one type of discretization approach. For an example of $df/dt = g + h$, we refer to fractional-step methods, in this book, as those that perform time stepping for each right-hand side term at a time, i.e., $f^F = f^n + \Delta t g$ and $f^{n+1} = f^F + \Delta t h$. Others may refer to fractional-step methods as those that employ different time integration schemes for g and h regardless of performing time stepping one term at a time (also called splitting or hybrid schemes). The *projection method* [4] refers to the approach that *projects* the velocity field u^F from Eq. (3.11) onto a solution space that satisfies incompressibility to determine the velocity field u^{n+1} using Eq. (3.12). An illustration of the projection method is provided in Fig. 3.2.

3.3.2 Simplified MAC (SMAC) Method

The *simplified MAC (SMAC) method* [1] also is based on Eq. (3.9) but decomposes the time stepping in the following manner

$$u^P = u^n + \Delta t(A^n - \nabla P^n + B^n) \tag{3.14}$$

$$u^{n+1} = u^P - \Delta t \nabla \phi, \tag{3.15}$$

where ϕ is used instead of P^{n+1}. The following Poisson equation

$$\nabla^2 \phi = \frac{1}{\Delta t} \nabla \cdot u^P \tag{3.16}$$

is solved to determine ϕ and consequently the pressure through $P^{n+1} = P^n + \phi$. The variable ϕ is a scalar potential that is proportional to the time rate of change of the pressure field. In principle, this method is the same as the fractional-step method discussed above. The term *simplified* in the name of this method is used to represent how simplified this method is in comparison to the original MAC method [10]. The name is not meant to imply simplification over the fractional-step method.

Let us discuss the differences between the fractional-step and SMAC methods. The fractional-step method does not include the influence from the pressure gradient in computing u^F. Thus, this velocity u^F is not a prediction but an intermediate solution at a fractional step which is later updated by adding the effect from the pressure gradient to become the velocity field at the next time level. On the other hand, u^P is the prediction of the velocity field in the SMAC method. The gradient of ϕ is added as a correction to the velocity field. Therefore, the SMAC method can be viewed as a predictor-corrector scheme. The superscripts F (fractional step) and P (prediction) were utilized to highlight these differences. Since the SMAC method predicts the velocity field in its first substep, the boundary condition can be incorporated rather easily. Further details on the treatment of boundary conditions are provided in Sect. 3.8.

3.3.3 Highly Simplified MAC (HSMAC) Method and Semi-Implicit Method for Pressure Linked Equation (SIMPLE)

The fractional-step and SMAC methods are based on solving the pressure Poisson equation and updating the velocity field at each time step. In contrast to those methods, there are the HSMAC and SIMPLE methods that simultaneously solve for the velocity and pressure fields in an iterative manner.

The *Highly Simplified MAC (HSMAC) method* [11] iteratively corrects the velocity and pressure fields in the direction that satisfies the continuity equation. First, the velocity field is predicted using Eq. (3.14) in the same manner as the SMAC method. The results are then set as the starting point $u^{\langle 0 \rangle} = u^P$ and $P^{\langle 0 \rangle} = P^n$ for solving the velocity and pressure fields in a simultaneous fashion by performing the following three steps iteratively:

$$\psi = -\frac{\omega \nabla \cdot u^{\langle m \rangle}}{2\Delta t \left[\frac{1}{(\Delta x)^2} + \frac{1}{(\Delta y)^2} + \frac{1}{(\Delta z)^2} \right]} \tag{3.17}$$

$$u^{\langle m+1 \rangle} = u^{\langle m \rangle} - \Delta t \nabla \psi \tag{3.18}$$

$$P^{\langle m+1 \rangle} = P^{\langle m \rangle} + \psi, \tag{3.19}$$

where the iteration number $m = 0, 1, 2, \ldots$. Here, the grid spacing in each direction is denoted as Δx, Δy, and Δz and the relaxation parameter ω is generally chosen to be around 1.7. We use Eqs. (3.17)–(3.19) instead of solving Eq. (3.16). Equation (3.17) is a gross diagonalization of the pressure Poisson equation and would not lead to the immediate satisfaction of the continuity equation by $u^{\langle 1 \rangle}$. However, the error in mass conservation would be reduced as the solution goes through the iterations with Eqs. (3.17)–(3.19). When $(\nabla \cdot u^{\langle m \rangle})$ becomes sufficiently small, we can let $u^{n+1} = u^{\langle m \rangle}$ and $P^{n+1} = P^{\langle m \rangle}$. Note that the error in the continuity equation $\nabla \cdot u^{\langle m \rangle}$ can be directly used as a convergence criterion in the HSMAC method. Additional details on the discretization and iterative solver can be found in Sect. 3.4.3.

The *Semi-Implicit Method for Pressure Linked Equation (SIMPLE)* [23] is often distinguished from the family of MAC methods but shares commonality with the MAC methods for discretizing the governing equations on the staggered grid. The SIMPLE method treats the advective and diffusive terms implicitly to determine the velocity and pressure variables through iterations. In that respect, this method is quite different from the HSMAC method that explicitly treats the advective and diffusive terms, which are used to compute the initial guess for the subsequent iterations. For details on the SIMPLE method, readers should consult with [22].

Since the purpose of this book is not to cover all solution techniques, we do not cover HSMAC, SIMPLE, and other methods in depth. We proceed our discussion below with the assumption that the SMAC method is selected as a time-advancement scheme unless it is noted otherwise.

3.3.4 Accuracy and Stability of Time Stepping

Up to this point, we have used the first-order explicit Euler scheme for the sake of simplicity. However, this scheme not only lacks in numerically stability but also in accuracy to predict the behavior of unsteady fluid flows. For the incompressible flow solver to be practically useful, we must increase the order of accuracy and enhance the numerical stability properties of the solver. Some of the time-stepping schemes employed to address these points are the Adams–Bashforth methods, the Crank–Nicolson method, and the Runge–Kutta methods.

For the predictor steps in the SMAC method, we can use the second and third-order Adams–Bashforth methods (see Sect. 2.4.2) for the advective and diffusive terms, and rewrite Eq. (3.14) as

$$\boldsymbol{u}^P = \boldsymbol{u}^n + \Delta t \left[-\nabla P^n + \frac{3}{2}(\boldsymbol{A}^n + \boldsymbol{B}^n) - \frac{1}{2}(\boldsymbol{A}^{n-1} + \boldsymbol{B}^{n-1}) \right] \tag{3.20}$$

$$\boldsymbol{u}^P = \boldsymbol{u}^n + \Delta t \left[-\nabla P^n + \frac{23}{12}(\boldsymbol{A}^n + \boldsymbol{B}^n) \right.$$
$$\left. - \frac{16}{12}(\boldsymbol{A}^{n-1} + \boldsymbol{B}^{n-1}) + \frac{5}{12}(\boldsymbol{A}^{n-2} + \boldsymbol{B}^{n-2}) \right], \tag{3.21}$$

respectively. The second-order accurate Adams–Bashforth method is often employed and the third-order accurate time integration scheme has good balance of stability and computation cost. Formulating the pressure term to be more accurate than first order is not common because P^{n+1} acts merely as a scalar potential, as discussed with Eq. (3.9) to satisfy the continuity equation of $\nabla \cdot \boldsymbol{u}^{n+1} = 0$ for every instantaneous flow field [24].

To improve numerical stability, it is effective to implicitly treat the viscous term. The influence of viscous diffusion is especially important near the walls where fine grid resolution is often employed. For constant viscosity ν, the viscous term becomes

$B = \nu\nabla^2 u$, which can be easily handled in an implicit manner. We can time step
the nonlinear advection term with the second-order Adams–Bashforth method and
change the viscous integration to be performed with the Crank–Nicolson method
(see Sect. 2.4.1). For the above choices of time stepping, the prediction step in the
SMAC method can be expressed as

$$u^P - \Delta t \frac{\nu}{2}\nabla^2 u^P = u^n + \Delta t\left(-\nabla P^n + \frac{3A^n - A^{n-1}}{2} + \frac{\nu}{2}\nabla^2 u^n\right). \tag{3.22}$$

By treating the viscous term implicitly, we must now solve this parabolic PDE for
the velocity in addition to an elliptic PDE, Eq. (3.16), to determine pressure. As we
will discuss later, the use of a staggered grid will define the variables at different
locations on the grid resulting in different finite-difference formula for the different
components of the velocity and pressure variables. Thus, programming may be more
laborious. Even if we treat the viscous term implicitly, the computational effort does
not significantly increase for each time step. As the velocity field is corrected with
$u^{n+1} = u^P - \Delta t\nabla\phi$, we must solve the Poisson equation, Eq. (3.16), to satisfy
$\nabla \cdot u^{n+1} = 0$, similar to how previous schemes have been formulated. Then the
pressure variable is updated with

$$P^{n+1} = P^n + \phi - \frac{\nu}{2}\Delta t\nabla^2\phi, \tag{3.23}$$

where the Laplacian ∇^2 in the last term should be discretized in the same manner
as in the pressure Poisson equation. What we have described above is based on
the algorithm by Kim and Moin [16] with appropriate modification for the SMAC
method.

We have presented an example of increasing the order of accuracy and enhancing
the numerical stability of the time advancement. For most engineering applications,
the second-order temporary accuracy should be sufficient. The choice to implicitly
time step the viscous term is dependent on how the grid is concentrated near the wall.
While it is difficult to say what the criteria is for implicitly treating the viscous term, it
would be wise to adopt the implicit approach if the overall amount of computation can
be saved by increasing the size of the time step even with the increase in computation
from the implicit solver. There is an option to treat the nonlinear advection term semi-
implicitly to allow the use of larger time steps. However, it may be more effective to
select a low storage Runge–Kutta method [33] for larger time steps while achieving
higher temporal accuracy.

When we select a numerical scheme, we should evaluate the overall work efficiency instead of just the computational efficiency. If the objective is to develop a
program that is to be used routinely, it would be worth investing the time to achieve
computational efficiency to the best we can. For a program that is to be used only
once, it may be sensible to choose a scheme that is easy to code, even if the efficiency
is somewhat sacrificed. The choice of schemes will depend on how the program is
to be used.

3.3.5   Summary of Time Stepping for Incompressible Flow Solvers

To analyze the temporal accuracy of MAC-based incompressible flow solvers, let us summarize the time stepping approaches in matrix based on [24]. We start from the incompressible Navier–Stokes equations in the following discretized form

$$\frac{u^{n+1} - u^n}{\Delta t} = -GP^{n+1} + A^n + \nu Lu^n \tag{3.24}$$

$$Du^{n+1} = 0, \tag{3.25}$$

where we consider constant density and viscosity with no external force. Here, A is the advective term, G is the discrete gradient operator, L is the discrete Laplacian operator, and D is the discrete divergence operator. For the ease of discussion on time-stepping schemes, we assume that the enforcement of boundary conditions does not generate (nonzero) inhomogeneous vectors on the right-hand side (see discussions in [24]). The enforcement of boundary conditions is discussed later in Sect. 3.8. Observe that L is a square matrix but G and D are non-square matrices. The above discretization is based on the simple explicit Euler scheme, which will be replaced with second-order methods in what follows.

Fractional-Step (Projection) Method

We can express Eqs. (3.24) and (3.25) together in a matrix equation form

$$\begin{bmatrix} I & \Delta t G \\ D & 0 \end{bmatrix} \begin{bmatrix} u^{n+1} \\ P^{n+1} \end{bmatrix} = \begin{bmatrix} u^n + \Delta t(A^n + \nu Lu^n) \\ 0 \end{bmatrix}, \tag{3.26}$$

where I is the identity matrix. Performing a block LU decomposition on the left-hand side matrix, we can split the above equation into two equations

$$\begin{bmatrix} I & 0 \\ D & -\Delta t DG \end{bmatrix} \begin{bmatrix} u^F \\ P^* \end{bmatrix} = \begin{bmatrix} u^n + \Delta t(A^n + \nu Lu^n) \\ 0 \end{bmatrix} \tag{3.27}$$

$$\begin{bmatrix} I & \Delta t G \\ 0 & I \end{bmatrix} \begin{bmatrix} u^{n+1} \\ P^{n+1} \end{bmatrix} = \begin{bmatrix} u^F \\ P^* \end{bmatrix}, \tag{3.28}$$

where $P^* = P^{n+1}$. Expanding the equations out, we arrive at the fractional-step (projection) method expressed in the traditional manner

$$u^F = u^n + \Delta t(A^n + \nu Lu^n), \tag{3.29}$$

$$DGP^{n+1} = \frac{1}{\Delta t} Du^F, \tag{3.30}$$

$$u^{n+1} = u^F - \Delta t GP^{n+1}. \tag{3.31}$$

It should be noted that for the discrete operators $DG \neq L$ (the discrete Laplacian operators for pressure and velocity variables are of different sizes).

SMAC Method (Predictor–Corrector Method)

By setting $\delta P = P^{n+1} - P^n$, Eqs. (3.24) and (3.25) can be expressed as

$$\begin{bmatrix} I & \Delta t G \\ D & 0 \end{bmatrix} \begin{bmatrix} u^{n+1} \\ \delta P \end{bmatrix} = \begin{bmatrix} u^n + \Delta t(-GP^n + A^n + \nu L u^n) \\ 0 \end{bmatrix}. \qquad (3.32)$$

With the application of the LU decomposition, we find

$$\begin{bmatrix} I & 0 \\ D & -\Delta t DG \end{bmatrix} \begin{bmatrix} u^P \\ \delta P^* \end{bmatrix} = \begin{bmatrix} u^n + \Delta t(-GP^n A^n + \nu L u^n) \\ 0 \end{bmatrix} \qquad (3.33)$$

$$\begin{bmatrix} I & \Delta t G \\ 0 & I \end{bmatrix} \begin{bmatrix} u^{n+1} \\ \delta P \end{bmatrix} = \begin{bmatrix} u^P \\ \delta P^* \end{bmatrix}, \qquad (3.34)$$

where $\delta P^* = \delta P$. This decomposition yields the time-stepping algorithm for the SMAC method

$$u^P = u^n + \Delta t(-GP^n + A^n + \nu L u^n), \qquad (3.35)$$

$$DG\delta P = \frac{1}{\Delta t} D u^P, \qquad (3.36)$$

$$u^{n+1} = u^P - \Delta t G(\delta P) \qquad (3.37)$$

$$P^{n+1} = P^n + \delta P. \qquad (3.38)$$

As discussed earlier, the pressure gradient term GP^n is used to find the prediction u^P.

Higher-Order Accurate Time Stepping

Let us use the second-order Adams–Bashforth method for the advection term A since it is nonlinear. On the other hand, we choose the second-order Crank–Nicolson method for the viscous term Lu, because it is linear and viscosity ν is constant, which makes it easy to treat implicitly. For these choices of time-stepping methods, the momentum equation can be discretized in time in a second-order manner

$$\frac{u^{n+1} - u^n}{\Delta t} = -GP^{n+1} + \frac{1}{2}\left(3A^n - A^{n-1}\right) + \frac{1}{2}\nu L\left(u^{n+1} + u^n\right). \qquad (3.39)$$

Let us now use the following notation for simplicity

$$R = I - \frac{\Delta t}{2}\nu L, \quad S = I + \frac{\Delta t}{2}\nu L \qquad (3.40)$$

and re-express Eq. (3.39) as

$$Ru^{n+1} + \Delta t G P^{n+1} = Su^n + \frac{\Delta t}{2} \left(3A^n - A^{n-1} \right). \tag{3.41}$$

We can combine Eqs. (3.41) and (3.25) into the matrix equation below

$$\begin{bmatrix} R & \Delta t G \\ D & 0 \end{bmatrix} \begin{bmatrix} u^{n+1} \\ P^{n+1} \end{bmatrix} = \begin{bmatrix} Su^n + \frac{\Delta t}{2}(3A^n - A^{n-1}) \\ 0 \end{bmatrix}. \tag{3.42}$$

Performing an LU decomposition of the left-hand side matrix, we find that

$$\begin{bmatrix} R & \Delta t G \\ D & 0 \end{bmatrix} = \begin{bmatrix} R & 0 \\ D & -\Delta t D R^{-1} G \end{bmatrix} \begin{bmatrix} I & \Delta t R^{-1} G \\ 0 & I \end{bmatrix}, \tag{3.43}$$

which allows us to arrive at the algorithm for the fractional-step method

$$Ru^F = Su^n + \frac{\Delta t}{2} \left(3A^n - A^{n-1} \right), \tag{3.44}$$

$$DR^{-1}GP^{n+1} = \frac{1}{\Delta t} Du^F, \tag{3.45}$$

$$u^{n+1} = u^F - \Delta t R^{-1} G P^{n+1}. \tag{3.46}$$

We can determine the fractional-step velocity u^F using a matrix solver for Eq. (3.44). However, the determination of the pressure field P^{n+1} is made cumbersome due to the appearance of R^{-1} in the left-hand side operator $DR^{-1}G$ in Eq. (3.45). Hence, it is impractical to directly solve Eq. (3.45) as a nested iteration is required.

In response to the above point, Kim and Moin [16] proposed the following fractional-step method

$$Ru^F = Su^n + \frac{\Delta t}{2} \left(3A^n - A^{n-1} \right), \tag{3.47}$$

$$DG\varphi = \frac{1}{\Delta t} Du^F, \tag{3.48}$$

$$u^{n+1} = u^F - \Delta t G\varphi, \tag{3.49}$$

$$P^{n+1} = R\varphi. \tag{3.50}$$

From Eq. (3.50), we have $\varphi = R^{-1} P^{n+1}$, which can be substituted into Eqs. (3.48) and (3.49) to yield

$$DGR^{-1}P^{n+1} = \frac{1}{\Delta t} Du^F, \tag{3.51}$$

$$u^{n+1} = u^F - \Delta t G R^{-1} P^{n+1}. \tag{3.52}$$

Comparing the above two equations with Eqs. (3.45) and (3.46), we observe that the operator acting on the pressure variable has been changed from $R^{-1}G$ to GR^{-1}. It should be noted that R^{-1} and G are not commutative (i.e., $R^{-1}G \neq GR^{-1}$). In the discrete form, GR^{-1} cannot be performed due to dimensional inconsistency [24]. While Eq. (3.47) appears to preserve second-order accuracy in time, the existence of the commutativity error leads to the loss of second-order accuracy of the overall method. This issue has been widespread in many variants of the fractional-step method and should be treated with care.

To avoid the loss of temporal accuracy and have compatible matrix operations, Perot [24] expands R^{-1} in Eq. (3.43) (or equivalently in Eqs. (3.44) and (3.46)) with a series expansion of

$$R^{-1} = \left(I - \frac{\Delta t}{2}\nu L\right)^{-1} = I + \frac{\Delta t}{2}\nu L + \left(\frac{\Delta t}{2}\nu L\right)^2 + \ldots \qquad (3.53)$$

By retaining two or more terms in the above expansion, second-order temporal accuracy can be achieved.

Another approach to maintain the desired temporal accuracy is to utilize the pressure difference (delta form) $\delta P = P^{n+1} - P^n$ in Eq. (3.42) to obtain

$$\begin{bmatrix} R & \Delta t G \\ D & 0 \end{bmatrix}\begin{bmatrix} u^{n+1} \\ \delta P \end{bmatrix} = \begin{bmatrix} -\Delta t G P^n + S u^n + \frac{\Delta t}{2}(3A^n - A^{n-1}) \\ 0 \end{bmatrix}. \qquad (3.54)$$

Let us premultiply the top right entry of the left-hand side matrix by R and perform an LU decomposition

$$\begin{bmatrix} R & \Delta t R G \\ D & 0 \end{bmatrix} = \begin{bmatrix} R & 0 \\ D & -\Delta t D G \end{bmatrix}\begin{bmatrix} I & \Delta t G \\ 0 & I \end{bmatrix} \qquad (3.55)$$

which produces the following solution algorithm

$$Ru^P = -\Delta t G P^n + S u^n + \frac{\Delta t}{2}\left(3A^n - A^{n-1}\right), \qquad (3.56)$$

$$DG\delta P = \frac{1}{\Delta t}Du^P, \qquad (3.57)$$

$$u^{n+1} = u^P - \Delta t G \delta P, \qquad (3.58)$$

$$P^{n+1} = P^n + \delta P. \qquad (3.59)$$

The approximation made by Eq. (3.55) for Eq. (3.54) introduces an error of

$$(I - R)\delta P = \frac{\Delta t}{2}\nu L \delta P = \mathcal{O}(\Delta t^2) \qquad (3.60)$$

which can be expected to be second-order given $\delta P = \mathcal{O}(\Delta t)$. With the predictor-based formulation, the loss of temporal accuracy is avoided using difference of pressure, δP.

Delta Formulation

The implicit velocity term from the Crank–Nicolson method can be factorized to solve for the velocity field. The method of Dukowicz and Dvinsky [6] utilizes the delta formulation to reduce the loss of the temporal accuracy of the overall solver even with the use of factorization of the implicit operator R. Let us take Eq. (3.55) and express the unknown velocity field in terms of the delta formation $\delta u = u^{n+1} - u^n$

$$\begin{bmatrix} R & \Delta t RG \\ D & 0 \end{bmatrix} \begin{bmatrix} \delta u \\ \delta P \end{bmatrix} = \begin{bmatrix} -\Delta t G P^n + \Delta t \nu L u^n + \frac{\Delta t}{2}(3A^n - A^{n-1}) \\ -Du^n \end{bmatrix}. \tag{3.61}$$

Through an LU decomposition, the above equation can be decomposed into the following two matrix equations

$$\begin{bmatrix} R & 0 \\ D & -\Delta t DG \end{bmatrix} \begin{bmatrix} \delta u^* \\ \delta P^* \end{bmatrix} = \begin{bmatrix} -\Delta t G P^n + \Delta t \nu L u^n + \frac{\Delta t}{2}(3A^n - A^{n-1}) \\ -Du^n \end{bmatrix} \tag{3.62}$$

$$\begin{bmatrix} I & \Delta t G \\ 0 & I \end{bmatrix} \begin{bmatrix} \delta u \\ \delta P \end{bmatrix} = \begin{bmatrix} \delta u^* \\ \delta P^* \end{bmatrix}, \tag{3.63}$$

where $\delta u^* = u^* - u^n$ and $\delta P^* = P^* - P^n$. Equations (3.62) and (3.63) can be expanded as

$$R\delta u^* = -\Delta t G P^n + \Delta t \nu L u^n + \frac{\Delta t}{2}\left(3A^n - A^{n-1}\right), \tag{3.64}$$

$$u^* = u^n + \delta u^*, \tag{3.65}$$

$$DG\delta P = \frac{1}{\Delta t}Du^*, \tag{3.66}$$

$$u^{n+1} = u^* - \Delta t G\delta P, \tag{3.67}$$

$$P^{n+1} = P^n + \delta P. \tag{3.68}$$

The first two equations here are essentially the same as Eq. (3.56). The subtle but important difference of the approach taken by Dukowicz and Dvinsky is to determine the velocity difference δu^*. This treatment becomes useful when the equation with the implicit operator $R = I - \frac{\Delta t}{2}\nu L$ is solved using factorization in each direction. For a Cartesian grid, the factorization yields

$$I - \frac{\Delta t}{2}\nu L = \left(I - \frac{\Delta t}{2}\nu L_x\right)\left(I - \frac{\Delta t}{2}\nu L_y\right)\left(I - \frac{\Delta t}{2}\nu L_z\right) + \mathcal{O}(\Delta t^2) \tag{3.69}$$

with a truncation error of $\mathcal{O}(\Delta t^2)$. Denoting the right-hand side of Eq. (3.64) by E, the velocity field δu^* can be determined with a three-step algorithm

$$\left.\begin{aligned}
\left(I - \frac{\Delta t}{2}\nu L_x\right) \delta u'' &= E \\
\left(I - \frac{\Delta t}{2}\nu L_y\right) \delta u' &= \delta u'' \\
\left(I - \frac{\Delta t}{2}\nu L_z\right) \delta u^* &= \delta u'
\end{aligned}\right\}, \tag{3.70}$$

which can reduce the amount of computation. This method is able to maintain the error low from factorization because Eq. (3.64) solves for the velocity difference that is of order Δt, instead of velocity.

3.4 Spatial Discretization of Pressure Gradient Term

When the MAC-based method is utilized to solve for the incompressible flow field, it is desirable to use a staggered grid instead of the regular grid,[3] as illustrated in Fig. 3.3. We discuss why the staggered grid is advantageous later. In this section, we present our discussion for two-dimensional flow on a Cartesian grid of (x, y). Extension to three-dimensional cases is straightforward.

On a staggered grid, scalar quantities such as the pressure variable are placed on the cell center and the velocity vectors are defined on the cell face, as illustrated in Fig. 3.3b. For computer programs, we use the notation of $u_{i,j}$ and $v_{i,j}$, instead of $u_{i+\frac{1}{2},j}$ and $v_{i,j+\frac{1}{2}}$, because the indices need to be in integer format. While we use the fractional notation for presenting the spatial discretization schemes below, we ask readers to translate those in Fig. 3.3b to the notation without fractional indices for coding purpose, as shown in Fig. 3.4.

3.4.1 Pressure Poisson Equation

First, let us consider the case of using a uniform grid with Δx and Δy as the spacings in the x and y directions. The correctional step in the SMAC method (Eq. (3.15)) for a staggered grid is shown below. Employing the second-order accurate formulation for the four discrete velocities around the position for pressure $p_{i,j}$ (cell center) in Fig. 3.3b, we find

$$\left.\begin{aligned}
u^{n+1}_{i-\frac{1}{2},j} &= u^{P}_{i-\frac{1}{2},j} - \Delta t\frac{-\phi_{i-1,j} + \phi_{i,j}}{\Delta x} \\
u^{n+1}_{i+\frac{1}{2},j} &= u^{P}_{i+\frac{1}{2},j} - \Delta t\frac{-\phi_{i,j} + \phi_{i+1,j}}{\Delta x}
\end{aligned}\right\} \tag{3.71}$$

---

[3] We make a distinction between the regular and collocated grids. The regular grid places all variables at the same location. On the other hand, the collocated grid places the fluxes between where u, v, and p are positioned. For further details, see Sect. 4.4.

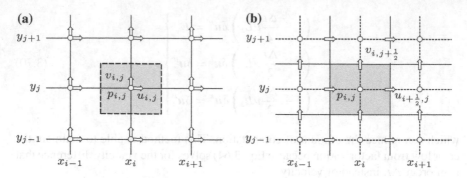

Fig. 3.3 Cartesian grid discretization. The gray box corresponds to the control volume for (x_i, y_j). **a** Regular grid. **b** Staggered grid

Fig. 3.4 Staggered grid notation without using 1/2 indices

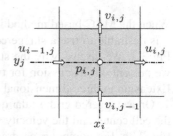

$$v^{n+1}_{i,j-\frac{1}{2}} = v^P_{i,j-\frac{1}{2}} - \Delta t \frac{-\phi_{i,j-1} + \phi_{i,j}}{\Delta y} \left.\vphantom{\begin{matrix} a \\ a \end{matrix}}\right\}$$
$$\left. v^{n+1}_{i,j+\frac{1}{2}} = v^P_{i,j+\frac{1}{2}} - \Delta t \frac{-\phi_{i,j} + \phi_{i,j+1}}{\Delta y} \right\}. \qquad (3.72)$$

Substituting these velocity values into the continuity equation,

$$\frac{-u^{n+1}_{i-\frac{1}{2},j} + u^{n+1}_{i+\frac{1}{2},j}}{\Delta x} + \frac{-v^{n+1}_{i,j-\frac{1}{2}} + v^{n+1}_{i,j+\frac{1}{2}}}{\Delta y} = 0, \qquad (3.73)$$

we obtain the pressure Poisson equation

$$\frac{\phi_{i-1,j} - 2\phi_{i,j} + \phi_{i+1,j}}{\Delta x^2} + \frac{\phi_{i,j-1} - 2\phi_{i,j} + \phi_{i,j+1}}{\Delta y^2}$$
$$= \frac{1}{\Delta t} \left(\frac{-u^P_{i-\frac{1}{2},j} + u^P_{i+\frac{1}{2},j}}{\Delta x} + \frac{-v^P_{i,j-\frac{1}{2}} + v^P_{i,j+\frac{1}{2}}}{\Delta y} \right), \qquad (3.74)$$

where the left-hand side represents the second-order central difference (Eq. (2.51)) of

Fig. 3.5 Nonuniform
staggered grid

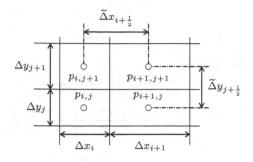

$$\nabla^2 \phi = \frac{\partial^2 \phi}{\partial x^2} + \frac{\partial^2 \phi}{\partial y^2}. \tag{3.75}$$

In general, it is not advisable to discretize $\nabla^2 \phi$ in an arbitrary manner. As an example of how issues can arise with arbitrary differencing, let us consider a nonuniform grid shown in Fig. 3.5. Letting the grid spacings be Δx_i and Δy_j for the (i, j)-th cell, the corresponding expressions for Eqs. (3.71) and (3.72) are

$$\left. \begin{aligned} u_{i-\frac{1}{2},j}^{n+1} &= u_{i-\frac{1}{2},j}^{P} - \Delta t \frac{-\phi_{i-1,j} + \phi_{i,j}}{\widetilde{\Delta x}_{i-\frac{1}{2}}} \\ u_{i+\frac{1}{2},j}^{n+1} &= u_{i+\frac{1}{2},j}^{P} - \Delta t \frac{-\phi_{i,j} + \phi_{i+1,j}}{\widetilde{\Delta x}_{i+\frac{1}{2}}} \end{aligned} \right\} \tag{3.76}$$

$$\left. \begin{aligned} v_{i,j-\frac{1}{2}}^{n+1} &= v_{i,j-\frac{1}{2}}^{P} - \Delta t \frac{-\phi_{i,j-1} + \phi_{i,j}}{\widetilde{\Delta y}_{j-\frac{1}{2}}} \\ v_{i,j+\frac{1}{2}}^{n+1} &= v_{i,j+\frac{1}{2}}^{P} - \Delta t \frac{-\phi_{i,j} + \phi_{i,j+1}}{\widetilde{\Delta y}_{j+\frac{1}{2}}} \end{aligned} \right\}, \tag{3.77}$$

where

$$\widetilde{\Delta x}_{i+\frac{1}{2}} = \frac{\Delta x_i + \Delta x_{i+1}}{2}, \quad \widetilde{\Delta y}_{j+\frac{1}{2}} = \frac{\Delta y_j + \Delta y_{j+1}}{2} \tag{3.78}$$

are the distances between the cell centers (where the discrete pressure is positioned).[4]
Substituting these velocity expressions into the continuity equation

---

[4]Let us consider an interval of $-1 \le x \le 1$, discretized nonuniformly into N cells with velocity defined on

$$x_j^u = -\cos \frac{\pi(j-1/2)}{N}, \quad j = \frac{1}{2}, \frac{3}{2}, \dots, N + \frac{1}{2}$$

In this case, there are two ways to place the pressure variables [22]. One way is to use the midpoints of where the velocity variables are positioned

$$x_j^p = \frac{1}{2}(x_{j-\frac{1}{2}}^u + x_{j+\frac{1}{2}}^u), \quad j = 1, 2, \dots, N.$$

$$\frac{-u_{i-\frac{1}{2},j}^{n+1} + u_{i+\frac{1}{2},j}^{n+1}}{\Delta x_i} + \frac{-v_{i,j-\frac{1}{2}}^{n+1} + v_{i,j+\frac{1}{2}}^{n+1}}{\Delta y_j} = 0, \tag{3.79}$$

we obtain

$$-\frac{-\phi_{i-1,j} + \phi_{i,j}}{\Delta x_i \tilde{\Delta} x_{i-\frac{1}{2}}} + \frac{-\phi_{i,j} + \phi_{i+1,j}}{\Delta x_i \tilde{\Delta} x_{i+\frac{1}{2}}}$$
$$-\frac{-\phi_{i,j-1} + \phi_{i,j}}{\Delta y_j \tilde{\Delta} y_{j-\frac{1}{2}}} + \frac{-\phi_{i,j} + \phi_{i,j+1}}{\Delta y_j \tilde{\Delta} y_{j+\frac{1}{2}}} \tag{3.80}$$
$$= \frac{1}{\Delta t} \left(\frac{-u_{i-\frac{1}{2},j}^{P} + u_{i+\frac{1}{2},j}^{P}}{\Delta x_i} + \frac{-v_{i,j-\frac{1}{2}}^{P} + v_{i,j+\frac{1}{2}}^{P}}{\Delta y_j} \right).$$

Observe that the differencing for $\nabla^2 \phi$ on the left-hand side is different from the second-order finite-difference formula found in Eq. (2.35) for nonuniform grids. This implies that ∇^2 on the left-hand side of the pressure Poisson equation, Eq. (3.16), should not be discretized arbitrarily. In principle, one should derive the pressure Poisson equation by inserting the discrete pressure gradient into the discrete continuity equation, as we have performed to arrive at Eq. (3.80). Otherwise the numerical solution to the Poisson equation does not satisfy the discrete continuity equation. This difference can appear not only for nonuniform grids but also for non-orthogonal grids and high-order finite-differencing schemes. To emphasize the importance of the compatibility of discretization, we express the discrete version of $\nabla^2 \phi$ as $\delta_x(\delta_x \phi) + \delta_y(\delta_y \phi)$, instead of $\delta_x^2 \phi + \delta_y^2 \phi$.

Next, let us discuss why the use of a staggered grid is recommended. Consider discretizing the pressure Poisson equation on a regular grid. The continuity equation about point (x_i, y_j) in Fig. 3.6 becomes

$$\frac{-u_{i-1,j}^{n+1} + u_{i+1,j}^{n+1}}{2\Delta x} + \frac{-v_{i,j-1}^{n+1} + v_{i,j+1}^{n+1}}{2\Delta y} = 0. \tag{3.81}$$

Take a close look at the control volume used for momentum balance shown in Fig. 3.6. Equation (3.73) performs a flux balance for a region boxed in by points

---

(Footnote 4 continued)

In this case, the distance between discrete pressure $\tilde{\Delta}_{j+\frac{1}{2}}$ is described in the text above. On the other hand, one can use

$$x_j^P = -\cos \frac{\pi(j - 1/2)}{N}, \quad j = 1, 2, \ldots, N$$

as a mapping function. This approach allows for a smooth distribution of

$$\tilde{\Delta}_{j+\frac{1}{2}} = -x_j^P + x_{j+1}^P$$

but such grid distribution based on a functional description is limited to spacial cases and is not common.

Fig. 3.6 Velocity and pressure coupling on a regular grid

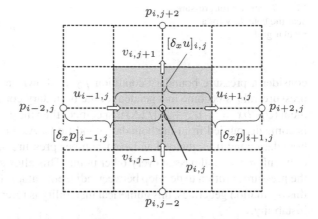

$(\pm\Delta x/2, \pm\Delta y/2)$. On the other hand, Eq. (3.81) considers a region, bound by points $(\pm\Delta x, \pm\Delta y)$ that is double in size in both directions, as the control volume. Because Eq. (3.81) does not represent a discretization scheme based on a control volume that shares a cell face with adjacent cells, it is not suitable for flux balancing.

In the final step of the SMAC method, we add the gradient of ϕ to values of $u_{i\pm1,j}$ and $u_{i,j\pm1}$ in Eq. (3.81):

$$\left.\begin{aligned}
u_{i-1,j}^{n+1} &= u_{i-1,j}^{P} - \Delta t\,\frac{-\phi_{i-2,j} + \phi_{i,j}}{2\Delta x} \\
u_{i+1,j}^{n+1} &= u_{i+1,j}^{P} - \Delta t\,\frac{-\phi_{i,j} + \phi_{i+2,j}}{2\Delta x}
\end{aligned}\right\} \tag{3.82}$$

$$\left.\begin{aligned}
v_{i,j-1}^{n+1} &= v_{i,j-1}^{P} - \Delta t\,\frac{-\phi_{i,j-2} + \phi_{i,j}}{2\Delta y} \\
v_{i,j+1}^{n+1} &= v_{i,j+1}^{P} - \Delta t\,\frac{-\phi_{i,j} + \phi_{i,j+2}}{2\Delta y}
\end{aligned}\right\} \tag{3.83}$$

Now, we substitute the above velocity expressions, Eqs. (3.82) and (3.83), into Eq. (3.81) to solve for ϕ that provides the gradient correction ($\nabla\phi$) to make velocity at next time step satisfy incompressibility:

$$\frac{\phi_{i-2,j} - 2\phi_{i,j} + \phi_{i+2,j}}{(2\Delta x)^2} + \frac{\phi_{i,j-2} - 2\phi_{i,j} + \phi_{i,j+2}}{(2\Delta y)^2}$$
$$= \frac{1}{\Delta t}\left(\frac{-u_{i-1,j}^{P} + u_{i+1,j}^{P}}{\Delta x} + \frac{-v_{i,j-1}^{P} + v_{i,j+1}^{P}}{\Delta y}\right). \tag{3.84}$$

Note that the second-derivative differencing scheme on the left-hand side of Eq. (3.84) consists of the values ($\phi_{i\pm2,j},\ \phi_{i,j\pm2}$) skipping every other point about $\phi_{i,j}$. If we

Fig. 3.7 Stencil for pressure
near the boundary for a
regular grid

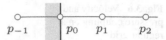

consider a pressure boundary condition p_0 as shown in Fig. 3.7, the odd-number
pressure values become independent of the boundary condition. If the pressure gradient $(\partial p/\partial x = [-p_{-1}+p_1]/2\Delta)$ is specified, then the even-number pressure points
become decoupled from the boundary conditions. As a result for either case of the
boundary condition, the odd and even-number pressure values are not coupled and
only enforce smoothness at every other point. This allows for spatial oscillations in
the pressure solution to develop between adjacent values. Because of the pattern that
this oscillation generates, this numerical instability is referred to as the checkerboard
instability.

If we try to suppress the pressure oscillation by replacing the left-hand side of
Eq. (3.84) with the left-hand side of Eq. (3.74) on a regular grid, that equation would
create incompatibility between the differencing of the continuity equation and the
pressure gradient. The use of a solution from such discrete pressure Poisson equation does not satisfy mass balance, which leads to error accumulation and eventual
deterioration of the computation.

The use of a staggered grid on the other hand, resolves this problem by shifting
the locations for discrete velocities by half a mesh width from the pressure locations.
In summary, the use of the staggered grid leads to discretizing the relationship

$$\frac{\partial^2 \phi}{\partial x^2} = \frac{\partial}{\partial x}\frac{\partial \phi}{\partial x} \tag{3.85}$$

with a compatible finite-difference scheme

$$\frac{\partial^2 \phi}{\partial x^2} = \delta_x(\delta_x \phi) \tag{3.86}$$

and simultaneously preventing the spatial oscillation in pressure to appear.

3.4.2 Iterative Method for the Pressure Poisson Equation

There are numerous solution techniques for elliptic partial differential equations.
The development of elliptic solvers by itself constitutes a field of research. For the
purpose of our discussion, we only discuss basic iterative methods.

First, let us rewrite Eq. (3.80) as

$$B_{y,j}^- \phi_{i,j-1} + B_{x,i}^- \phi_{i-1,j} - B_{i,j}^0 \phi_{i,j} + B_{x,i}^+ \phi_{i+1,j} + B_{y,j}^+ \phi_{i,j+1} = \psi_{i,j}, \tag{3.87}$$

where the coefficients are

$$
\begin{aligned}
B_{x,i}^- &= 1/(\Delta x_i \tilde{\Delta} x_{i-\frac{1}{2}}), \quad B_{x,i}^+ = 1/(\Delta x_i \tilde{\Delta} x_{i+\frac{1}{2}}), \\
B_{y,j}^- &= 1/(\Delta y_j \tilde{\Delta} y_{j-\frac{1}{2}}), \quad B_{y,j}^+ = 1/(\Delta y_j \tilde{\Delta} y_{j+\frac{1}{2}}), \\
B_{i,j}^0 &= B_{x,i}^- + B_{x,i}^+ + B_{y,j}^- + B_{y,j}^+
\end{aligned}
\tag{3.88}
$$

and ψ represents the right-hand side of the Poisson equation.

We denote the iteration number by $\langle m \rangle$. As an example, consider an initial guess of $\phi_{i,j}^{\langle 0 \rangle} = 0$ everywhere (although we can use a better initial guess if available), which yields Eq. (3.87) to have an error associated with that guess. Based on the error, we can gradually adjust the solution ϕ. This procedure is repeated to produce $\phi_{i,j}^{\langle 1 \rangle}$, $\phi_{i,j}^{\langle 2 \rangle}$, $\phi_{i,j}^{\langle 3 \rangle}$, ... and once convergence is achieved, the solution is found for Eq. (3.87). This is referred to as an *iterative method* (*relaxation method*).

The *Jacobi method* solves for $\phi_{i,j}$ in Eq. (3.87) by using solution from the previous iteration to compute the right-hand side:

$$
\phi_{i,j}^{\langle m+1 \rangle} = \frac{B_{y,j}^- \phi_{i,j-1}^{\langle m \rangle} + B_{x,i}^- \phi_{i-1,j}^{\langle m \rangle} + B_{x,i}^+ \phi_{i+1,j}^{\langle m \rangle} + B_{y,j}^+ \phi_{i,j+1}^{\langle m \rangle} - \psi_{i,j}}{B_{i,j}^0}
\tag{3.89}
$$

This equation updates $\phi_{i,j}^{\langle m+1 \rangle}$ using values of $\phi_{i\pm1,j}^{\langle m \rangle}$ and $\phi_{i,j\pm1}^{\langle m \rangle}$. The above equation can be expressed as

$$
\phi_{i,j}^{\langle m+1 \rangle} = \phi_{i,j}^{\langle m \rangle} + \frac{E_{i,j}^{\langle m \rangle}}{B_{i,j}^0},
\tag{3.90}
$$

where

$$
\begin{aligned}
E_{i,j}^{\langle m \rangle} &= B_{y,j}^- \phi_{i,j-1}^{\langle m \rangle} + B_{x,i}^- \phi_{i-1,j}^{\langle m \rangle} - B_{i,j}^0 \phi_{i,j}^{\langle m \rangle} \\
&\quad + B_{x,i}^+ \phi_{i+1,j}^{\langle m \rangle} + B_{y,j}^+ \phi_{i,j+1}^{\langle m \rangle} - \psi_{i,j}
\end{aligned}
\tag{3.91}
$$

represents the residual from the iterative solver after the m-th iteration (for satisfying Eq. (3.87)). Hence, we can regard Eq. (3.90) as a finite-difference approximation of

$$
\phi_{i,j}^{\langle m+1 \rangle} = \phi_{i,j}^{\langle m \rangle} + \frac{1}{B_{i,j}^0} (\nabla^2 \phi_{i,j} - \psi_{i,j})^{\langle m \rangle}.
\tag{3.92}
$$

When the iterative solution satisfies $\phi_{i,j}^{\langle m+1 \rangle} = \phi_{i,j}^{\langle m \rangle}$, the term $E_{i,j}^{\langle m \rangle}$ becomes zero and tells us that Eq. (3.91) has been solved. Let us examine the Jacobi iteration from the viewpoint of the Newton–Raphson method. Fixing $\phi_{i\pm1,j}^{\langle m \rangle}$ and $\phi_{i,j\pm1}^{\langle m \rangle}$ and treating E_{ij} as a function of ϕ_{ij}, we approximate

$$
E_{i,j}^{\langle m+1 \rangle} - E_{i,j}^{\langle m \rangle} = \frac{\partial F_{i,j}}{\partial \phi_{i,j}} \left(\phi_{i,j}^{\langle m+1 \rangle} - \phi_{i,j}^{\langle m \rangle} \right).
\tag{3.93}
$$

To achieve $E_{i,j}^{\langle m+1 \rangle} = 0$ with the above equation, we should update $\phi_{i,j}$ in the following manner:

$$\phi_{i,j}^{\langle m+1 \rangle} = \phi_{i,j}^{\langle m \rangle} - \frac{\partial E_{i,j}^{\langle m \rangle}}{\partial E_{i,j}/\partial \phi_{i,j}}. \tag{3.94}$$

Because $\partial E_{i,j}/\partial \phi_{i,j} = -B_{i,j}^0$, we can observe that Eq. (3.94) is equivalent to Eq. (3.90). Once $\phi_{i,j}$ is determined satisfying $E_{i,j} = 0$, we move to the next point to update $\phi_{i+1,j}$ enforcing $E_{i+1,j} = 0$, which then offsets $E_{i,j}$ to become nonzero. Hence, it is necessary to iteratively update all values until the error at all points approach zero.

If we consider $1/B_{i,j}^0$ to be an artificial time step $\Delta\tau$, Eq. (3.92) can be thought of as an approximation to

$$\frac{\partial \phi}{\partial \tau} = \nabla^2 \phi - \psi, \tag{3.95}$$

where τ is a virtual time. This implies that the iterative scheme adds a time-varying term to the Poisson equation to make the elliptic PDE into a parabolic PDE. The converged solution (steady-state solution with respect to the virtual time τ) becomes the solution to the original elliptic PDE.

A sample program for Eq. (3.90) is provided below.

```
DO J = 1,NY
    DO I = 1,NX
        R(I,J) = BYM(J)*P(I,J-1) + BXM(I)*P(I-1,J) &
               - B0(I,J)*P(I,J) &
               + BXP(I)*P(I+1,J) + BYP(J)*P(I,J+1) &
               - Q(I,J)
    END DO
END DO
DO J = 1,NY
    DO I = 1,NX
        P(I,J) = P(I,J) + R(I,J)/B0(I,J)
    END DO
END DO
```

In the above snippet of the code, the residual is stored in R inside the first loop and P ($= \phi$) is updated in the second loop. Suppose we combine the two loops and modify the above code in the following fashion:

```
DO J = 1,NY
    DO I = 1,NX
        P(I,J) = P(I,J) &
               + (BYM(J)*P(I,J-1) + BXM(I)*P(I-1,J) &
               - B0(I,J)*P(I,J) &
               + BXP(I)*P(I+1,J) + BYP(J)*P(I,J+1) &
               - Q(I,J) ) / B0(I,J)
    END DO
END DO
```

By the time $P(I, J)$ is to be calculated, $P(I - 1, J)$ and $P(I, J - 1)$ are already updated. This algorithm allows not only for faster convergence but is simpler in programming and uses less memory. This method is called the *Gauss–Seidel method* and can be expressed as

$$\phi_{i,j}^{\langle m+1 \rangle} = \phi_{i,j}^{\langle m \rangle} + \frac{E_{i,j}^{\langle m* \rangle}}{B_{i,j}^0}, \tag{3.96}$$

where

$$E_{i,j}^{\langle m* \rangle} = B_{y,j}^- \phi_{i,j-1}^{\langle m+1 \rangle} + B_{x,i}^- \phi_{i-1,j}^{\langle m+1 \rangle} - B_{i,j}^0 \phi_{i,j}^{\langle m \rangle}$$

$$+ B_{x,i}^+ \phi_{i+1,j}^{\langle m \rangle} + B_{y,j}^+ \phi_{i,j+1}^{\langle m \rangle} - \psi_{i,j}. \tag{3.97}$$

When the residual $E_{i,j}$ is sufficiently small compared to the norm of $\psi_{i,j}$, the solution ϕ is considered to be converged. Assuming that the difference between the solutions from the m-th and the $(m + 1)$-th iterations is small, we can use the norm of $E_{i,j}^{\langle m* \rangle}$ to assess convergence.

There is also the well-known *SOR (successive over-relaxation) method*, which uses an over-relaxation parameter β $(1 < \beta < 2)$:

$$\phi_{i,j}^{\langle m+1 \rangle} = \phi_{i,j}^{\langle m \rangle} + \beta + \frac{E_{i,j}^{\langle m* \rangle}}{B_{i,j}^0}. \tag{3.98}$$

The optimal value of β is problem dependent but usually lies between 1.5 and 1.7.

While the rate of convergence depends on the uniformity of the grid and the type of boundary conditions, the aforementioned methods should converge to the correct solution. If convergence is extremely slow, we may have to consider the use of another solver. When the updated velocity u^{n+1} based on ϕ does not satisfy the continuity equation, we should revisit the finite-difference scheme utilized for the pressure Poisson equation, even if convergence is achieved. We should make sure that the boundary condition is correctly implemented and the finite-difference scheme for the pressure gradient is indeed inserted into the discrete continuity equation. That is, we must verify that the discretization schemes are compatible.

Convergence Criteria

For Eq. (3.89), we can consider the solution to be converged when the difference $\| \phi^{\langle m+1 \rangle} - \phi^{\langle m \rangle} \|$ becomes several orders of magnitude smaller than $\| \phi^{\langle m+1 \rangle} \|$. At that stage, we can terminate the iteration. Thus the convergence criteria can be set as

$$\frac{\| \phi^{\langle m+1 \rangle} - \phi^{\langle m \rangle} \|}{\| \phi^{\langle m+1 \rangle} \|} < \varepsilon', \tag{3.99}$$

where

$$\| f \| = \left[\frac{1}{N} \sum_{i=1}^{N} f_i^2 \right]^{1/2} \tag{3.100}$$

is the norm used above (known as the L_2 norm). The convergence tolerance ε' in Eq. (3.99) can be set to appropriate values such as 10^{-2} or 10^{-5} depending on the problem. The normalization by $\|\phi^{\langle m+1 \rangle}\|$ should be avoided if its value is supposed to be zero or very small. In such case, the unnormalized difference $\|\phi^{\langle m+1 \rangle} - \phi^{\langle m \rangle}\|$ can be used to examine the convergence.

With the criteria described by Eq. (3.99), it is not obvious how much error is left in Eq. (3.87). We therefore can use an alternative criteria that terminates the iteration when the residual $\|E\| = \|\nabla^2 \phi - \psi\|$ becomes smaller than the right-hand side of the Poisson equation $\|\psi\|$ by a few orders of magnitude

$$\|E\| / \|\psi\| < \varepsilon. \tag{3.101}$$

Generally speaking, there is no standard value for ε, but we can consider the solution to be converged when $\nabla \cdot \boldsymbol{u}$ is several orders of magnitude smaller than $|\boldsymbol{u}|/\Delta$ (where the grid size is Δ). If the flow becomes steady, the denominators in Eqs. (3.99) and (3.101) can become small. Moreover, as the grid uniformity and boundary conditions can affect the rate of convergence, it is strongly advised to set a limit on the number of iterations inside the solver.

When the flow being solved for is not too sensitive to errors, the MAC method is forgiving to the residual from the pressure equation. If the MAC method is correctly implemented, the error in $\nabla \cdot \boldsymbol{u}$ does not accumulate since it is projected out of the solution. Iterating the solution excessively is not required to achieve stable time advancement. Note however that convergence to high accuracy is required when the flow of interest is sensitive to perturbations, such as in turbulent flow.

Every computer suffers from round-off error and such error is referred to as machine epsilon. It is rare to force the solution to converge with residual reaching machine epsilon. The residual being larger than single-precision round-off error does not imply that single-precision computation is faster. Double-precision computation often allows for faster convergence.

Notes on Programming

Note that the aforementioned program uses the values of

$$\texttt{P(I,J-1), P(I-1,J), P(I,J), P(I+1,J), P(I,J+1)}$$

to evaluate the residual. This ordering is intended for programming languages that store data or order arrays *column-wise* (*column-major*), such as FORTRAN. For programming languages that order arrays *row-wise* (*row-major*), such as C/C++, the order for evaluating the above residual should be along the row. If reversed, after the data P(I, J) is referred, the pointer goes to the end of the memory and returns to the front of the memory to then call P(I − 1, J). This causes increase in memory access time. Some compilers can identify this issue and can correct the order during compilation. However, this point should be kept in mind while programming.

In this book, we express the differencing as $(-u_{i-\frac{1}{2},j} + u_{i+\frac{1}{2},j})$ in the order of increasing index (in the order of memory storage) instead of $(u_{i+\frac{1}{2},j} - u_{i-\frac{1}{2},j})$. It is a good habit to express the difference formula in this manner, instead of just implementing them in such order in the program.

For the same reason, rather than writing the do loop as

```
DO I = 1,NX
    DO J = 1,NY
        P(I,J) = ...
    END DO
END DO
```

it would be better to implement

```
DO J = 1,NY
    DO I = 1,NX
        P(I,J) = ...
    END DO
END DO
```

with the do loop for j on the outside. This would allow for quicker access to the memory locations for column-major data such as in Fortran. The order of the do loops needs to be reversed for row-major data structure, such as in C/C++.

The SOR method cannot compute $\phi_{i,j}^{\langle m+1 \rangle}$ until $\phi_{i-1,j}^{\langle m+1 \rangle}$ is computed and stored on memory. Since this results in recursive referencing, the algorithm in its current form is not suitable for a vector processor. To vectorize the algorithm, the most inner-loop can be separated into even and odd number indices.

```
DO J = 1,NY
    DO L = 1,2
        DO I = L,NX,2
            P(I,J) = P(I,J) + BETA* &
            ( BYM(J)*P(I,J-1) + BXM(I)*P(I-1,J) &
            - B0(I,J)*P(I,J) &
            + BXP(I)*P(I+1,J) + BYP(J)*P(I,J+1) &
            - Q(I,J) ) / B0(I,J)
        END DO
    END DO
END DO
```

Because the above program computes results for index $I - 1$ based on a separate loop from index I, it would not result in recursive referencing. In some cases, it is necessary to force vector instructions. There are also multi-color methods that are more effective for vectorized computation.

Fast Solution Algorithm Using FFT

When periodicity can be assumed for at least one of the directions for a variable, the use of Fast Fourier Transform (FFT) allows for an efficient solution algorithm. Here, we require an orthonormal coordinate system with a uniform grid in that periodic direction.[5] If the pressure field is periodic in the x and y directions, it is possible to use

---

[5]Grid stretching can be incorporated for certain cases. See Cain et al. [2] for details.

the two-dimensional Fourier transform. The pressure Poisson equation discretized with the second-order central difference, Eq. (3.74), can be transformed to become

$$
-\left[\frac{2(1-\cos k_x \Delta x)}{\Delta x^2} + \frac{2(1-\cos k_y \Delta y)}{\Delta y^2}\right]\widetilde{\phi}(k_x, k_y) = \widetilde{\psi}(k_x, k_y) \qquad (3.102)
$$

for wave numbers of k_x and k_y. The variable ψ represents the right-hand side of Eq. (3.74). Solving Eq. (3.102) in Fourier space is a simple algebraic operation (division) for each wave number. We need to perform a Fourier transform of the right-hand side to obtain $\widetilde{\psi}$, solve for $\widetilde{\phi}$ using Eq. (3.102), and perform the inverse transform to find the solution ϕ. Note that we must be careful for the wave number combination of $k_x = k_y = 0$. This algorithm requires forward and inverse discrete Fourier transforms but can be more efficient than using an iterative scheme to converge the solution with error level down to machine epsilon.

The FFT algorithm can be useful even when periodicity is applicable for only one of the directions (say x here). Performing Fourier transform in the periodic x direction, we find

$$
\frac{\widetilde{\phi}_{k_x, j-1}}{\Delta y^2} - \left[\frac{2(1-\cos k_x \Delta x)}{\Delta x^2} + \frac{2}{\Delta y^2}\right]\widetilde{\phi}_{k_x, j} + \frac{\widetilde{\phi}_{k_x, j+1}}{\Delta y^2} = \widetilde{\psi}(k_x, j). \qquad (3.103)
$$

In this case, we end up with the usual finite-difference approach in the y-direction (index j). While it is possible to employ an iterative solver, we can directly solve this equation since the equation reduces to a tridiagonal matrix.

For higher-order finite-difference methods, the finite-difference stencil becomes wider in real space and increases the amount of computation. By using the discrete Fourier transform, the amount of computation remains effectively the same while the effective wave number is changed. Although the above approach is limited for cases with periodicity, the use of FFT becomes valuable for simplifying the finite-difference calculation and accelerating the computation.

There are methods that seek further acceleration of the computation. If we do not overemphasize efficiency and try to keep programming simple, SOR would be sufficient for most applications. Since SOR requires recursive referencing during iterations, one must be careful in balancing the reduction in computation time and amount of memory access. Other useful methods with accelerated convergence include the *Biconjugate gradient stabilized (Bi-CGSTAB) method* [28, 32] and the *residual cutting method* [30].

3.4.3 Iterative Method for HSMAC Method

Let us discuss an iterative approach for the HSMAC method. In the above discussion on the Poisson equation, we had Eqs. (3.76) and (3.77) utilizing four velocity components surrounding point (i, j), which necessitated five scalar potential values of $(\phi_{i,j}, \phi_{i\pm1,j}, \phi_{i,j\pm1})$. Here, we consider using only the value at (i, j)

$$\widehat{u}_{i-\frac{1}{2},j} = u^P_{i-\frac{1}{2},j} - \Delta t \frac{\varphi_{i,j}}{\Delta x_{i-\frac{1}{2}}}, \quad \widehat{u}_{i+\frac{1}{2},j} = u^P_{i+\frac{1}{2},j} + \Delta t \frac{\varphi_{i,j}}{\Delta x_{i+\frac{1}{2}}}, \tag{3.104}$$

$$\widehat{v}_{i,j-\frac{1}{2}} = v^P_{i,j-\frac{1}{2}} - \Delta t \frac{\varphi_{i,j}}{\Delta y_{j-\frac{1}{2}}}, \quad \widehat{v}_{i,j+\frac{1}{2}} = v^P_{i,j+\frac{1}{2}} + \Delta t \frac{\varphi_{i,j}}{\Delta y_{j+\frac{1}{2}}}. \tag{3.105}$$

Because these equations are based on different procedures from those of Eqs. (3.76) and (3.77), we used $\widehat{u}$ and $\widehat{v}$ in place of u^{n+1} and v^{n+1}, respectively. Consequently ϕ is altered by these velocity values, which leads us to instead denote the scalar variable with φ. Similar to Eq. (3.79), the continuity equation can be expressed as

$$\frac{-\widehat{u}_{i-\frac{1}{2},j} + \widehat{u}_{i+\frac{1}{2},j}}{\Delta x_i} + \frac{-\widehat{v}_{i,j-\frac{1}{2}} + \widehat{v}_{i,j+\frac{1}{2}}}{\Delta y_j} = 0, \tag{3.106}$$

which yields

$$\varphi_{i,j} = -\frac{1}{\Delta t B^0_{i,j}} \left(\frac{-u^P_{i-\frac{1}{2},j} + u^P_{i+\frac{1}{2},j}}{\Delta x_i} + \frac{-v^P_{i,j-\frac{1}{2}} + v^P_{i,j+\frac{1}{2}}}{\Delta y_j} \right). \tag{3.107}$$

An iterative scheme can be constructed using the above set of equations. First, we set $\widehat{u}^{(0)} = u^P$, $\widehat{v}^{(0)} = v^P$, and $\widehat{P}^{(0)} = P^n$. We then solve for the solution iteratively through

$$\varphi^{(m+1)}_{i,j} = -\frac{\beta}{\Delta t B^0_{i,j}} \left(\frac{-\widehat{u}^{(m+\frac{1}{2})}_{i-\frac{1}{2},j} + \widehat{u}^{(m)}_{i+\frac{1}{2},j}}{\Delta x_i} + \frac{-\widehat{v}^{(m+\frac{1}{2})}_{i,j-\frac{1}{2}} + \widehat{v}^{(m)}_{i,j+\frac{1}{2}}}{\Delta y_j} \right) \tag{3.108}$$

$$\widehat{u}^{(m+1)}_{i-\frac{1}{2},j} = \widehat{u}^{(m+\frac{1}{2})}_{i-\frac{1}{2},j} - \Delta t \frac{\varphi^{(m+1)}_{i,j}}{\Delta x_{i-\frac{1}{2}}}, \quad \widehat{u}^{(m+\frac{1}{2})}_{i+\frac{1}{2},j} = \widehat{u}^{(m)}_{i+\frac{1}{2},j} + \Delta t \frac{\varphi^{(m+1)}_{i,j}}{\Delta x_{i+\frac{1}{2}}} \tag{3.109}$$

$$\widehat{v}^{(m+1)}_{i,j-\frac{1}{2}} = \widehat{v}^{(m+\frac{1}{2})}_{i,j-\frac{1}{2}} - \Delta t \frac{\varphi^{(m+1)}_{i,j}}{\Delta y_{j-\frac{1}{2}}}, \quad \widehat{v}^{(m+\frac{1}{2})}_{i,j+\frac{1}{2}} = \widehat{v}^{(m)}_{i,j+\frac{1}{2}} + \Delta t \frac{\varphi^{(m+1)}_{i,j}}{\Delta y_{j+\frac{1}{2}}} \tag{3.110}$$

$$\widehat{P}^{(m+1)}_{i,j} = \widehat{P}^{(m)}_{i,j} + \varphi^{(m+1)}_{i,j} \tag{3.111}$$

for $m = 0, 1, 2, \cdots$. The parameter β is an over-relaxation parameter, similar to the one that appears in the SOR method. When the above iterative scheme converges, the solutions reach the velocity and pressure values at the next time step. Hence, we can set $u^{n+1} = \widehat{u}$, $u^{n+1} = \widehat{v}$, $P^{n+1} = \widehat{P}$. This method is known as the HSMAC method [11].

For Eqs. (3.108)–(3.111), we note that the velocity values are updated twice as the algorithm steps through i and j between iteration number m and $m + 1$. For instance for $\widehat{u}_{i+\frac{1}{2},j}$ and $\widehat{v}_{i,j+\frac{1}{2}}$, we have

$$\widehat{u}^{\langle m+1\rangle}_{i+\frac{1}{2},j} = \widehat{u}^{\langle m+\frac{1}{2}\rangle}_{i+\frac{1}{2},j} - \Delta t \frac{\varphi^{\langle m+1\rangle}_{i+1,j}}{\widetilde{\Delta}x_{i+\frac{1}{2}}} = \widehat{u}^{\langle m\rangle}_{i+\frac{1}{2},j} - \Delta t \frac{-\varphi^{\langle m+1\rangle}_{i,j} + \varphi^{\langle m+1\rangle}_{i+1,j}}{\widetilde{\Delta}x_{i+\frac{1}{2}}}, \tag{3.112}$$

$$\widehat{v}^{\langle m+1\rangle}_{i,j+\frac{1}{2}} = \widehat{v}^{\langle m+\frac{1}{2}\rangle}_{i,j+\frac{1}{2},j} - \Delta t \frac{\varphi^{\langle m+1\rangle}_{i,j+1}}{\widetilde{\Delta}y_{j+\frac{1}{2}}} = \widehat{v}^{\langle m\rangle}_{i,j+\frac{1}{2}} - \Delta t \frac{-\varphi^{\langle m+1\rangle}_{i,j} + \varphi^{\langle m+1\rangle}_{i,j+1}}{\widetilde{\Delta}y_{j+\frac{1}{2}}}. \tag{3.113}$$

From Eqs. (3.111), (3.112), and (3.113), the variable $\phi (= P^{n+1} - P^n)$ from the SMAC method and the variable φ from the HSMAC method are related by

$$\phi = \sum_m \varphi^{\langle m\rangle}. \tag{3.114}$$

Assuming that the boundary conditions are implemented correctly (discussed later), the SMAC and HSMAC methods are in principal equivalent [31]. There can however be difference in the convergence process, depending on whether the velocity field is simultaneously relaxed or not.

3.5   Spatial Discretization of Advection Term

The use of a staggered grid to discretize the pressure Poisson equation allows us to achieve mass conservation and prevent spurious pressure oscillations. This is the central idea behind the MAC method and the basis for stable time advancement of the incompressible Navier–Stokes equations. The discretization of the advective term in the momentum equation also influences the accuracy and stability of the computation. The discretization of the advective term is also closely related to energy conservation.

As discussed in Sect. 1.3.5, the internal and kinetic energies are decoupled for incompressible flows. The conservation equation for kinetic energy is derived from the momentum equation. In other words, the kinetic energy needs to be implicitly accounted for while simultaneously satisfying the mass and momentum conservations. When kinetic energy conservation is grossly violated, the simulated flow physics can be excessively damped or can lead to blowup of the solution.

3.5.1 Compatibility and Conservation

We showed two forms of the nonlinear advection term in the momentum equation in Sect. 1.3.3. Here, we consider the advective term for the incompressible Navier–Stokes equations. The nonlinear term $\nabla \cdot (\boldsymbol{uu})$ in the momentum equation is referred to as being in the *divergence form*. This form can also be expressed as $\boldsymbol{u} \cdot \nabla \boldsymbol{u}$ with the use of the incompressibility $\nabla \cdot \boldsymbol{u} = 0$. The later form of this term is called the *gradient (advective) form*. The divergence and the gradient forms are also called the *conservative* and *non-conservative forms*, respectively, but such nomenclature can be misleading for the reason explained below.

In two-dimensional Cartesian coordinates, the divergence form is

$$\frac{\partial(u^2)}{\partial x} + \frac{\partial(uv)}{\partial y}, \quad \frac{\partial(uv)}{\partial x} + \frac{\partial(v^2)}{\partial y} \tag{3.115}$$

and the gradient form is

$$u\frac{\partial u}{\partial x} + v\frac{\partial u}{\partial y}, \quad u\frac{\partial v}{\partial x} + v\frac{\partial v}{\partial y}. \tag{3.116}$$

Although these two forms are equivalent through the use of the continuity equation, they oftentime leads to different numerical solutions. This implies that the derivative relation in Eq. (2.1) does not hold at the discrete level. Let us examine this paradox with examples of finite-difference approximations.

Finite-Difference Approximation of the Divergence Form

For the first expression of the divergence form (Eq. (3.115)), we consider the finite-difference approximation for the location of $u_{i+\frac{1}{2},j}$ on a uniform grid. For the stencil illustrated in Fig. 3.8a, it would be appropriate to write

$$\left[(u^2)_x + (uv)_y\right]_{i+\frac{1}{2},j}$$
$$= \frac{1}{\Delta x}\left[-\left(\frac{u_{i-\frac{1}{2},j}+u_{i+\frac{1}{2},j}}{2}\right)^2 + \left(\frac{u_{i+\frac{1}{2},j}+u_{i+\frac{3}{2},j}}{2}\right)^2\right]$$
$$+ \frac{1}{\Delta y}\left[-\frac{v_{i,j-\frac{1}{2}}+v_{i+1,j-\frac{1}{2}}}{2}\frac{u_{i+\frac{1}{2},j-1}+u_{i+\frac{1}{2},j}}{2}\right.$$
$$\left. + \frac{v_{i,j+\frac{1}{2}}+v_{i+1,j+\frac{1}{2}}}{2}\frac{u_{i+\frac{1}{2},j}+u_{i+\frac{1}{2},j+1}}{2}\right]. \tag{3.117}$$

For the second term in (3.115), we have for the location of $v_{i,j+\frac{1}{2}}$,

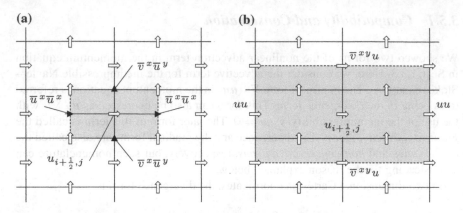

Fig. 3.8 The second-order accurate central-difference stencil for $(i + \frac{1}{2}, j)$ and its control volume for the divergence form (conservative form) of $[\partial(u^2)/\partial x + \partial(uv)/\partial y]_{i+\frac{1}{2},j}$. **a** Using the difference operator δ. **b** Using the difference operator δ' (inappropriate)

$$\left[(uv)_x + (v^2)_y\right]_{i,j+\frac{1}{2}}$$
$$= \frac{1}{\Delta x}\left[-\frac{u_{i-\frac{1}{2},j}+u_{i-\frac{1}{2},j+1}}{2}\frac{v_{i-1,j+\frac{1}{2}}+v_{i,j+\frac{1}{2}}}{2}\right.$$
$$\left.+\frac{u_{i+\frac{1}{2},j}+u_{i+\frac{1}{2},j+1}}{2}\frac{v_{i,j+\frac{1}{2}}+v_{i+1,j+\frac{1}{2}}}{2}\right] \qquad (3.118)$$
$$+\frac{1}{\Delta y}\left[-\left(\frac{v_{i,j-\frac{1}{2}}+v_{i,j+\frac{1}{2}}}{2}\right)^2+\left(\frac{v_{i,j+\frac{1}{2}}+v_{i,j+\frac{3}{2}}}{2}\right)^2\right].$$

In the above Eqs. (3.117) and (3.118), products to be differentiated are interpolated from the neighboring two points and these interpolated values are differenced using the values from half-grid points in the respective x or y directions to find the approximation of the divergence form. We can denote the *difference* and *interpolation* operations over the $\pm\frac{1}{2}$ stencil by δ_x and $\overline{}^x$, respectively. The subscripts and superscripts in these operators represent the direction in which the operations are being performed. Correspondingly, Eqs. (3.117) and (3.118) can be expressed as

$$\left[(u^2)_x + (uv)_y\right]_{i+\frac{1}{2},j} = \left[\delta_x(\overline{u}^x\overline{u}^x) + \delta_y(\overline{v}^x\overline{u}^y)\right]_{i+\frac{1}{2},j} \qquad (3.119)$$

$$\left[(uv)_x + (v^2)_y\right]_{i,j+\frac{1}{2}} = \left[\delta_x(\overline{u}^y\overline{v}^x) + \delta_y(\overline{v}^y\overline{v}^y)\right]_{i,j+\frac{1}{2}} \qquad (3.120)$$

in a short-hand notation. On the other hand, if we use the difference across the two grid cells as shown in Fig. 3.8b, we would need values of v and u at positions where u and v are respectively defined. This necessitates two-dimensional interpolations from four points to find $\overline{v}^{xy}$ and $\overline{u}^{xy}$. In this case, we obtain

$$\left[(u^2)_x + (uv)_y\right]_{i+\frac{1}{2},j} = \left[\delta'_x(uu) + \delta'_y(\overline{v}^{xy}u)\right]_{i+\frac{1}{2},j} \qquad (3.121)$$

$$[(uv)_x + (v^2)_y]_{i,j+\frac{1}{2}} = [\delta'_x(\overline{u}^{xy}v) + \delta'_y(vv)]_{i,j+\frac{1}{2}}, \tag{3.122}$$

which is described by a stencil δ' over ± 1 grid points. This difference scheme results in different accuracy depending on the spatial direction and is not independent of the spatial directions.

Let us examine the two different stencils discussed above from the point of view of the momentum balance in the framework of finite volume methods. Considering Fig. 3.8, Eqs. (3.119) and (3.120) balance the momentum flux in a region of $(\pm \Delta x/2, \pm \Delta y/2)$. The discretization described by Eqs. (3.121) and (3.122) performs the momentum flux balance over a region of $(\pm \Delta x, \pm \Delta y)$. The problem with the later discretization is that overlapping stencil occurs even when points are not positioned adjacent to each other.

Finite-Difference Approximation of the Gradient Form

Based on Fig. 3.9, let us consider the discretization of the gradient (advective) form of the advection term, Eq. (3.116), for $u_{i+\frac{1}{2},j}$. With the stencil for δ' being over ± 1, the different formula at location for $u_{i+\frac{1}{2},j}$ becomes

$$[uu_x + vu_y]_{i+\frac{1}{2},j} = u_{i+\frac{1}{2},j} \frac{-u_{i-\frac{1}{2},j} + u_{i+\frac{3}{2},j}}{2\Delta x}$$
$$+ \frac{v_{i,j-\frac{1}{2}} + v_{i+1,j-\frac{1}{2}} + v_{i,j+\frac{1}{2}} + v_{i+1,j+\frac{1}{2}}}{4} \tag{3.123}$$
$$\times \frac{-u_{i+\frac{1}{2},j-1} + u_{i+\frac{1}{2},j+1}}{2\Delta y},$$

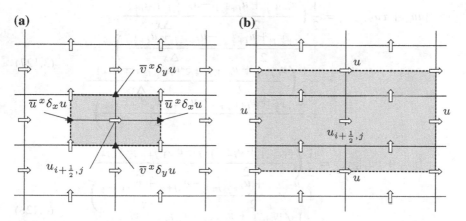

Fig. 3.9 The second-order accurate central-difference stencil for $(i + \frac{1}{2}, j)$ and its control volume for the gradient form (advection form) of $[u\partial u/\partial x + v\partial u/\partial y]_{i+\frac{1}{2},j}$. **a** Using the difference operator δ. **b** Using the difference operator δ' (inappropriate)

as shown in Fig. 3.9b. Similarly, we find that the position for $v_{i,j+\frac{1}{2}}$, the discretization of the gradient form of the advective terms yields

$$
[uv_x + vv_y]_{i,j+\frac{1}{2}} = \frac{u_{i-\frac{1}{2},j} + u_{i+\frac{1}{2},j} + u_{i-\frac{1}{2},j+1} + u_{i+\frac{1}{2},j+1}}{4}
$$
$$
\times \frac{-v_{i-1,j+\frac{1}{2}} + u_{i+1,j+\frac{1}{2}}}{2\Delta x} + v_{i,j+\frac{1}{2}} \frac{-v_{i,j-\frac{1}{2}} + v_{i,j+\frac{3}{2}}}{2\Delta x}. \tag{3.124}
$$

Here, we utilize the two-dimensional interpolation $\overline{v}^{xy}$ and $\overline{u}^{xy}$ to obtain the values of v and u at locations where u and v, respectively, reside on the staggered grid. Note that the differencing is performed with δ and not with δ'.

We can alternatively express Eqs. (3.123) and (3.124) as

$$
[uu_x + vu_y]_{i+\frac{1}{2},j} = [u\delta'_x u + \overline{v}^{xy}\delta'_y u]_{i+\frac{1}{2},j} \tag{3.125}
$$

$$
[uv_x + vv_y]_{i,j+\frac{1}{2}} = [\overline{u}^{xy}\delta'_x v + v\delta'_y v]_{i,j+\frac{1}{2}}, \tag{3.126}
$$

respectively. Because spatial accuracy depends on the direction if we use u, v, $\overline{u}^{xy}$, and $\overline{v}^{xy}$, choosing these discretization schemes are inappropriate. This is the same reason why Eqs. (3.121) and (3.122) of the divergence form are inappropriate. While most MAC methods show the divergence form, Eqs. (3.123) and (3.124) can be derived if the gradient form is naively discretized. Some of the earlier literature utilize these discretizations. It should however be noted that these forms of differencing are different from Eqs. (3.117) and (3.118) and result in larger error in momentum conservation.

For the gradient form, we should use

$$
[uu_x + vu_y]_{i+\frac{1}{2},j} = \frac{1}{2} \left(\frac{u_{i-\frac{1}{2},j} + u_{i+\frac{1}{2},j}}{2} \frac{-u_{i-\frac{1}{2},j} + u_{i+\frac{1}{2},j}}{\Delta x} \right.
$$
$$
+ \frac{u_{i+\frac{1}{2},j} + u_{i+\frac{3}{2},j}}{2} \frac{-u_{i+\frac{1}{2},j} + u_{i+\frac{3}{2},j}}{\Delta x} \right)
$$
$$
+ \frac{1}{2} \left(\frac{v_{i,j-\frac{1}{2}} + v_{i+1,j-\frac{1}{2}}}{2} \frac{-u_{i+\frac{1}{2},j-1} + u_{i+\frac{1}{2},j}}{\Delta y} \right. \tag{3.127}
$$
$$
+ \frac{v_{i,j+\frac{1}{2}} + v_{i+1,j+\frac{1}{2}}}{2} \frac{-u_{i+\frac{1}{2},j} + u_{i+\frac{1}{2},j+1}}{\Delta y} \right)
$$

$$
[uv_x + vv_y]_{i,j+\frac{1}{2}} = \frac{1}{2} \left(\frac{u_{i-\frac{1}{2},j} + u_{i-\frac{1}{2},j+1}}{2} \frac{-v_{i-1,j+\frac{1}{2}} + v_{i,j+\frac{1}{2}}}{\Delta x} \right.
$$
$$
+ \frac{u_{i+\frac{1}{2},j} + u_{i+\frac{1}{2},j+1}}{2} \frac{-v_{i,j+\frac{1}{2}} + v_{i+1,j+\frac{1}{2}}}{\Delta x} \right)
$$
$$
+ \frac{1}{2} \left(\frac{v_{i,j-\frac{1}{2}} + v_{i,j+\frac{1}{2}}}{2} \frac{-v_{i,j-\frac{1}{2}} + v_{i,j+\frac{1}{2}}}{\Delta y} \right. \tag{3.128}
$$
$$
+ \frac{v_{i,j+\frac{1}{2}} + v_{i,j+\frac{3}{2}}}{2} \frac{-v_{i,j+\frac{1}{2}} + v_{i,j+\frac{3}{2}}}{\Delta y} \right)
$$

as the finite-difference approximation. For Eq. (3.127), finite differencing is performed half grid cell off the velocity vector position in each advective direction and then interpolated at the position for $u_{i+\frac{1}{2},j}$, as shown in Fig. 3.9a. For this reason, we call this scheme the *advective interpolation* scheme. Using the short-hand notation, we can rewrite these formulas as

$$[uu_x + vu_y]_{i+\frac{1}{2},j} = [\overline{\overline{u^x}\delta_x u}^x + \overline{\overline{v^x}\delta_y u}^y]_{i+\frac{1}{2},j} \tag{3.129}$$

$$[uv_x + vv_y]_{i,j+\frac{1}{2}} = [\overline{\overline{u^y}\delta_x v}^x + \overline{\overline{v^y}\delta_y v}^y]_{i,j+\frac{1}{2}}. \tag{3.130}$$

Note that the treatment of the terms are the same for the x and y directions.

Compatibility and Conservation of Momentum

The difference between Eqs. (3.117) and (3.127) is

$$\frac{u_{i+\frac{1}{2},j}}{2}\left[\left(\frac{-u_{i-\frac{1}{2},j}+u_{i+\frac{1}{2},j}}{\Delta x} + \frac{-v_{i,j-\frac{1}{2}}+v_{i,j+\frac{1}{2}}}{\Delta y}\right)\right.$$
$$\left. + \left(\frac{-u_{i+\frac{1}{2},j}+u_{i+\frac{3}{2},j}}{\Delta x} + \frac{-v_{i+1,j-\frac{1}{2}}+v_{i+1,j+\frac{1}{2}}}{\Delta y}\right)\right] \tag{3.131}$$

which can be simply expressed as $[u(\overline{\delta_x u + \delta_y v}^x)]_{i+\frac{1}{2},j}$ by noticing that

$$\left[\overline{\delta_x u + \delta_y v}^x\right]_{i+\frac{1}{2},j} = \frac{[\delta_x u + \delta_y v]_{i,j} + [\delta_x u + \delta_y v]_{i+1,j}}{2}. \tag{3.132}$$

Similarly, the difference between Eqs. (3.118) and (3.128) is $[v(\overline{\delta_x u + \delta_y v}^y)]_{i,j+\frac{1}{2}}$. When the discretization is performed appropriately, we observe that the momentum is conserved to the level of mass conservation even for the gradient form (note that the value of $\delta_x u + \delta_y v$ is the numerical error in conserving mass). Thus, the gradient and divergence forms have compatibility and momentum conservation properties. For this reason, it is misleading to refer to the gradient form as a non-conservative form.

Kinetic Energy Conservation

Let us consider the divergence form of the advective terms shown below

$$\left[\delta_x\left(\overline{u}^x\frac{\widetilde{u^2}}{2}\right) + \delta_y\left(\overline{v}^x\frac{\widetilde{u^2}}{2}\right)\right]_{i+\frac{1}{2},j}$$
$$\left[\delta_x\left(\overline{u}^y\frac{\widetilde{v^2}}{2}\right) + \delta_y\left(\overline{v}^y\frac{\widetilde{v^2}}{2}\right)\right]_{i,j+\frac{1}{2}} \tag{3.133}$$

defined at locations where velocities $u_{i+\frac{1}{2},j}$ and $v_{i,j+\frac{1}{2}}$ are positioned. The variables $\widetilde{u^2}$ and $\widetilde{v^2}$ are

$$\widetilde{u^2}^x_{i,j} = u_{i-\frac{1}{2},j}u_{i+\frac{1}{2},j}, \quad \widetilde{v^2}^y_{i,j} = v_{i,j-\frac{1}{2}}v_{i,j+\frac{1}{2}} \tag{3.134}$$

positioned at the cell centers and are known as the *quadratic quantities*. On the other hand,

$$\widetilde{u^2}^y_{i+\frac{1}{2},j+\frac{1}{2}} = u_{i+\frac{1}{2},j}u_{i+\frac{1}{2},j+1}, \quad \widetilde{v^2}^x_{i+\frac{1}{2},j+\frac{1}{2}} = v_{i,j+\frac{1}{2}}v_{i+1,j+\frac{1}{2}} \tag{3.135}$$

are placed at the cell edges. The quantity $(\widetilde{u^2} + \widetilde{v^2})/2$ uses velocity values from different positions for evaluation and thus should be treated as a pseudo-energy. Multiplying $u_{i+\frac{1}{2},j}$ to Eqs. (3.117) and (3.127) and taking a difference with the first expression in Eq. (3.133) returns $\pm[u^2(\overline{\delta_x u + \delta_y v}^x)]_{i+\frac{1}{2},j}/2$ (an error of the same magnitude with opposite sign). Multiplying $u_{i+\frac{1}{2},j}$ to Eqs. (3.118) and (3.128) and taking a difference with the second equation in Eq. (3.133) returns $\pm[v^2(\overline{\delta_x u + \delta_y v}^y)]_{i+\frac{1}{2},j}/2$. With these error expressions in mind, we can conceive the use of the *mixed form* (*skew-symmetric form*, [21]). That is, by evaluating the following expressions

$$\frac{1}{2}\left[\{\delta_x(\overline{u}^x\overline{u}^x) + \delta_y(\overline{v}^x\overline{u}^y)\} + (\overline{\overline{u}^x\delta_x u}^x + \overline{\overline{v}^x\delta_y u}^y)\right]_{i+\frac{1}{2},j} \tag{3.136}$$

$$\frac{1}{2}\left[\{\delta_x(\overline{u}^y\overline{v}^x) + \delta_y(\overline{v}^y\overline{v}^y)\} + (\overline{\overline{u}^y\delta_x v}^x + \overline{\overline{v}^y\delta_y y}^y)\right]_{i,j+\frac{1}{2}} \tag{3.137}$$

at positions for $u_{i+\frac{1}{2},j}$ and $v_{i,j+\frac{1}{2}}$, respectively, the quadratic quantity can be conserved. This formulation was proposed by Piacsek and Williams [25] and is referred to as the *quadratic conservation form*. Strictly speaking, this quadratic quantity is not the kinetic energy. The quadratic quantity evaluates $\widetilde{u^2}/2$ and $\widetilde{v^2}/2$ at different points for $u_{i+\frac{1}{2},j}$ and $v_{i,j+\frac{1}{2}}$, respectively, and cannot be used for local conservation of energy. Nonetheless, the conservation of this pseudo-energy is thought to provide numerical stability and can be in fact an important property for calculations.

The description of "energy conservative form $\frac{1}{2}(u_j\partial u_i/\partial x_j + \partial(u_i u_j)/\partial x_j)$ being utilized" is oftentimes seen in various studies. More precisely, we should refer to the scheme as quadratic conservative. This conservation property is satisfied only for cases where the difference scheme of the two terms have compatibility, such as in the case of Eqs. (3.136) and (3.137). All of these points become meaningless when any one of the terms are discretized inappropriately or when upwinding (described later) is employed. One of the most frequently introduced incompatibilities is to use discretizations based on the formulation similar to Eqs. (3.123) and (3.124) yielding

$$\frac{1}{2}\left[\{\delta_x(\overline{u}^x\overline{u}^x) + \delta_y(\overline{v}^x\overline{u}^y)\} + (u\delta'_x u + \overline{v}^{xy}\delta'_y u)\right]_{i+\frac{1}{2},j} \tag{3.138}$$

$$\frac{1}{2}\left[\{\delta_x(\overline{u}^y\overline{v}^x) + \delta_y(\overline{v}^y\overline{v}^y)\} + (\overline{u}^{xy}\delta'_x v + v\delta'_y v)\right]_{i,j+\frac{1}{2}} \tag{3.139}$$

at the locations for $u_{i+\frac{1}{2},j}$ and $v_{i,j+\frac{1}{2}}$, respectively. The other incompatibility often seen is the use of upwinding for the second terms.

To summarize, the divergence forms, Eqs. (3.117) and (3.118) conserve the momentum but not the quadratic quantity. On the other hand, the mixed forms of Eqs. (3.136) and (3.137) conserve the quadratic quantity but not the momentum. The gradient forms, Eqs. (3.127) and (3.128) do not conserve either quantities. Nonetheless, the conservation error for all of these cases are proportional to the error in the mass conservation $\nabla \cdot \boldsymbol{u}$. If the mass conservation error is sufficiently small, the errors in the conservation of momentum and quadratic quantity are made small as well. Therefore, when appropriate discretization is implemented and the continuity equation is satisfied to tolerable error, we can say we have conservation of momentum and the quadratic quantity. In actual calculations, one can hardly notice any difference between the results obtained from Eqs. (3.117) and (3.118) and those from Eqs. (3.127) and (3.128). However, the numerical results from Eqs. (3.123) and (3.124) cannot reach the same results [12].

Both the MAC-based approach by Harlow and Welch [10] and the quadratic conserving form by Piacsek and Williams [25] are methods that are capable of maintaining numerical stability while permitting some level of unavoidable error. These influential methods were inspired to perform stable calculations at a time when computational resources to spend on the Poisson solver were limited. They continue to empower researchers to simulate incompressible flows.

3.5.2 Discretization on Nonuniform Grids

Let us expand finite-difference schemes to nonuniform rectangular grids as shown in Fig. 3.10. For nonuniform grids, there is some flexibility in how we can perform interpolation. However, a naive implementation of interpolation in what we have presented in Sect. 3.5.1 does not provide compatibility between the divergence and gradient forms.

To resolve this issue, we need to revisit the governing equations on a generalized coordinate system [14]. Here we only present a brief description, since a detailed discussion is offered later. We consider transforming (mapping) a nonuniform grid $x_j(x, y)$ onto a uniform grid $\xi^k(\xi, \eta)$. The difference operation in the mapped space for the divergence form is then

$$\frac{\partial u_i u_j}{\partial x_j} = \frac{1}{J} \delta_{\xi_k} (\overline{JU^k}^{\xi^i} \overline{u_i}^{\xi^k}) \tag{3.140}$$

and the advective interpolation used for the gradient form in the mapped space becomes

$$u_j \frac{\partial u_i}{\partial x_j} = \frac{1}{J} \overline{\overline{JU^k}^{\xi^i} \delta_{\xi^k} u_i}^{\xi^k} \tag{3.141}$$

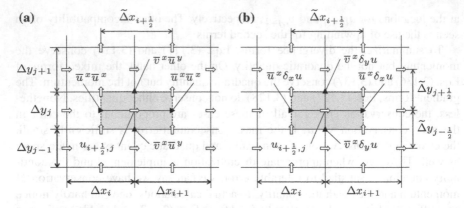

Fig. 3.10 The second-order accurate central-difference stencil for $(i + \frac{1}{2}, j)$ and its control volume for the advective term in the x-direction momentum equation for nonuniform grid. **a** Divergence form $[\partial(u^2)/\partial x + \partial(uv)/\partial y]_{i+\frac{1}{2},j}$. **b** Gradient form $[u\partial u/\partial x + v\partial u/\partial y]_{i+\frac{1}{2},j}$

Note that in the above expressions, we do not sum over the indices appearing in the interpolation (*e.g.*, k in $\overline{\cdot}^{\xi^k}$) unless that index appears twice excluding the interpolation index.

Details on this matter will be presented in Sect. 4.2. Here, $J = |\partial x^j/\partial \xi^k|$ represents the transformation Jacobian and $U^k = u_i \partial \xi^k/\partial x_i$ is the contravariant component of the advection velocity. For a two-dimensional rectangular grid, we can choose the grid width in the mapped space to be $\Delta \xi = 1$ and $\Delta \eta = 1$. The corresponding Jacobian then becomes

$$J_{i,j} = \Delta x_i \Delta y_j, \quad J_{i+\frac{1}{2},j} = \widetilde{\Delta} x_{i+\frac{1}{2}} \Delta y_j, \quad J_{i,j+\frac{1}{2}} = \Delta x_i \widetilde{\Delta} y_{j+\frac{1}{2}} \quad (3.142)$$

with the contravariant velocity components being provided by

$$U_{i+\frac{1}{2},j} = u_{i+\frac{1}{2}}/\widetilde{\Delta} x_{i+\frac{1}{2}}, \quad V_{i,j+\frac{1}{2}} = v_{i,j+\frac{1}{2}}/\widetilde{\Delta} y_{j+\frac{1}{2}}. \quad (3.143)$$

Because we are mapping the physical grid onto a uniform grid with unit mesh size, the interpolation and gradient can be simply computed as the average and difference of the values, respectively, between the two points. With this mapping approach, Eq. (3.140) for divergence form becomes

$$[\delta_x (\overline{u}^x \overline{u}^x) + \delta(\overline{v}^x \overline{u}^y)]_{i+\frac{1}{2},j}$$
$$= \{-[\overline{JU^\xi \overline{u}^\xi}]_{i,j} + [\overline{JU^\xi \overline{u}^\xi}]_{i+1,j}\}/J_{i+\frac{1}{2},j} \quad (3.144)$$
$$+\{-[\overline{JV^\xi \overline{u}^\eta}]_{i+\frac{1}{2},j-\frac{1}{2}} + [\overline{JV^\xi \overline{u}^\eta}]_{i+\frac{1}{2},j+\frac{1}{2}}\}/J_{i+\frac{1}{2},j}$$

$$[\,\delta_x(\overline{u}^y\overline{v}^x) + \delta_y(\overline{v}^y\overline{v}^y)]_{i,j+\frac{1}{2}}$$
$$= \{-[\overline{JU}^\eta\overline{v}^\xi]_{i-1,j+\frac{1}{2}} + [\overline{JU}^\eta\overline{v}^\xi]_{i+\frac{1}{2},j+\frac{1}{2}}\}/J_{i,j+\frac{1}{2}} \qquad (3.145)$$
$$+\{-[\overline{JV}^\eta\overline{v}^\eta]_{i,j} + [\overline{JV}^\eta\overline{v}^\eta]_{i,j+1}\}/J_{i,j+\frac{1}{2}}.$$

Equation (3.141) for the gradient form becomes

$$[\overline{\overline{u}^x\delta_x u}^x + \overline{\overline{v}^x\delta_y u}^y]_{i+\frac{1}{2},j}$$
$$= \{[\overline{JU}^\xi\delta_\xi u]_{i,j} + [\overline{JU}^\xi\delta_\xi u]_{i+1,j}\}/(2J_{i+\frac{1}{2},j}) \qquad (3.146)$$
$$+\{[\overline{JV}^\xi\delta_\eta u]_{i+\frac{1}{2},j-\frac{1}{2}} + [\overline{JV}^\xi\delta_\eta u]_{i+\frac{1}{2},j+\frac{1}{2}}\}/(2J_{i+\frac{1}{2},j})$$

$$[\overline{\overline{u}^y\delta_x v}^x + \overline{\overline{v}^y\delta_y v}^y]_{i,j+\frac{1}{2}}$$
$$= \{[\overline{JU}^\eta\delta_\xi v]_{i-\frac{1}{2},j+\frac{1}{2}} + [\overline{JU}^\eta\delta_\xi v]_{i+\frac{1}{2},j+\frac{1}{2}}\}/(2J_{i,j+\frac{1}{2}}) \qquad (3.147)$$
$$+\{[\overline{JV}^\eta\delta_\eta v]_{i,j} + [\overline{JV}^\eta\delta_\eta v]_{i,j+1}\}/(2J_{i,j+\frac{1}{2}}).$$

Interpolations simply become the following averaging operations

$$\left.\begin{array}{l} [\overline{JU}^\xi]_{i,j} = \dfrac{\Delta y_j u_{i-\frac{1}{2},j} + \Delta y_j u_{i+\frac{1}{2},j}}{2} \\[2mm] [\overline{JV}^\xi]_{i+\frac{1}{2},j+\frac{1}{2}} = \dfrac{\Delta x_i v_{i,j+\frac{1}{2}} + \Delta x_{i+1} v_{i+1,j+\frac{1}{2}}}{2} \end{array}\right\} \qquad (3.148)$$

$$\left.\begin{array}{l} [\overline{JV}^\eta]_{i,j} = \dfrac{\Delta x_i v_{i,j-\frac{1}{2}} + \Delta x_i v_{i,j+\frac{1}{2}}}{2} \\[2mm] [\overline{JU}^\eta]_{i+\frac{1}{2},j+\frac{1}{2}} = \dfrac{\Delta y_j u_{i+\frac{1}{2},j} + \Delta y_{j+1} u_{i+\frac{1}{2},j+1}}{2} \end{array}\right\} \qquad (3.149)$$

$$[\overline{u}^\xi]_{i,j} = \frac{u_{i-\frac{1}{2},j} + u_{i+\frac{1}{2},j}}{2}, \quad [\overline{u}^\eta]_{i+\frac{1}{2},j+\frac{1}{2}} = \frac{u_{i+\frac{1}{2},j} + u_{i+\frac{1}{2},j+1}}{2} \qquad (3.150)$$

$$[\overline{v}^\eta]_{i,j} = \frac{v_{i,j-\frac{1}{2}} + v_{i,j+\frac{1}{2}}}{2}, \quad [\overline{v}^\xi]_{i+\frac{1}{2},j+\frac{1}{2}} = \frac{v_{i,j+\frac{1}{2}} + v_{i+1,j+\frac{1}{2}}}{2} \qquad (3.151)$$

and the first-derivative approximations as the difference between two values are

$$[\delta_\xi u]_{i,j} = -u_{i-\frac{1}{2},j} + u_{i+\frac{1}{2},j}, \quad [\delta_\eta u]_{i+\frac{1}{2},j+\frac{1}{2}} = -u_{i+\frac{1}{2},j} + u_{i+\frac{1}{2},j+1} \qquad (3.152)$$

$$[\delta_\eta v]_{i,j} = -v_{i,j-\frac{1}{2}} + v_{i,j+\frac{1}{2}}, \quad [\delta_\xi v]_{i+\frac{1}{2},j+\frac{1}{2}} = -v_{i,j+\frac{1}{2}} + v_{i+1,j+\frac{1}{2}} \qquad (3.153)$$

We can note that Eqs. (3.144) and (3.145) and Eqs. (3.146) and (3.147) are compatible, as discussed in the previous section. Taking the difference between Eqs. (3.144) and (3.146), we have

$$
\begin{aligned}
&\frac{1}{J}\,\delta_{\xi^k}(\overline{JU^k}^{\xi^i}\,\overline{u_i}^{\xi^k}) - \frac{1}{J}\overline{\overline{JU^k}^{\xi^i}\,\delta_{\xi^k}u_i}^{\xi^k} \\
&= \frac{u_{i+\frac{1}{2}}}{2\widetilde{\Delta x}_{i+\frac{1}{2}}\Delta y_j} \times
\left\{
\begin{array}{l}
\Delta y_j(-u_{i-\frac{1}{2},j} + u_{i+\frac{1}{2},j}) \\
+\Delta x_i(-v_{i,j-\frac{1}{2}} + v_{i,j+\frac{1}{2}}) \\
+\Delta y_j(-u_{i+\frac{1}{2},j} + u_{i+\frac{3}{2},j}) \\
+\Delta x_{i+1}(-v_{i+1,j-\frac{1}{2}} + v_{i+1,j+\frac{1}{2}})
\end{array}
\right\} \\
&= \left[\frac{u}{J}\delta_\xi(JU) + \delta_\eta(JV) \right]^{\xi}_{i+\frac{1}{2},j}.
\end{aligned}
\tag{3.154}
$$

Correspondingly, when the continuity equation at the two cells adjacent to $u_{i+\frac{1}{2},j}$

$$
\frac{\partial u}{\partial x} + \frac{\partial v}{\partial y} = \frac{1}{J}\left[\frac{\partial(JU)}{\partial \xi} + \frac{\partial(JV)}{\partial \eta} \right] = 0
\tag{3.155}
$$

satisfies the discrete relation for

$$
\frac{1}{J}\left[\delta_\xi(JU) + \delta_\eta(JV) \right] = 0,
\tag{3.156}
$$

we can observe that there is no difference between the two discretizations. This is true for the difference between Eqs. (3.145) and (3.147) as well.

3.5.3 Upwinding Schemes

It is known that numerical diffusion can provide added stability for numerical simulations of fluid flow. In the momentum equation, this would correspond to numerical viscosity. Such effect would generally be introduced to the system by upwinding[6] the advection term. Note that the use of upwinding cannot only reduce the effective Reynolds number, as discuss in Sect. 2.3.6, but also bring in higher-order numerical viscosity that behaves differently from the second-derivative viscous term. Since there are reported cases that upwinding may introduce unphysical vortices into the simulated flow, the use of these concepts must be considered carefully. We refer to the numerical viscosity that is intentionally introduced to the numerical solver as *artificial viscosity*.

Upwinding should be limited in its use for regions where grid resolution cannot be further refined or for situations where the use of a numerically unstable model equation is unavoidable. Nonetheless, upwinding appears to be utilized often just because the simulation diverges due to the use of central differencing. Naively taming numerical instability with upwinding usually implies that the solution is computed with an inappropriate central-difference scheme and is overly smoothed by artificial viscosity. Such practice cannot produce reliable results. With these cautionary notes

---

[6]The term upwind is commonly used. However, the use of the term upstream may be more appropriate so that applications are not limited to air.

in mind, let us discuss a few appropriate upwinding schemes suitable for the staggered grid.

Upwinding the Divergence Form of the Advective Term

Approximating the flux term $\partial(uf)/\partial x$ using the divergence form of the advective term with the second-order central-difference scheme, we have

$$[(uf)_x]_{i+\frac{1}{2}} = \frac{-[uf]_i + [uf]_{i+1}}{\Delta x}. \tag{3.157}$$

Without the use of upwinding, we can provide the average value for f from both neighboring points

$$[uf]_i = u_i \frac{f_{i-\frac{1}{2}} + f_{i+\frac{1}{2}}}{2} = [u\overline{f}^x]_i. \tag{3.158}$$

We can alternatively use an interpolation for f that weighs the upstream value more heavily. The resulting scheme amounts to addicting numerical viscosity. Two well-known methods for upwinding are the *donor cell method* and the *QUICK method*, which we describe below.

The donor cell method uses

$$[uf]_i = \begin{cases} u_i f_{i-\frac{1}{2}} & u_i \geq 0 \\ u_i f_{i+\frac{1}{2}} & u_i < 0 \end{cases} \tag{3.159}$$

where only the upstream value is referenced. Rewriting the above expression without the need to classify cases based on the sign of u_i, we have

$$\begin{aligned} [uf]_i &= u_i \frac{f_{i-\frac{1}{2}} + f_{i+\frac{1}{2}}}{2} - |u_i| \frac{-f_{i-\frac{1}{2}} + f_{i+\frac{1}{2}}}{2} \\ &= [u\overline{f}^x]_i - \frac{|u_i|\Delta x}{2}[\delta_x f]_i, \end{aligned} \tag{3.160}$$

where $\overline{f}^x$ and $\delta_x f$ are the two-point interpolation and central-difference operations, respectively, that use $f_{i\pm\frac{1}{2}}$.

The QUICK (Quadratic Upstream Interpolation for Convective Kinematics) method by Leonard [18]

$$[uf]_i = \begin{cases} u_i \dfrac{-f_{i-\frac{3}{2}} + 6f_{i-\frac{1}{2}} + 3f_{i+\frac{1}{2}}}{8} & u_i \geq 0 \\ u_i \dfrac{3f_{i-\frac{1}{2}} + 6f_{i+\frac{1}{2}} - f_{i+\frac{3}{2}}}{8} & u_i < 0 \end{cases} \tag{3.161}$$

uses information from downstream as well but weighs the upstream values higher. The above equation can be written without an if statement by

$$[uf]_i = u_i \frac{-f_{i-\frac{3}{2}} + 9f_{i-\frac{1}{2}} + 9f_{i+\frac{1}{2}} - f_{i+\frac{3}{2}}}{16}$$

$$+ |u_i| \frac{-f_{i-\frac{3}{2}} + 3f_{i-\frac{1}{2}} - 3f_{i+\frac{1}{2}} + f_{i+\frac{3}{2}}}{16} \qquad (3.162)$$

$$= [u\overline{f}^x]_i + \frac{|u_i|(\Delta x)^3}{16} [\delta_x^3 f]_i,$$

where $\overline{f}^x$ and $\delta_x^3 f$ are the four-point interpolation and central differencing for the third derivative, respectively, based on $f_{i\pm\frac{1}{2}}$ and $f_{i\pm\frac{3}{2}}$.

In general, the right-hand side advective flux uf in Eq. (3.157) is weighed more upstream and is provided by the form consisting of the m-point interpolation $\overline{f}^x$ and the difference approximation for the $(m - 1)$-th derivative multiplied by $(\Delta x)^{m-1}$

$$uf = u\overline{f}^x + (-1)^{\frac{m}{2}} \alpha (\Delta x)^{m-1} |u| \delta_x^{m-1} f, \qquad (3.163)$$

where m is even, α is a positive constant, and δ_x^{m-1} is an m-point central differencing operator for the $(m - 1)$-th derivative over a stencil of $\pm\frac{1}{2}, \pm\frac{3}{2}, \cdots, \pm\frac{1}{2}(m - 1)$. Usually, the constant α is the inverse of the numerator in the interpolation $\overline{f}^x$. The added term in the above relation undergoes an extra differencing in Eq. (3.157) to result in the m-th derivative (even derivative) to act as artificial viscosity. For $m = 2$ and $\alpha = 1/2$, we have the donor cell method (second-order artificial viscosity) and for $m = 4$ and $\alpha = 1/16$, we have the QUICK method (fourth-order artificial viscosity).

Based on numerical experiments, it is reported that the first-order accurate donor cell method overdamps solutions but the QUICK method provides relatively better results. While the interpolation scheme (Eq. (3.162)) in the QUICK method is third-order accurate, the evaluation of the advective term by taking the divergence (Eq. (3.157)) leads to second-order accuracy. Even if higher-order artificial viscosity (with higher-order interpolation) is introduced, the accuracy stays at second order. Perhaps for that reason, the attempt to combine Eqs. (3.157) and (3.163) for the divergence form has not been observed.

Upwinding the Gradient Form of the Advective Term

Upwind finite differencing is often used with the gradient form. The first-order upwind difference is given by

$$[uf_x]_i = \begin{cases} u_i \dfrac{-f_{i-1} + f_i}{\Delta x} & u_i \geq 0 \\ u_i \dfrac{-f_i + f_{i+1}}{\Delta x} & u_i < 0 \end{cases} \qquad (3.164)$$

which can be alternatively written without classifying into cases (as seen above)

$$[u f_x]_i = u_i \frac{-f_{i-1} + f_{i+1}}{2\Delta x} - |u_i| \frac{f_{i-1} - 2 f_i + f_{i+1}}{2\Delta x}$$
$$= [u \delta_x' f]_i - \frac{|u_i| \Delta x}{2} [\delta_x'^2 f]_i, \tag{3.165}$$

where δ_x' and $\delta_x'^2$ are the first and second-derivative central differencing, respectively, over three points $i, i \pm 1$. From this expression, we can observe that the upwinding has added viscosity of $|u|\Delta x/2$. The first term in Eq. (3.165) is the second-order accurate central differencing. Due to the addition of the second term, the overall accuracy is however reduced to first order. This term gives rise to artificial viscosity in the momentum equation. What the simulation solves for with artificial viscosity is the flow field with effective viscosity of $\nu + |u|\Delta x/2$. When the artificial viscosity is of significant magnitude relative to the physical viscosity, the flow field cannot be predicted correctly, as discussed in Sect. 2.3.6.

Thus, there are attempts to consider reducing the amount of artificial viscosity or to limit its use only in regions where the gradients in the flow field are large. One example is the scheme by Kawamura and Kuwahara [15]

$$[u f_x]_i = \begin{cases} u_i \dfrac{2 f_{i-2} - 10 f_{i-1} + 9 f_i - 2 f_{i+1} + f_{i+2}}{6\Delta x} & u_i \geq 0 \\[3mm] u_i \dfrac{-f_{i-2} + 2 f_{i-1} - 9 f_i + 10 f_{i+1} - 2 f_{i+2}}{6\Delta x} & u_i < 0 \end{cases} \tag{3.166}$$

As opposed to the first-order upwind formula, this scheme uses downstream information as well. The weights are selected to weigh the upstream values more, compared to those from downstream. This can equivalently be expressed as

$$[u f_x]_i = u_i \frac{f_{i-2} - 8 f_{i-1} + 8 f_{i+1} - f_{i+2}}{12\Delta x}$$
$$+ 3|u_i| \frac{f_{i-2} - 4 f_{i-1} + 6 f_i - 4 f_{i+1} + f_{i+2}}{12\Delta x}$$
$$= [u \delta_x' f]_i + \frac{3|u_i|(\Delta x)^3}{12} [\delta_x'^4 f]_i, \tag{3.167}$$

where δ_x' and $\delta_x'^4$ are the five-point stencil central-difference operators for the first and fourth derivatives, respectively, using $i, i \pm 1, i \pm 2$. In Eq. (3.167), the first term is the fourth-order central differencing. The addition of the second term makes this a third-order accurate upwind method with artificial viscosity added through the fourth derivative.

Use of higher derivatives to introduce artificial viscosity tends to suppress high wavenumber fluctuations. If we consider an oscillation of $f = \exp(ikx)$, we have $d^m f/dx^m = (ik)^m f$ which implies that with large m (derivative order), high wavenumber components can be greatly affected by the artificial viscosity. While it appears that high-order artificial viscosity can remove high-frequency oscillations often related to numerical instability, there is some report of unphysical behavior of fluid flow appearing due to the addition of high-order artificial viscosity.

In general, the implementation of artificial viscosity in the gradient form of the advective terms leads to the upwind difference of

$$uf_x = u\delta'_x f + (-1)^{\frac{m}{2}}\alpha(\Delta x)^{m-1}|u|\delta'^m_x f, \tag{3.168}$$

where m is even and α is a positive constant. The right-hand side of the above expression weighs the upstream values in the finite-difference expression. Here δ' is the central-difference $(m + 1)$-point scheme with stencil over $0, \pm 1, \pm 2, \cdots, \pm\frac{m}{2}$. When $m = 4$, we have the third-order accurate upwind scheme and when $m = 6$ we have the fifth-order accurate upwind scheme. The scheme by Kawamura and Kuwahara chooses $m = 4$ and $\alpha = 3/12$. If we consider α to be the inverse of the integer multiplied to Δx on the denominator of the difference scheme $\delta_x f$, it may be more common to select $\alpha = 1/12$. There is also the fifth-order upwind scheme by Rai and Moin [27] that uses $m = 6$ and $\alpha = 1/60$.

Let us focus on the first term on the right-hand side of Eq. (3.168). This term corresponds to the inappropriate discretization of the advective term (gradient form) on a staggered grid, as previously discussed. Even if we are to increase the accuracy to third or fifth order, this type of discretization would not be appropriate. In cases of artificial viscosity being necessary, we should at least use an appropriate discretization for the central difference. For example, if we are to perform upwind differencing of the advective term in the x-direction momentum equation in two dimension, we would not want to use

$$\begin{aligned}[uu_x + vu_y]_{i+\frac{1}{2},j} &= [u\delta'_x u + \overline{v}^{xy}\delta'_y u]_{i+\frac{1}{2},j} \\ &+ (-1)^{\frac{m}{2}}\alpha[(\Delta x)^{m-1}|u|\delta'^m_x u + (\Delta y)^{m-1}|\overline{v}^{xy}|\delta'^m_y u]_{i+\frac{1}{2},j}\end{aligned} \tag{3.169}$$

but would want to choose

$$\begin{aligned}[uu_x + vu_y]_{i+\frac{1}{2},j} &= [\overline{\overline{u}^x\delta_x u}^x + \overline{\overline{v}^x\delta_y u}^y]_{i+\frac{1}{2},j} \\ &+ (-1)^{\frac{m}{2}}\alpha[(\Delta x)^{m-1}|u|\delta'^m_x u + (\Delta y)^{m-1}|\overline{v}^{xy}|\delta'^m_y u]_{i+\frac{1}{2},j}\end{aligned}. \tag{3.170}$$

The above discretization is suggested as one of the corrected upwinding schemes [13].

While there are numerous reports of simulations that necessitated the use of upwinding for stability, there may be cases where the inappropriate choice of central differencing have caused numerical problems. The utilization of upwinding may become unnecessary if central differencing is performed with care.

3.6 Spatial Discretization of Viscous Term

The viscous term in the momentum equation represents the effect of viscous diffusion. As we have seen in Chap. 2, the diffusion term acts to smooth out the solution profile.

This implies that the viscous term smears out the velocity fluctuations in the flow. Here, let us discuss the role of viscosity from the perspective of kinetic energy, $k = u_i u_i / 2$. Assuming ρ and ν are constant, the inner product of the momentum equation with velocity yields

$$u_i \frac{\partial u_i}{\partial t} = -u_i u_j \frac{\partial u_i}{\partial x_j} - \frac{u_i}{\rho} \frac{\partial p}{\partial x_i} + \nu u_i \frac{\partial^2 u_i}{\partial x_j \partial x_j}, \tag{3.171}$$

which can be further transformed into the kinetic energy equation

$$\frac{\partial k}{\partial t} + \frac{\partial}{\partial x_j} \left(u_j k + \frac{u_j p}{\rho} - \nu \frac{\partial k}{\partial x_j} \right) = -\nu \frac{\partial u_i}{\partial x_j} \frac{\partial u_i}{\partial x_j}. \tag{3.172}$$

The term in Eq. (3.172) represented by $\partial / \partial x_j (\cdots)$ on the left-hand side is in divergence form and can lead to the conservation of kinetic energy. However, the last term on the right-hand side

$$\Phi = \nu \frac{\partial u_i}{\partial x_j} \frac{\partial u_i}{\partial x_j} \tag{3.173}$$

is always positive and acts to dissipate kinetic energy. The term Φ denotes the *dissipation rate of kinetic energy*. The existence of the viscous diffusion term in the momentum equation does not alter momentum conservation. However, there is a non-conservative dissipative term in the kinetic energy equation.

For the discretization of the viscous term, most finite-difference schemes for the second derivative suppress velocity fluctuations and dissipate energy. Hence, the solutions do not appear to be sensitive to the discretization of the second derivative for most problems.

However, for higher-fidelity simulations, the compatibility and high-order accuracy of schemes become important. For example, when solving for the velocity profile within the boundary layer in which viscous effects dominate the flow physics, high resolution becomes necessary which can be achieved with the finer grid size and higher spatial accuracy. Post-processing of turbulence simulation data to accurately calculate each term in the kinetic energy equation also requires care. Since the computational results are based on the continuity and momentum equations but not explicitly on the kinetic energy equation, the satisfaction of compatibility properties becomes important in computing the terms in the kinetic energy equation for correctly analyzing the results.

With the above discussion in mind, let us present finite-difference schemes for the viscous term. A general expression is based on differencing the derivative of the stress

$$\frac{\partial}{\partial x_j}\left[\nu\left(\frac{\partial u_i}{\partial x_j}+\frac{\partial u_j}{\partial x_i}\right)\right]=\delta_{x_j}\left[\nu\left(\delta_{x_j}u_i+\delta_{x_i}u_j\right)\right]. \tag{3.174}$$

If viscosity ν is constant, the viscous term becomes $\nu\nabla^2 u_i$ and leads to

$$\nu\frac{\partial^2 u_i}{\partial x_j\partial x_j}=\nu\delta_{x_j}(\delta_{x_j}u_i). \tag{3.175}$$

As we have considered in the pressure Poisson equation, we discretize the Laplacian by performing the finite difference of the velocity gradient, rather than directly computing the second derivative. Based on this idea, the relation corresponding to transforming the differential equation, Eq. (3.171)–(3.172),

$$u_i\frac{\partial^2 u_i}{\partial x_j\partial x_j}=\frac{\partial^2 k}{\partial x_j\partial x_j}-\frac{\partial u_i}{\partial x_j}\frac{\partial u_i}{\partial x_j} \tag{3.176}$$

should hold also in the context of the finite difference

$$u_i\delta_{x_j}(\delta_{x_j}u_i)=\delta_{x_j}(\delta_{x_j}k)-\overline{\delta_{x_j}u_i\delta_{x_j}u_i}^{x_j}. \tag{3.177}$$

Now, let us examine the above expressions for nonuniform grids. For the viscous term, what we considered in discretizing the advection term with the generalized coordinate system becomes useful. With Fig. 3.11, we examine the viscous term for the x-direction momentum equation. For the divergence form of the stress from Eq. (3.174), we have

Fig. 3.11 The second order accurate central-difference stencil and control volume for the viscous term in the x-direction momentum equation for nonuniform grid

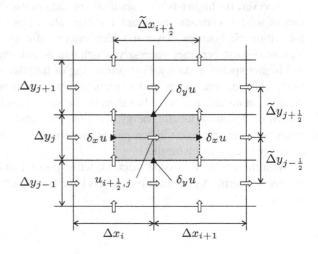

$$\left[(2\nu u_x)_x + \{\nu(u_y + v_x)\}_y\right]_{i+\frac{1}{2},j}$$

$$= \frac{2}{\widetilde{\Delta}x_{i+\frac{1}{2}}}\left[-\nu\frac{-u_{i-\frac{1}{2},j} + u_{i+\frac{1}{2},j}}{\Delta x_i} + \nu\frac{-u_{i+\frac{1}{2},j} + u_{i+\frac{3}{2},j}}{\Delta x_{i+1}}\right]$$

$$+\frac{1}{\Delta y_j}\left[-\nu\left\{\frac{-u_{i+\frac{1}{2},j-1} + u_{i+\frac{1}{2},j}}{\widetilde{\Delta}y_{j-\frac{1}{2}}} + \frac{-v_{i,j-\frac{1}{2}} + v_{i+1,j-\frac{1}{2}}}{\widetilde{\Delta}x_{i+\frac{1}{2}}}\right\}\right.$$

$$\left.+\nu\left\{\frac{-u_{i+\frac{1}{2},j} + u_{i+\frac{1}{2},j+1}}{\widetilde{\Delta}y_{j+\frac{1}{2}}} + \frac{-v_{i,j+\frac{1}{2}} + v_{i+1,j+\frac{1}{2}}}{\widetilde{\Delta}x_{i+\frac{1}{2}}}\right\}\right]. \tag{3.178}$$

If ν is constant, we have $\nu\nabla^2 u_i$ which yields

$$\nu\left[(u_x)_x + (u_y)_y\right]_{i+\frac{1}{2},j}$$

$$= \frac{\nu}{\widetilde{\Delta}x_{i+\frac{1}{2}}}\left[-\frac{-u_{i-\frac{1}{2},j} + u_{i+\frac{1}{2},j}}{\Delta x_i} + \frac{-u_{i+\frac{1}{2},j} + u_{i+\frac{3}{2},j}}{\Delta x_{i+1}}\right]$$

$$+\frac{\nu}{\Delta y_j}\left[-\frac{-u_{i+\frac{1}{2},j-1} + u_{i+\frac{1}{2},j}}{\widetilde{\Delta}y_{j-\frac{1}{2}}} + \frac{-u_{i+\frac{1}{2},j} + u_{i+\frac{1}{2},j+1}}{\widetilde{\Delta}y_{j+\frac{1}{2}}}\right]. \tag{3.179}$$

For the above discretization, let us consider the effect of discretization on the energy equation. With finite-differencing approximation of $\partial^2/\partial x^2$ in Eq. (3.179), the left-hand side and the first term on the right-hand side in Eq. (3.176) for $i = 1$ become

$$[u(u_x)_x]_{i+\frac{1}{2},j}$$

$$= \frac{u_{i+\frac{1}{2},j}}{\widetilde{\Delta}x_{i+\frac{1}{2}}}\left[-\frac{-u_{i-\frac{1}{2},j} + u_{i+\frac{1}{2},j}}{\Delta x_i} + \frac{-u_{i+\frac{1}{2},j} + u_{i+\frac{3}{2},j}}{\Delta x_{i+1}}\right] \tag{3.180}$$

$$\left[\{(u^2/2)_x\}_x\right]_{i+\frac{1}{2},j}$$

$$= \frac{1}{2\widetilde{\Delta}x_{i+\frac{1}{2}}}\left[-\frac{-u_{i-\frac{1}{2},j}^2 + u_{i+\frac{1}{2},j}^2}{\Delta x_i} + \frac{-u_{i+\frac{1}{2},j}^2 + u_{i+\frac{3}{2},j}^2}{\Delta x_{i+1}}\right]. \tag{3.181}$$

The difference of these expressions is

$$[u_x u_x]_{i+\frac{1}{2},j} = \frac{1}{2\widetilde{\Delta}x_{i+\frac{1}{2}}}\left[\Delta x_i\left(\frac{-u_{i-\frac{1}{2},j} + u_{i+\frac{1}{2},j}}{\Delta x_i}\right)^2\right.$$

$$\left.+\Delta x_{i+1}\left(\frac{-u_{i+\frac{1}{2},j} + u_{i+\frac{3}{2},j}}{\Delta x_{i+1}}\right)^2\right] \tag{3.182}$$

$$= \frac{1}{2J_{i+\frac{1}{2}}}\left([J(\delta_x u)^2]_i + [J(\delta_x u)^2]_{i+1}\right),$$

which is the average value of the dissipation rate weighted by the Jacobian (cell volume). Similar to the case of the advective term from Sect. 3.5.2, the weighing is applied in an opposite manner to distance-based interpolation and is not the simple arithmetic mean.

3.7 Summary of the Staggered Grid Solver

Let us summarize the finite-difference technique to solve for incompressible flows on a Cartesian grid. We will also briefly mention how one can extend the above discussions to three-dimensional flows. For simplicity, we represent the nonlinear advection term here for a uniform grid discretization but we refer readers to Sect. 3.5.2 for nonuniform grids.

Consider a rectangular grid shown in Fig. 3.12. The placement of variables on a staggered grid should be as the following:

C_p	(cell center)	$p, \phi, \overline{u_i}^{x_i}, \delta_{x_i} u_i$ (no summation implied)
C_i	(cell face)	u_i
C_{ij}	(cell edge)	$\overline{u_i}^{x_j}, \overline{u_j}^{x_i}, \delta_{x_j} u_i, \delta_{x_i} u_j$

where p and u_i (u, v, and w) are the fundamental variables and all others are variables used during the calculation.

For calculations involving turbulence models, the eddy viscosity ν_T (which is a function of space and time) can be introduced in addition to the kinematic viscosity ν. There can also be cases where ν is not a constant. For these cases, the scalar (eddy) viscosity can be placed at C_p and use the interpolated value at C_{ij} when the shear stress is needed (discussed further in Sect. 8.7.1).

Explicit Method

The treatment of the advective and viscous terms with the second-order Adams-Bashforth method and implicitly incorporating the pressure gradient and incompressibility leads to the following discretizations:

Fig. 3.12 Variable placements on a three-dimensional staggered grid

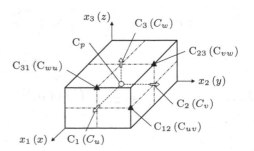

$$\delta_{x_i} u_i^{n+1} = 0 \tag{3.183}$$

$$\frac{u_i^{n+1} + u_i^n}{\Delta t} = -\delta_{x_i} P^{n+1} + \frac{3 (A_i + B_i)^n - (A_i + B_i)^{n-1}}{2} \tag{3.184}$$

$$A_i = -\delta_{x_j} \left(\overline{u_j}^{x_i} \overline{u_i}^{x_j} \right) \quad \text{or} \quad A_i = -\overline{\overline{u_j}^{x_i} \delta_{x_j} u_i}^{x_j} \tag{3.185}$$

$$B_i = \delta_{x_j} \left[\nu \left(\delta_{x_j} u_i + \delta_{x_i} u_j \right) \right], \tag{3.186}$$

where we do not invoke the summation for the index used for interpolation (e.g., for the index k that appears in the interpolation $\overline{}^{x_k}$) unless three of the same indices appear. The viscous term becomes $B_i = \nu \delta_{x_j} (\delta_{x_j} u_i)$ for constant ν. We can also adopt the average of the above two advective term discretizations to yield a quadratic conservation form. The time advancement scheme can be described by the following steps:

1. Predict the velocity at cell face, C_i, explicitly.

$$u_i^P = u_i^n - \Delta t \, \delta_{x_i} P^n + \Delta t \, \frac{3 (A_i + B_i)^n - (A_i + B_i)^{n-1}}{2} \tag{3.187}$$

2. Solve the Poisson equation to determine the scalar potential ϕ (pressure correction) at cell center, C_p.

$$\delta_{x_j} \left(\delta_{x_j} \phi \right) = \frac{1}{\Delta t} \delta_{x_i} u_i^P \tag{3.188}$$

3. Correct the velocity at C_i and update the pressure at C_p.

$$u_i^{n+1} = u_i^P - \Delta t \, \delta_{x_i} \phi \tag{3.189}$$

$$P^{n+1} = P^n + \phi \tag{3.190}$$

Starting from an appropriate initial condition, the flow field can evolve in time with time step Δt. If the solution ceases to change over time, the steady solution is found.

Implicit Treatment of Viscous Term

When viscosity ν is constant, the linear viscous term can be treated implicitly in a straightforward manner. This implicit treatment is especially useful when the stability limit from viscous time integration is restrictive with an explicit formulation.[7] Here, let us consider changing the viscous integration to the second-order Crank–Nicolson method

$$\frac{u_i^{n+1} - u_i^n}{\Delta t} = -\delta_{x_i} P^{n+1} + \frac{3 A_i^n - A_i^{n-1}}{2} + \nu \delta_{x_j} \delta_{x_j} \frac{u_i^n + u_i^{n+1}}{2} \tag{3.191}$$

---

[7] As in cases where fine grids are required near wall boundaries (e.g., near-wall turbulence).

The time stepping routine becomes threefold:

1. Solve the momentum equation to predict the velocity at C_i.

$$
\begin{aligned}
\left[1 - \Delta t \, \tfrac{\nu}{2} \delta_{x_j} \delta_{x_j}\right] u_i^P &= u_i^n + \Delta t \, \tfrac{\nu}{2} \delta_{x_j} \delta_{x_j} u_i^n \\
&\quad - \Delta t \, \delta_{x_i} P^n + \Delta t \, \tfrac{3A_i^n - A_i^{n-1}}{2}
\end{aligned}
\tag{3.192}
$$

2. Solve the Poisson equation to determine the scalar potential ϕ at C_p.

$$
\delta_{x_j} \left(\delta_{x_j} \phi \right) = \frac{1}{\Delta t} \delta_{x_i} u_i^P
\tag{3.193}
$$

3. Correct the velocity at C_i and update pressure at C_p.

$$
u_i^{n+1} = u_i^P - \Delta t \, \delta_{x_i} \phi
\tag{3.194}
$$

$$
P^{n+1} = P^n + \phi - \frac{\nu}{2} \Delta t \, \delta_{x_j} \delta_{x_j} \phi
\tag{3.195}
$$

By substituting Eq. (3.194) into Eq. (3.192) to eliminate u_i^P, we find that P^{n+1} should be determined with Eq. (3.195) instead of $P^{n+1} = P^n + \phi$. Nonetheless, the magnitude of the last term $\frac{\nu}{2} \Delta t \, \delta_{x_j} \delta_{x_j} \phi$ in Eq. (3.195) is usually very small.

3.8   Boundary and Initial Conditions

3.8.1   Boundary Setup

Let us consider various types of flows that are often simulated. Shown as examples in Fig. 3.13 are (a) external flow around a body, (b) internal flow, (c) boundary-layer flow, and (d) a jet in semi-infinite space. There are cases of flows in which bodies can be in motion. We can also encounter flows with a moving gas–liquid interface. If we neglect the influence from the gas, the interface can be treated as a free surface for the liquid.

Because we cannot generate a finite-size grid for an infinitely large domain, we are forced to truncate the computational space to be a finite region in many cases (unless the flow over interest is within a closed space). This spatial truncation necessitates the specification of inflow, outflow, or far-field boundary conditions. In general, these conditions are difficult to specify exactly to match the flow physics described by the governing equations. If we are to simulate unsteady turbulent flow, we would need an inflow condition that contains the correct turbulence that is a function of time and also an outflow condition that allows for the vortices to exit the computational domain without creating unphysical reflections back into the inner domain. Furthermore, the specification of the far-field boundary should be treated with care since it can

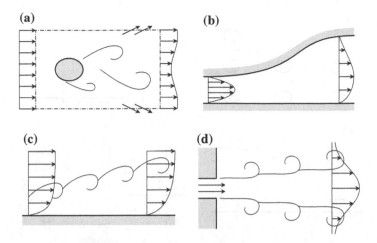

Fig. 3.13 Examples of computational domains for spatially-developing flows: **a** flow around a body, **b** internal flow in a channel, **c** boundary-layer flow, and **d** jet in semi-infinite space

affect the temporal and spatial evolution of flow around bodies. At the moment, there is not a universal artificial boundary condition that provides correct physics for all of the inlet, outlet, and far-field boundaries. On the other hand, solid walls are physical boundaries and do not usually affect the flow in unphysical manner if prescribed correctly. However, it should be kept in mind that boundary layers have large velocity gradients near the wall. In order to capture the flow field accurately, the mesh must be highly refined, especially for high-Reynolds number flows.

For some special cases with spatial periodicity as shown in Fig. 3.14, the specification of inflow, outflow, or far-field boundary conditions may be facilitated with the use of periodic boundary conditions. For Case (a), we can invoke periodicity around the shown box. The size of the box can be increased to cover more than one object if flow structures of interest are larger than the box shown in Fig. 3.14a. For the internal flow of Case (b), it may be possible to specify periodic boundary conditions. For spatially developing flows in Cases (c) and (d), where there is flow profile similarity between the inlet and outlet, we may use a periodic boundary condition that takes the growth into account. For example, if we can allow for time development (i.e., $x = Ut$), the downstream boundary condition can be transformed as a spatial development. If we restrict the temporal development at the boundary, we would restrict some of the spatial development. Prescribing artificial boundary conditions should be dealt with care with potential influence on the flow in mind.

To assess the influence of the position of the artificial boundary, let us consider the flow around a circular cylinder in an infinitely large domain in the reference frame of the body, as illustrated in Fig. 3.15. For boundaries that are placed sufficiently far from the cylinder, the upstream and downstream flow would be uniform. However, in numerical simulations, we truncate the spatial domain and consider only the finite domain close to the body. For low-Reynolds number flow, the viscous effect can

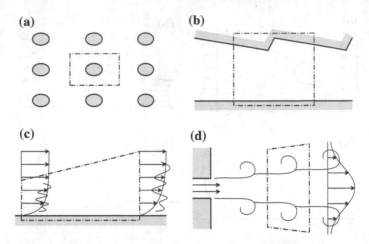

Fig. 3.14 Examples for which periodic boundary conditions may be applied. **a** Flow around an array of objects. **b** Flow through a channel with expansions and contractions. **c** Self-similar boundary-layer flow. **d** Jet in self-similarity region

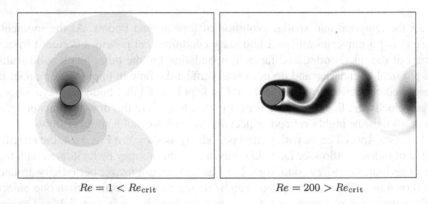

$Re = 1 < Re_{\mathrm{crit}}$ $Re = 200 > Re_{\mathrm{crit}}$

Fig. 3.15 Vorticity fields around a circular cylinder for flows with Re below and above the critical Reynolds number Re_{crit} for shedding. Same contour levels are shown for both figures

affect the flow field over a large distance, especially in the cross-stream direction. On the other hand, for higher Reynolds number flows, the viscous effect is confined to the neighborhood of the body and in its large wake from the shedding of the vortices. Note that the velocity boundary condition is not likely to become uniform at the outlet in most cases.

For simulations that use a finite domain with uniform flow being inappropriate, we must provide far-field or outflow boundary conditions. In such cases, we must ensure that the inner flow field is not influenced significantly by the artificial boundary conditions by considering the following points. First, the computational domain should be chosen as large as possible. Second, the boundary condition should not

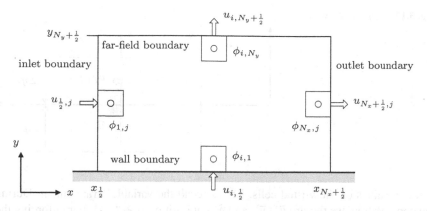

Fig. 3.16 Cells adjacent to the boundaries of the computational domain

allow non-physical reflections to propagate back into the inner field. Since artificial boundary conditions usually do not strictly satisfy the Navier–Stokes equations, the motion of fluid may be affected near the artificial boundary. Third, we can remove the region near the artificial boundary from what we consider a reliable numerical solution for the aforementioned reason.

In this chapter, we restrict our discussion to simulations using a Cartesian coordinate system. Thus, let us consider a flow field such as the one presented in Fig. 3.16. Here, we denote the streamwise direction with x, which extends from the inflow to the outflow boundaries. The wall-normal direction is represented with y which covers the domain from the wall to the artificial far-field boundary. This far-field boundary refers to the computational boundary where the influence from the body (or wall) is sufficiently small. For the solid wall, the velocity boundary condition is usually specified. For inflow, outflow, and far-field boundary conditions, there are cases where the velocity is specified or the pressure is prescribed.

3.8.2 Solid Wall Boundary Condition

To apply the boundary condition at a solid wall, we place a virtual cell (ghost cell) inside the wall as shown by the dashed lines in Fig. 3.17. We can let the width of the virtual cell (length normal to the surface) to be equal to the cell size adjacent to the wall (*i.e.*, $\Delta y_0 = \Delta y_1$). Along the solid wall, we often apply the *no-slip boundary condition* of $u = 0$ or the *slip boundary condition* of $\partial u / \partial y = 0$, in addition to the no-penetration (permeable) boundary condition $v = 0$.

For these boundary conditions, we can respectively set $u_{i+\frac{1}{2},0} = \mp u_{i+\frac{1}{2},1}$ in Fig. 3.17. Within the actual program, we do not need to allocate an array for the

Fig. 3.17 Stencil at the wall

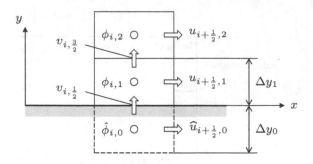

velocity values on the virtual cells. Let us recall the variable arrangement. During the computation, we require $\bar{u}^y$, $\bar{v}^x$, and $\delta_y u$ at position $(i + \frac{1}{2}, \frac{1}{2})$. For enforcing the no-slip condition, we have

$$[\bar{u}^y]_{i+\frac{1}{2},\frac{1}{2}} = 0, \quad [\delta_y u]_{i+\frac{1}{2},\frac{1}{2}} = \frac{u_{i+\frac{1}{2},1}}{\Delta y_1/2} \qquad (3.196)$$

with the velocity at the virtual cell being taken into account. It is sometimes desirable to increase the accuracy of $[\delta_y u]_{i+\frac{1}{2},\frac{1}{2}}$ for the no-slip condition. In case of slip boundary, we can instead use

$$[\bar{u}^y]_{i+\frac{1}{2},\frac{1}{2}} = u_{i+\frac{1}{2},1}, \quad [\delta_y u]_{i+\frac{1}{2},\frac{1}{2}} = 0. \qquad (3.197)$$

In general, if the velocity of the wall $u_{i+\frac{1}{2},\text{wall}}$ is provided, we can set

$$[\bar{u}^y]_{i+\frac{1}{2},\frac{1}{2}} = u_{i+\frac{1}{2},\text{wall}}, \quad [\delta_y u]_{i+\frac{1}{2},\frac{1}{2}} = \frac{u_{i+\frac{1}{2},1} - u_{i+\frac{1}{2},\text{wall}}}{\Delta y_1/2}. \qquad (3.198)$$

For the normal components, we provide

$$[\bar{v}^x]_{i+\frac{1}{2},\frac{1}{2}} = \frac{v_{i,\frac{1}{2}} + v_{i+1,\frac{1}{2}}}{2}, \quad [\delta_x v]_{i+\frac{1}{2},\frac{1}{2}} = \frac{-v_{i,\frac{1}{2}} + v_{i+1,\frac{1}{2}}}{\tilde{\Delta}x_{i+\frac{1}{2}}}. \qquad (3.199)$$

In many cases, the *non-permeable (no-penetration) boundary condition* is prescribed for the wall-normal velocity at the wall boundary, which is zero if the body is stationary.

There are multiple pressure boundary conditions. Based on the momentum equation at the wall, boundary conditions can be derived for $\partial p/\partial n$, $\partial p/\partial s_1$, and $\partial p/\partial s_2$, where n is the wall-normal direction and s_1 and s_2 are the two independent wall-tangential directions. These boundary conditions should be satisfied, but with discretized equations that may not be necessarily true.

To determine the pressure boundary condition in conjunction with the time advancement scheme, we should utilize the pressure gradient normal to boundary in the wall-normal momentum equation. There is usually no penetrating flow at a solid wall. However, in the case of a porous wall or when suction/blowing is enforced at the wall, we can have non-zero flow into or out of the wall. In either case, we assume that the wall-normal velocity v_{wall} is specified in some fashion. The simplest treatment of the boundary condition is to set

$$v_{i,\frac{1}{2}}^P = v_{i,\text{wall}}^{n+1} \tag{3.200}$$

in the prediction step of the SMAC method for the wall-normal velocity v^P, instead of time advancing the boundary velocity through the Navier–Stokes equation, Eq. (3.14). Here we have denoted the normal velocity with superscript $(n + 1)$ to represent the boundary condition to be from the new time step. For the subscript, Fig. 3.17 should be referred. Since we cannot alter the boundary value during the correction step, Eq. (3.15), the boundary condition for ϕ must be $[\partial\phi/\partial y]_{\text{wall}} = 0$. To prescribe $\partial\phi/\partial y$ at the boundary, we make use of a virtual ϕ inside the wall. In this case, we have

$$v_{i,\frac{1}{2}}^{n+1} = v_{i,\frac{1}{2}}^P - \Delta t \frac{-\widehat{\phi}_{i,0} + \phi_{i,1}}{\widetilde{\Delta y}_{\frac{1}{2}}}, \tag{3.201}$$

where

$$\widetilde{\Delta y}_{\frac{1}{2}} = \frac{\Delta y_0 + \Delta y_1}{2} = \Delta y_1. \tag{3.202}$$

Because $v_{i,\frac{1}{2}}^P$ should not be altered as we have already provided the boundary condition with Eq. (3.200), the following must hold

$$\widehat{\phi}_{i,0} = \phi_{i,1}. \tag{3.203}$$

The extrapolated value of $\widehat{\phi}_{i,0}$ can then be inserted into Eq. (3.87) at the boundary cell for the y direction finite difference

$$\left.\frac{\partial^2 \phi}{\partial y^2}\right|_{i,1} = B_{y,1}^- \widehat{\phi}_{i,0} - (B_{y,1}^- + B_{y,1}^+)\phi_{i,1} + B_{y,1}^+ \phi_{i,2}. \tag{3.204}$$

Note that the virtual $\widehat{\phi}_{i,0}$ does not need to be allocated since Eq. (3.203) can be combined with Eq. (3.204) to yield

$$\left.\frac{\partial^2 \phi}{\partial y^2}\right|_{i,1} = B_{y,1}^+ (-\phi_{i,1} + \phi_{i,2}) \tag{3.205}$$

saving some memory allocation.

We should carefully examine the implication of prescribing the above boundary condition. That is, there is an issue of whether the variable P determined from the above algorithm is equivalent to the physical pressure. When there is no slip and no penetration for the velocity at the wall ($u = v = w = 0$), the wall-normal pressure gradient becomes nonzero with

$$\frac{\partial P}{\partial y} = \nu \frac{\partial^2 v}{\partial y^2} \tag{3.206}$$

based on the Navier–Stokes equations. We however $\partial p/\partial y = 0$ is specified as long as we use Eq. (3.203). This incompatibility is generated by providing Eq. (3.200) for the prediction velocity v^P along the boundary in the SMAC method, instead of time integrating the boundary velocity with the Navier–Stokes equations. Because we utilized the velocity from the next time step where the viscous term and the pressure term are in balance, we have to set the gradient of the corrective ϕ to be zero in the numerical algorithm.

There are instances where the boundary condition for the next time step is provided in the intermediate step of the fractional-step method, without using Eq. (3.11). As we have warned in Sect. 3.3.2, the fractional-step velocity v^F contains the influence of viscosity but not the pressure gradient and thus is not yet a physical velocity. Let us consider the consequence of applying a physical boundary condition at this step. We add the influence of the pressure gradient to the intermediate velocity field to advance the flow field to the next time step in the inner domain. At the boundary, the boundary condition at the next time step is already provided for the intermediate velocity and no modification is performed along the boundary. There clearly is a difference in how the velocity is fractionally advanced between the inner domain and the boundary. Therefore, the treatment of ∇P along the boundary becomes different from that of the interior domain.

The variable P that is solved for in the fractional-step method and the SMAC method corresponds to the scalar potential that is the sum of the pressure and the potential function needed in numerical computation. What is meant by "needed in numerical computation" is to let the potential to ensure the compatibility between the boundary condition and the inner flow condition, and to eliminate the error in the continuity equation at the present time step.[8]

When the physical pressure is needed, we can provide the divergence-free velocity field to the right-hand side of Eq. (1.40)

$$\nabla^2 P = -\nabla \cdot \nabla \cdot (uu) + \nabla \cdot f. \tag{3.207}$$

(where we have set $P = p/\rho$ for constant ρ) and prescribe the physical solid wall boundary condition, such as Eq. (3.206). The above pressure Poisson equation can be solved to find the physical pressure field. To obtain a good approximation for the physical pressure field, we should predict $v^P_{i,\frac{1}{2}}$ even along the boundary with Eq. (3.11) that includes the viscous term. We can correct this boundary velocity such that the velocity at the next step becomes $v^{n+1}_{i,\text{wall}}$ by employing the appropriately

---

[8]Here, we have not specified a pressure boundary condition for p but utilized the zero-gradient boundary condition for ϕ. To determine the divergence-free (solenoidal) velocity field, the pressure boundary condition for p is not needed.

extrapolated $\widehat{\phi}_{i,0}$ via the boundary condition on ϕ,

$$v_{i,\text{wall}}^{n+1} = v_{i,\frac{1}{2}}^{P} - \Delta t \frac{-\widehat{\phi}_{i,0} + \phi_{i,1}}{\widehat{\Delta y}_{\frac{1}{2}}} \qquad (3.208)$$

The issue here is the numerical accuracy of $\nu \partial^2 v / \partial y^2$ at the wall. As mentioned in Sect. 2.3.1, the use of one-sided difference reduces the accuracy. Hence it would be necessary to employ a stencil at the boundary that is wider than the stencil in the inner domain to maintain the desired accuracy.

Treatment of Flow Around a Corner

Consider flow around a stationary wall with a corner at $(x_{i-\frac{1}{2}}, y_{j-\frac{1}{2}})$, as shown in Fig. 3.18a. From the boundary condition, $u_{i-\frac{1}{2},j-1} = v_{i-1,j-\frac{1}{2}} = 0$. Using the approach described in Sect. 3.5, let us consider determining the velocity $u_{i-\frac{1}{2},j}$ and $v_{i,j-\frac{1}{2}}$ near the corner. When casting the advection term in the divergence or gradient form, we need to evaluate $[\overline{u}^y \overline{v}^x]_{i-\frac{1}{2},j-\frac{1}{2}}$ or $[\overline{u}^y \delta_x v]_{i-\frac{1}{2},j-\frac{1}{2}}$, respectively, at the corner. If we construct the discrete operators that satisfy compatibility and conservation properties in accordance to Sect. 3.5, we must have

$$[\overline{u}^y]_{i-\frac{1}{2},j-\frac{1}{2}} = \frac{u_{i-\frac{1}{2},j-1} + u_{i-\frac{1}{2},j}}{2}, \qquad [\delta_y u]_{i-\frac{1}{2},j-\frac{1}{2}} = \frac{-u_{i-\frac{1}{2},j-1} + u_{i-\frac{1}{2},j}}{\Delta y}$$

$$[\overline{v}^x]_{i-\frac{1}{2},j-\frac{1}{2}} = \frac{v_{i-1,j-\frac{1}{2}} + v_{i,j-\frac{1}{2}}}{2}, \qquad [\delta_x v]_{i-\frac{1}{2},j-\frac{1}{2}} = \frac{-v_{i-1,j-\frac{1}{2}} + v_{i,j-\frac{1}{2}}}{\Delta x} \qquad (3.209)$$

However, to enforce the boundary condition of $u = 0$ and $v = 0$ at the corner, we should have

$$[\overline{u}^y]_{i-\frac{1}{2},j-\frac{1}{2}} = 0, \qquad [\delta_y u]_{i-\frac{1}{2},j-\frac{1}{2}} = \frac{2u_{i-\frac{1}{2},j}}{\Delta y}$$

$$[\overline{v}^x]_{i-\frac{1}{2},j-\frac{1}{2}} = 0, \qquad [\delta_x v]_{i-\frac{1}{2},j-\frac{1}{2}} = \frac{2v_{i,j-\frac{1}{2}}}{\Delta x} \qquad (3.210)$$

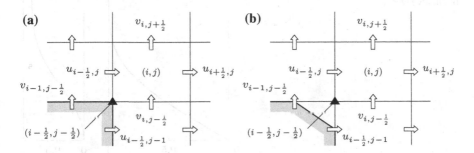

Fig. 3.18 Staggered grid arrangement near a corner. **a** Variable arrangement near a corner. **b** Geometry representation with staggered grid

Thus, we observe that the desired properties and the boundary conditions cannot be satisfied simultaneously. Equation (3.209) does not satisfy the velocity boundary conditions at the corner and Eq. (3.210) does not meet the compatibility and conservation properties discussed in Sect. 3.5. Since compatibility and conservation are important, we cannot blindy adopt Eq. (3.210). This dilemma can be resolved, if we consider the corner to be the shaven off as illustrated in Fig. 3.18b. For this reason, the boundary condition at solid wall corners cannot be implemented exactly within the framework of staggered grids while maintaining compatibility.

3.8.3 Inflow and Outflow Boundary Conditions

Inflow Condition

For the inflow condition, the velocity $u_{\frac{1}{2},j}$ at $x_{\frac{1}{2}}$ is often prescribed (see Fig. 3.16). For simulating flow in pipes, ducts, and channels, we can specify the inflow condition with the fully developed velocity profile using an analytical solution, numerical solution, or experimental measurements. For cases of simulating turbulent flow, we need to also prescribe the turbulent stress at the inlet. Otherwise, the flow can behave in a non-physical manner downstream of the inlet due to the imbalance in the momentum equation.

Specifying the inflow boundary condition for a semi-infinite domain with a boundary layer should be dealt with care. For a desired boundary layer thickness, we can use the *Blasius profile* for a laminar boundary layer over a flat plat as shown in Fig. 3.19. Here, U is the freestream velocity, x is the streamwise distance from the leading edge, and $\mathrm{Re}_x = Ux/\nu$. This solution is a self-similar solution for a growing boundary layer [29]. Note that the Blasius profile also has a vertical component v as shown in Fig. 3.19, which also needs to be provided at the inlet. Also see Problem 3.4.

Fig. 3.19 Velocity profile for laminar flat-plate boundary layer

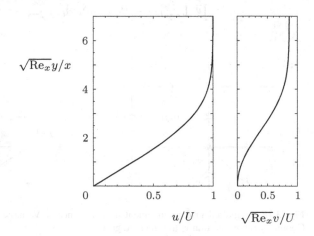

In many cases, only the streamwise velocity profile is specified using the Blasius solution and the vertical component is set to zero. However, specifying only the streamwise velocity creates an imbalance in the governing equations and can lead to the solution behaving incorrectly near the inlet boundary. Even if the vertical velocity profile is specified, one should specify the far field boundary condition with care to ensure that there is no mass imbalance between the flow entering and leaving the computational domain.

Specifying an inlet condition can be a challenge for many problems. To ensure that the flow of interest is predicted correctly, sufficient run-up region should be provided upstream to allow for the flow to relax prior to reaching the region of the flow being examined.

Advective (Convective) Outflow Condition

The outflow boundary condition for velocity $u_{N_x+\frac{1}{2},j}$ is generally unknown at $x_{N_x+\frac{1}{2}}$. Outflow condition should be time varying to allow for vortical structures to cleanly exit the computational domain without reflecting back into the domain or disturbing the solution in the inner domain.

To determine the outflow velocity profile in Fig. 3.16, the advective outflow condition can be utilized

$$\frac{\partial u_i}{\partial t} + u_m \frac{\partial u_i}{\partial x} = 0, \qquad (3.211)$$

where u_m is a characteristic advective velocity for the outlet. This condition is an advection equation for u_i. The boundary velocity is computed by advecting the velocity profile u_i by $\Delta t u_m$ in the x direction to the boundary location.

Referring to Fig. 3.20a, we can approximate the advective outflow condition, Eq. (3.211), with the first-order upwind difference

$$u^{n+1}_{N_x+\frac{1}{2},j} = u^n_{N_x+\frac{1}{2},j} - \Delta t u_m \frac{-u^n_{N_x-\frac{1}{2},j} + u^n_{N_x+\frac{1}{2},j}}{\Delta x_{N_x}}, \qquad (3.212)$$

(a)   **(b)**

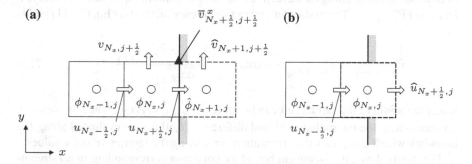

Fig. 3.20 Cells and finite-difference stencil near the outflow boundary for the case of specifying **a** normal velocity and **b** pressure

which corresponds to linear interpolation of $u^{n+1}(x_{N_{x+\frac{1}{2}}}, y) = u^n(x_{N_{x+\frac{1}{2}}} - \Delta t u_m, y)$.
To maintain global conservation (mass conservation over the entire computational
domain), the advective velocity u_m can be taken to be constant [34]. There is arbitrariness to the choice of u_m. For instance, u_m can be set to the average streamwise
velocity across the outlet boundary, the average between the minimum and maximum
value, or a characteristic advection velocity.

We can also consider the outflow condition that includes the influence of viscosity

$$\frac{\partial u_i}{\partial t} + u_m \frac{\partial u_i}{\partial x} = \nu \left(\frac{\partial^2 u_i}{\partial x^2} + \frac{\partial^2 u_i}{\partial y^2} \right), \tag{3.213}$$

which appears to provide approximation to the outflow dynamics at the boundary.
The derivative $\partial^2 u_i/\partial y^2$ can be evaluated with the central-difference scheme, but
$\partial^2 u_i/\partial x^2$ needs to be computed with an upwinding scheme. Note that further mim-
icking the Navier–Stokes equations with the above convective boundary condition
does not appear to improve the simulated results. In some cases, the overall conservation can be violated. For example, if we consider the use of instantaneous velocity
values (u, v) in

$$\frac{\partial u_i}{\partial t} + u \frac{\partial u_i}{\partial x} + v \frac{\partial u_i}{\partial y} = \nu \left(\frac{\partial^2 u_i}{\partial x^2} + \frac{\partial^2 u_i}{\partial y^2} \right), \tag{3.214}$$

the solution near the outlet becomes unphysical [20, 34].

Next, let us consider the convective boundary condition for the velocity v that is
tangential to the outlet boundary. Due to the arrangement of the staggered grid, the
velocity v is not positioned along the outlet boundary. There are candidates of $v_{N_x, j+\frac{1}{2}}^{n+1}$
(inside the domain), $\overline{v}_{N_x+\frac{1}{2}, j+\frac{1}{2}}^x$ (interpolated at the boundary), and $\widehat{v}_{N_x+1, j+\frac{1}{2}}^{n+1}$ (exterior
virtual value) to set the convective boundary condition. Velocity values inside the
computational domain should be determined by the momentum equation. When we
compute $v_{N_x, j+\frac{1}{2}}$, velocity values $\overline{v}_{N_x+\frac{1}{2}, j+\frac{1}{2}}^x$ and $[\delta_x v]_{N_x+\frac{1}{2}, j+\frac{1}{2}}$ at the boundary are
necessary. Thus, we specify the boundary condition at the computational boundary
or downstream of it. Using the advective boundary condition, Eqs. (3.211) or (3.213),
let us find $\widehat{v}_{N_x+1, j+\frac{1}{2}}^{n+1}$. The first-order upwind-difference method for Eq. (3.211) yields

$$\widehat{v}_{N_x+1, j+\frac{1}{2}}^{n+1} = \widehat{v}_{N_x+1, j+\frac{1}{2}}^n - \Delta t u_m \frac{-v_{N_x, j+\frac{1}{2}}^n + \widehat{v}_{N_x+1, j+\frac{1}{2}}^n}{\widehat{\Delta x}_{N_x+\frac{1}{2}}}, \tag{3.215}$$

where the width of the virtual cell can be set to $\widehat{\Delta x}_{N_x+\frac{1}{2}} = \Delta x_{N_x}$. Once the velocity
$\widehat{v}$ is computed, the interpolation $\overline{v}^x$ and difference $[\delta_x v]$ can be evaluated along the
boundary which completes the preparation for solving the interior velocity values.

For steady flows, the Neumann boundary condition corresponding to no streamwise gradient

$$\frac{\partial u_i}{\partial x} = 0 \tag{3.216}$$

is often employed to suggest that the flow exits the computational domain in a fully developed manner. This boundary condition however should not be used at the outlet boundary for unsteady flow. We consider that steady flow is attained when temporal change diminishes in unsteady calculations and hence do not use Eq. (3.216) when the flow is developing. When steady state is reached, Eq. (3.211) reduces to Eq. (3.216).

Pressure Boundary Condition with Prescribed Velocity

Specifying the normal velocity condition at the inlet or outlet is similar to how the velocity at the solid wall is prescribed. For the SMAC method we can provide the inflow and outflow conditions at the next time step u^{n+1} for the intermediate velocity u^P. In such cases, we can impose the normal gradient for ϕ to be zero such that the velocity at the next time level would not be affected.

The actual pressure gradient at the inflow and outflow boundary is not necessarily zero. This is similar to the discussion regarding the condition along a solid wall boundary raised in Sect. 3.8.2. The zero-pressure gradient condition results from the use of two different equations to predict the velocity due to the boundary condition and due to the evolution through the momentum equation. In the SMAC method, this is caused by how P includes pressure as well as the scalar potential.

In order to make P close to be the actual pressure in the SMAC method, we perform the following. First, we predict the velocity $u^P_{N_x+\frac{1}{2},j}$ using the Navier–Stokes equations at the outlet. Then we select a boundary condition such that the gradient of ϕ lets the velocity $u^{n+1}_{N_x+\frac{1}{2},j}$ satisfy the advective/convective boundary condition Eqs. (3.211) and (3.213). That is, the equation below provides an extrapolating formula for $\widehat{\phi}_{N_x+1,j}$:

$$u^{n+1}_{N_x+\frac{1}{2},j} = u^P_{N_x+\frac{1}{2},j} - \Delta t \frac{-\phi_{N_x,j} + \widehat{\phi}_{N_x+1,j}}{\widehat{\Delta x}_{N_x+\frac{1}{2}}}. \tag{3.217}$$

Even if the prediction is based on the Navier–Stokes equations, the computed results are somewhat different from the inner solution because the stencil used in computation is one-sided. For a realistic pressure solution, the pressure equation should be resolved with a velocity field satisfying incompressibility and a physical pressure boundary condition.

Another issue related to the used of the zero-gradient pressure condition is the reduction in the speed of convergence for the pressure solver. The use of Dirichlet condition allows for a faster convergence. From the momentum equation, the pressure gradient is

$$\frac{\partial P}{\partial x} = -\frac{\partial u}{\partial t} - u\frac{\partial u}{\partial x} - v\frac{\partial u}{\partial y} + \nu\left(\frac{\partial^2 u}{\partial x^2} + \frac{\partial^2 u}{\partial y^2}\right). \tag{3.218}$$

Substituting Eq. (3.213) into the above relation, we find

$$\frac{\partial P}{\partial x} = -(u - u_m)\frac{\partial u}{\partial x} - v\frac{\partial u}{\partial y}. \tag{3.219}$$

Now, let us consider using this as an outflow boundary condition. We can calculate the right-hand side using upwinding and let the left-hand side be $(-P_{N_x,j}^{n+1} + \widehat{P}_{N_x+1,j}^{n+1})/\widehat{\Delta}x_{N_x+\frac{1}{2}}$. Of course, $P_{N_x,j}^{n+1}$ is unknown when we solve for the inner pressure field. If this value can be predicted somehow, Eq. (3.219) would become an extrapolation formula for $\widehat{P}_{N_x+1,j}^{n+1}$ and we can use it as a Dirichlet condition for the Poisson equation. Miyauchi et al. [20] have proposed to use a transport equation model for pressure

$$\frac{\partial P}{\partial t} + u_m \frac{\partial P}{\partial x} = \frac{\nu}{2}\omega^2, \tag{3.220}$$

to predict the pressure value $P_{N_x,j}^{n+1}$. Here, ω is the vorticity. They have demonstrated that the use of this boundary condition provides reasonable solutions. In their work, they have employed a different type of grid and difference method, and the details of how the Dirichlet boundary condition is implemented differs.

Outflow Boundary Condition with Prescribed Pressure

The pressure can be prescribed along the boundary as a Dirichlet boundary condition. For such case, the pressure boundary condition is applied to P^{n+1} for the fractional-step method and $\phi = P^{n+1} - P^n$ for the SMAC method. The convergence of the numerical solution to the pressure Poisson equation is faster when a Dirichlet condition is applied (even along a portion of the boundary) compared to when a Neumann condition is employed along the entire boundary.

The specification of the Dirichlet condition can either be enforced at the N_x-th cell adjacent to the boundary as shown in Fig. 3.20a or at the pressure location aligned with the boundary as shown in Fig. 3.20b.

To specify the pressure at the outlet boundary as shown in Fig. 3.20a, we need to provide $\widehat{\phi}_{N_x+1,j}$ into the pressure equation such that $(\phi_{N_x,j} + \widehat{\phi}_{N_x+1,j})/2$ satisfies the desired boundary condition. Using extrapolation based on the advective boundary condition, Eq. (3.215), to calculate $\widetilde{v}_{N_x+1,j+\frac{1}{2}}^{n+1}$ and employing the continuity equation at virtual cell $(N_x + 1, j)$

$$\widehat{u}_{N_x+\frac{3}{2},j} = -u_{N_x+\frac{1}{2},j} - \widetilde{\Delta}x_{i+1} \frac{-\widetilde{v}_{N_x+1,j-\frac{1}{2}} + \widetilde{v}_{N_x+1,j+\frac{1}{2}}}{\Delta y_j}, \tag{3.221}$$

to compute $\widehat{u}_{N_x+\frac{3}{2},j}$, necessary values become available to perform the finite-difference calculations of the momentum equation for boundary value velocity $u_{N_x+\frac{1}{2},j}$.

For the arrangement shown in Fig. 3.20b, we have direct access to the value $\phi_{N_x,j}$. From the continuity equation of cell (N_x, j),

$$u_{N_x+\frac{1}{2},j} = -u_{N_x-\frac{1}{2},j} - \Delta x_i \frac{-v_{N_x,j-\frac{1}{2}} + v_{N_x,j+\frac{1}{2}}}{\Delta y_j} \tag{3.222}$$

can be extrapolated. If we also extrapolate $\widehat{v}^{n+1}_{N_x+1,j+\frac{1}{2}}$ based on the advective boundary condition, Eq. (3.215), we have all necessary ingredients to compute the boundary velocity $v_{N_x,j+\frac{1}{2}}$ through the momentum equation.

3.8.4 Far-Field Boundary Condition

The specification of uniform flow as far-field boundary condition is often used when flow over a body in uniform flow is simulated. As shown in Fig. 3.16, the velocity $v_{i,N_y+\frac{1}{2}}$ can be set to the flow velocity at infinity at the far-field boundary location $y_{N_y+\frac{1}{2}}$. However, when the far-field boundary $y_{N_y+\frac{1}{2}}$ in not placed sufficiently away from a body, the specifying uniform flow as a boundary condition may be inappropriate. In reality, there is nonzero flow across the far-field boundary ($v \neq 0$). For boundary layer flows, the boundary layer can displace fluid out of the computational domain (see Fig. 3.14c). Fluid entrainment due to jets can cause the flow to be drawn into the computational domain from the far field (see Fig. 3.14d). The flow crossing the far-field boundary can influence the global flow field. The far-field velocity profile in general is unknown and should be determined as a part of the computation.

A boundary condition that can accommodate far-field flows is the traction-free boundary condition [8, 9]. Let us consider how we can implement the traction-free condition using Fig. 3.21. The traction-free boundary condition assumes that there is no stress on the fluid element on the computational boundary. Denoting the outward normal unit vector along the far-field boundary with $\boldsymbol{n}$,

$$\boldsymbol{T} \cdot \boldsymbol{n} = 0, \quad \text{in indicial notation} \quad T_{ij}n_j = 0 \tag{3.223}$$

represents the traction-free boundary condition. For a boundary at $y = y_{N_y+\frac{1}{2}}$ in a two-dimensional domain shown in Fig. 3.21, Eq. (3.223) becomes

$$T_{xy} = \nu \left(\frac{\partial u}{\partial y} + \frac{\partial v}{\partial x} \right) = 0 \tag{3.224}$$

Fig. 3.21 Finite-difference grid adjacent to the far-field boundary

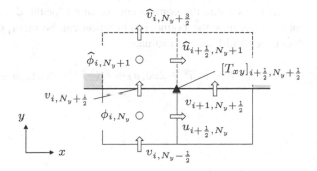

$$T_{yy} = -P + 2\nu \frac{\partial v}{\partial y} = 0 \tag{3.225}$$

since $n = (0, 1)$. For two-dimensional flow, we only have two boundary conditions while we have three variables (u, v, and P) and, for three-dimensional flow, we have three boundary conditions while we have four variables (u, v, w and P). Thus, it becomes necessary that one of the variables be specified separately. In most cases, the sufficiently far pressure value is set to be a constant value (e.g., zero).

It would be appropriate to apply Eq. (3.224) at the point for $[T_{xy}]_{i+\frac{1}{2},N_y+\frac{1}{2}}$ in Fig. 3.21. The discretization of Eq. (3.224) leads to

$$\frac{-u_{i+\frac{1}{2},N_y} + \widehat{u}_{i+\frac{1}{2},N_y+1}}{\widetilde{\Delta} y_{N_y+\frac{1}{2}}} + \frac{-v_{i,N_y+\frac{1}{2}} + v_{i+1,N_y+\frac{1}{2}}}{\widetilde{\Delta} x_{i+\frac{1}{2}}} = 0. \tag{3.226}$$

Equation (3.226) enables us to extrapolate for $\widehat{u}_{i+\frac{1}{2},N_y+1}$. For Eq. (3.225), there is flexibility in employing the condition at the position for $\widehat{P}_{i,N_y+1}$ outside the boundary or for $v_{i,N_y+\frac{1}{2}}$ along the boundary. The selection of the former position suggests

$$- \widehat{P}_{i,N_y+1} + 2\nu \frac{-v_{i,N_y+\frac{1}{2}} + \widehat{v}_{i,N_y+\frac{3}{2}}}{\widetilde{\Delta} y_{N_y+1}} = 0 \tag{3.227}$$

as the discretization. This relation takes either the specified $\widehat{P}_{i,N_y+1}$ or $\widehat{v}_{i,N_y+\frac{3}{2}}$ outside the boundary and extrapolates the other. The width of the virtual cell $\widehat{\Delta} y_{N_y+1}$ is somewhat arbitrary. In most cases, it would suffice to let $\widehat{\Delta} y_{N_y+1} = \Delta y_{N_y}$ and also in Eq. (3.226) to set $\widetilde{\Delta} y_{N_y+1} = \Delta y_{N_y}$. By having $\widehat{u}$ extrapolated with Eq. (3.226), we can utilize the continuity equation

$$\frac{-\widehat{u}_{i-\frac{1}{2},N_y+1} + \widehat{u}_{i+\frac{1}{2},N_y+1}}{\Delta x_i} + \frac{-v_{i,N_y+\frac{1}{2}} + \widehat{v}_{i,N_y+\frac{3}{2}}}{\widehat{\Delta} y_{N_y+1}} = 0 \tag{3.228}$$

at cell (i, N_y), which is outside of the computational domain, to determine $\widehat{v}_{i,N_y+\frac{3}{2}}$. Since the satisfaction of incompressibility on a virtual grid is not absolutely necessary, it is also possible to determine $\widehat{P}_{i,N_y+1}$ beforehand.

Let us examine the components required for the discrete momentum equations and discuss how the boundary condition can be incorporated. For the divergence form, we would need to evaluate

$$[\delta_x(-P - \overline{u}^x\overline{u}^x + 2\nu\delta_x u) + \delta_y(-\overline{u}^y\overline{v}^x + \nu\{\delta_y u + \delta_x v\})]_{i+\frac{1}{2},j} \tag{3.229}$$

$$[\delta_x(-\overline{u}^y\overline{v}^x + \nu\{\delta_y u + \delta_x v\}) + \delta_y(-P - \overline{v}^y\overline{v}^y + 2\nu\delta_y v)]_{i,j+\frac{1}{2}}, \qquad (3.230)$$

which necessitates $-\overline{u}^y\overline{v}^x + \nu\{\delta_y u + \delta_x v\}$ for the location of $[T_{xy}]_{i+\frac{1}{2},N_y+\frac{1}{2}}$ and $-P - \overline{v}^y\overline{v}^y + 2\nu\delta_y v$ for the location of P_{i,N_y+1}, which is outside of the domain. For the gradient form we would use

$$[-\delta_x P - \overline{\overline{u}^x\delta_x u}^x - \overline{\overline{v}^x\delta_y u}^y + \nu\{\delta_x(\delta_x u) + \delta_y(\delta_y u)\}]_{i+\frac{1}{2},j} \qquad (3.231)$$

$$[-\delta_y P - \overline{\overline{u}^y\delta_x v}^x - \overline{\overline{v}^y\delta_y v}^y + \nu\{\delta_x(\delta_x v) + \delta_y(\delta_y v)\}]_{i,j+\frac{1}{2}}, \qquad (3.232)$$

for which, we would need $\overline{u}^y$, $\overline{v}^x$, $\delta_y u$, and $\delta_x v$ for the location of $[T_{xy}]_{i+\frac{1}{2},N_y+\frac{1}{2}}$, and P, $\overline{v}^y$ and $\delta_y v$ for the location of P_{i,N_y+1}, which is outside of the domain. Note that the traction-free boundary condition can be implemented directly with the divergence form. In any case, when $\widehat{u}$, $\widehat{v}$, $\widetilde{P}$ are provided through Eqs. (3.226) and (3.227), all necessary ingredients to integrate the momentum equation should be available.

3.8.5 Initial Condition

For the initial condition, it is ideal to provide a flow field that is close to the desired solution that discretely satisfies the continuity equation. However, this in general can be difficult. Unfortunately, there is not a rule of thumb for choosing an initial condition. We can let the initial flow field be quiescent and instantaneously prescribe the boundary conditions or add external force to time advance the numerical solution. In most cases, the flow would reach the desired steady state solution. Even if the provided initial condition does not satisfy incompressibility, correctly implemented MAC type methods ensure that the error in continuity is removed through the projection steps. Such error can be suppressed due to the construction of the method.

Nonetheless, if we provide an unphysical initial condition with inappropriate mass or momentum balance, the solution would likely blow up before reaching the fully-developed flow field. It may be necessary to consider smaller time steps or increasing the Reynolds number gradually to allow for the numerical flow field to reach the steady flow (fully developed flow) in a stable manner. Starting the simulation in such fashion however would be disadvantageous since longer computation time would most likely be required before the numerical solution settles into the physical solution. In many cases, experience and knowledge of the flow physics can be useful in selecting an appropriate initial condition.

What is most important is that the final outcome of the simulation (steady-state or fully-developed flow) does not contain artifacts from the use of artificial initial conditions.

3.9   High-Order Accurate Spatial Discretization

3.9.1   High-Order Accurate Finite Difference

In this section, let us consider enhancing the spatial accuracy to numerically solve for incompressible flows by employing high-order accurate finite-difference schemes. The second-order interpolation and differencing are

$$\left[\overline{u}^x\right]_i = \frac{u_{i-\frac{1}{2}} + u_{i+\frac{1}{2}}}{2} \tag{3.233}$$

$$\left[\delta_x u\right]_i = \frac{-u_{i-\frac{1}{2}} + u_{i+\frac{1}{2}}}{\Delta x}. \tag{3.234}$$

For the fourth-order accurate interpolation and differencing, we have

$$\left[\overline{u}^x\right]_i = \frac{-u_{i-\frac{3}{2}} + 9u_{i-\frac{1}{2}} + 9u_{i+\frac{1}{2}} - u_{i+\frac{3}{2}}}{16} \tag{3.235}$$

$$\left[\delta_x u\right]_i = \frac{u_{i-\frac{3}{2}} - 27u_{i-\frac{1}{2}} + 27u_{i+\frac{1}{2}} - u_{i+\frac{3}{2}}}{24\Delta x}. \tag{3.236}$$

To achieve sixth-order accuracy, we can use

$$\left[\overline{u}^x\right]_i = \frac{3u_{i-\frac{5}{2}} - 25u_{i-\frac{3}{2}} + 150u_{i-\frac{1}{2}} + 150u_{i+\frac{1}{2}} - 25u_{i+\frac{3}{2}} + 3u_{i+\frac{5}{2}}}{256} \tag{3.237}$$

$$\left[\delta_x u\right]_i = \frac{-9u_{i-\frac{5}{2}} + 125u_{i-\frac{3}{2}} - 2250u_{i-\frac{1}{2}} + 2250u_{i+\frac{1}{2}} - 125u_{i+\frac{3}{2}} + 9u_{i+\frac{5}{2}}}{1920\Delta x}. \tag{3.238}$$

For staggered grids, the central-difference stencil uses midpoint data at $\pm\frac{1}{2}, \pm\frac{3}{2}$, $\pm\frac{5}{2}, \ldots$ as shown above (with even number of points). Use of the stencil with points from $\pm 1, \pm 2, \pm 3, \ldots$, for the advective term (with odd number of points; including $0 \times u_i$) with fourth-order accuracy

$$\left[\delta'_x u\right]_i = \frac{u_{i-2} - 8u_{i-1} + 8u_{i+1} - u_{i+2}}{12\Delta x} \tag{3.239}$$

and sixth-order accuracy

$$\left[\delta'_x u\right]_i = \frac{-u_{i-3} + 9u_{i-2} - 45u_{i-1} + 45u_{i+1} - 9u_{i+2} + u_{i+3}}{60\Delta x} \tag{3.240}$$

are not recommended.

With the above high-order formulas, we can increase the spatial order of accuracy to any desired level. It should however be realized that we encounter the following two issues. First, the compatibility of the discrete differences can be compromised. Although the global conservation may be maintained by treating the boundary suitably, local conservation on the other hand may not hold. Second, central-difference schemes shown above necessitate additional points outside of the computational domain. In what follows, we provide some discussions on these two issues.

3.9.2 Compatibility of High-Order Finite Differencing of Advective Term

Let us examine the compatibility (consistency and conservation) property of the advective term with high-order accurate discretization. Recall first the second-order central difference for the advective term with a uniform grid as presented in Eq. (3.185). For the divergence form, the discretization becomes

$$C_i = -\delta_{x_j}(\overline{u_j}^{x_i}\overline{u_i}^{x_j}) \qquad (3.241)$$

and for the gradient (advective) form, we have

$$C_i = -\overline{u_j}^{x_i}\delta_{x_j}u_i^{x_j}. \qquad (3.242)$$

In principle, the use of Eqs. (3.235) and (3.236) provides fourth-order accuracy and the utilization of Eqs. (3.237) and (3.238) leads to sixth-order accuracy. However, the compatibility and conservation properties discussed in Sect. 3.5.1 no longer hold locally.

A method that retains compatibility and conservation properties while achieving high-order accuracy is proposed by Morinishi [21]. To facilitate the discussion, let us define the following two-point interpolation and difference operators over an arbitrary stencil width as

$$[\overline{f}^{mx}]_j = \frac{f_{j-\frac{m}{2}} + f_{j+\frac{m}{2}}}{2}, \quad [\delta_{mx}f]_j = \frac{-f_{j-\frac{m}{2}} + f_{j+\frac{m}{2}}}{m\Delta x}. \qquad (3.243)$$

Using these notations, we can write the second-order operations simply as $\overline{f}^x = \overline{f}^{1x}$ and $\delta_x f = \delta_{1x} f$. However, for the fourth-order accurate formula, we have

$$\overline{f}^x = \frac{9}{8}\overline{f}^{1x} - \frac{1}{8}\overline{f}^{3x} \qquad (3.244)$$

$$\delta_x f = \frac{9}{8}\delta_{1x}f - \frac{1}{8}\delta_{3x}f \qquad (3.245)$$

and for the sixth-order accurate formula, we have

$$\overline{f}^x = \frac{150}{128}\overline{f}^{1x} - \frac{25}{128}\overline{f}^{3x} + \frac{3}{128}\overline{f}^{5x} \tag{3.246}$$

$$\delta_x f = \frac{150}{128}\delta_{1x} f - \frac{25}{128}\delta_{3x} f + \frac{3}{128}\delta_{5x} f. \tag{3.247}$$

For the divergence form, Eq. (3.241), the interpolation scheme for the transported quantity $\overline{u_i}^{x_j}$ is selected according to the chosen difference stencil δ_{x_j}. For the fourth-order scheme

$$\frac{\partial(u_j u_i)}{\partial x_j} = \frac{9}{8}\delta_{1x_j}\left(\overline{u_j}^{x_i}\overline{u_i}^{1x_j}\right) - \frac{1}{8}\delta_{3x_j}\left(\overline{u_j}^{x_i}\overline{u_i}^{3x_j}\right) \tag{3.248}$$

and for the sixth-order scheme

$$\frac{\partial(u_j u_i)}{\partial x_j} = \frac{150}{128}\delta_{1x_j}\left(\overline{u_j}^{x_i}\overline{u_i}^{1x_j}\right) - \frac{25}{128}\delta_{3x_j}\left(\overline{u_j}^{x_i}\overline{u_i}^{3x_j}\right)$$
$$+ \frac{3}{128}\delta_{5x_j}\left(\overline{u_j}^{x_i}\overline{u_i}^{5x_j}\right). \tag{3.249}$$

For the gradient form, Eq. (3.242), the difference of the transported variable $\delta_{x_j} u_i$ is chosen based on the selection of the interpolation method $\overline{(\,)}^j$. For the fourth-order scheme

$$u_j\frac{\partial u_i}{\partial x_j} = \frac{9}{8}\overline{\overline{u_j}^{x_i}\delta_{1x_j} u_i}^{1x_j} - \frac{1}{8}\overline{\overline{u_j}^{x_i}\delta_{3x_j} u_i}^{3x_j} \tag{3.250}$$

and for the sixth-order scheme

$$u_j\frac{\partial u_i}{\partial x_j} = \frac{150}{128}\overline{\overline{u_j}^{x_i}\delta_{1x_j} u_i}^{1x_j} - \frac{25}{128}\overline{\overline{u_j}^{x_i}\delta_{3x_j} u_i}^{3x_j} + \frac{3}{128}\overline{\overline{u_j}^{x_i}\delta_{5x_j} u_i}^{5x_j}. \tag{3.251}$$

In any of these forms, it should be observed that the interpolation of the advective velocity $\overline{u_j}^{x_i}$ remains the same. Use of these approaches can be extended to provide higher-order accurate central-difference schemes.

With the above schemes, the compatibility property can be satisfied. That is Eqs. (3.248) and (3.250) and Eqs. (3.9.2) and (3.251) provide the same results. Additionally, it is shown that quadratic conservation is achieved [21]. The desired level of accuracy is also reported to be achieved. Error assessments and actual error plots are shown in [21].

For high-order accurate discretizations on nonuniform grids, the methodology explained in Sect. 3.5.2 for the divergence and gradient forms (Eqs. (3.140) and (3.141)) can be utilized to discretize $\overline{u_i}^{\xi^k}$ and $\delta_{\xi^k} u_i$.

3.9.3 Boundary Conditions for High-Order Accurate Schemes

When we adopt high-order accurate schemes, the differencing stencil near the boundaries can extended outside of the computational domain (e.g., into the wall). Here, we discuss how we can modify finite-difference stencil near the computational boundary.

Let us consider an example illustrated in Fig. 3.22, where a six-point stencil along the y-axis intersects with the wall boundary. The velocity u along the boundary ($u_{i+\frac{1}{2},\frac{1}{2}}$) is not defined due to the arrangement of the staggered grid. The discrete momentum equation requires the evaluation of interpolation $[\overline{u}^y]_{i+\frac{1}{2},\frac{1}{2}}$ and the difference $[\delta_y u]_{i+\frac{1}{2},\frac{1}{2}}$. For computing these terms with high-order accurate interpolation and differencing, some of the points on the stencil can extend outside of the computational domain. However, placing virtual cells and extrapolating values for the variables using the inner solution face two difficulties. First, extrapolating inner values does not increase the degrees of freedom, even if the outer stencil size is widened. The number of independent values of the variable needs to be increased to improve the accuracy of the method. This can be seen from the polynomial analysis performed for the finite-difference derivation in Sect. 2.3.2. Second, there is arbitrariness in the width and shape of the virtual cell. We cannot easily determine whether there is an optimal arrangement of the virtual cells. Placing more than one virtual cell outside of the computational domain probably would not be beneficial.

Based on the above arguments, there are four choices for the stencil near the boundary as shown in Fig. 3.22a–d, depending on the usage of virtual cells and the extent of the stencil. For Fig. 3.22a, c, a single virtual cell is placed with its variable

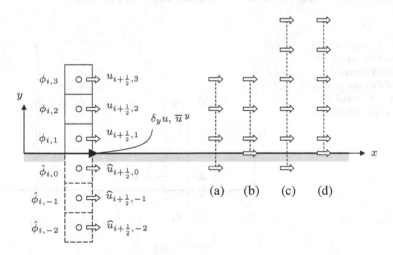

Fig. 3.22 Approaches to construct the stencil for high-order accurate finite-difference schemes (for cases of variables not located on the boundary)

value provided by extrapolating the boundary condition. In the cases of Fig. 3.22b, d, we adopt stencils that directly provide the boundary condition although the spacing becomes nonuniform. All stencil points except for the adjacent cell outside of the computational domain are omitted in Fig. 3.22a, b. For cases Fig. 3.22c, d, additional interior points are added to supplement the number of points outside of the domain (that are removed). We have used the stencil of Fig. 3.22b for the Dirichlet boundary condition in which $[u]_{\text{wall}}$ is prescribed at the solid wall and utilized the stencil in Fig. 3.22a for the Neumann boundary condition in which $[\partial u/\partial y]_{\text{wall}}$ is given at the slip-wall or far-field boundaries. The reason for such choice is to avoid the usage of finite-difference coefficients with large fluctuations for wide one-sided stencils. Based on experience, the choice of these stencils do not appear to deteriorate the accuracy significantly for fourth or higher-order accurate schemes, compared to the cases of Fig. 3.22c, d that keep the same stencil width.

Since the pressure $p_{i,j}$ is positioned at the cell center, there is no pressure defined along the boundary. When we are given the penetrating velocity as a boundary condition for an inlet, an outlet, or a wall, a Neumann pressure condition $[\partial p/\partial y]_{\text{wall}}$ can be specified. This leads to the use of the stencil in Fig. 3.22a. We can also use the same stencil to specify the Dirichlet boundary conditions for the pressure.

Next, let us consider interpolation and differencing needed for wall-normal velocity component v as illustrated in Fig. 3.23. Although this velocity component is located at the boundary, we need the interpolated $[\overline{v}^y]_{i,1}$ and the difference $[\delta_y v]_{i,1}$ at the cell centers for the momentum and continuity equations. Since the wall velocity $[v]_{\text{wall}}$ is directly utilized, there is no need to extrapolate virtual cell values. Thus, we have choices of stencils with the cell outside of the computational domain removed (Fig. 3.23a) or a one-sided stencil with the width unchanged (Fig. 3.23b). For the aforementioned reasons, we choose to use the stencil in Fig. 3.23a.

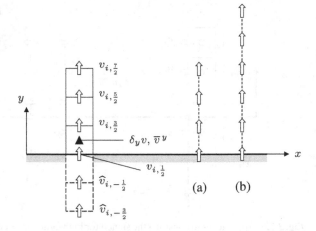

Fig. 3.23 Approaches to construct the stencil for high-order accurate finite-difference schemes (for cases of variables located on the boundary)

There is not an established universal rule for the treatment of high-order accurate finite differences near boundaries. The material described in this section should be taken as suggestions based on experience.

3.10 Remarks

We presented numerical simulation techniques for unsteady incompressible flow on Cartesian grid. The coupling of the continuity equation with the pressure term and the spatial discretization of the advective and diffusive terms were discussed. In order to avoid spatial oscillations of the pressure field while satisfying the continuity equation, the velocity components for the discrete continuity equation and pressure variables need to be arranged on a staggered grid and discretized in a compatible manner. For the advective term, which can trigger numerical instability from nonlinearity, discretization should be performed with care to satisfy compatibility without dependency on the form of the advective term and at the same time numerically conserve energy. Although the viscous term does not generate as many issues as the pressure and advective terms may, we should also ensure that the term is discretized appropriately to maintain compatibility for the purpose of, for example, evaluating kinetic energy budget. Compatibility properties essentially comes from the application of carefully constructed finite-difference schemes as discussed in Chap. 2, which correspond to the fundamental relations such as $f'' = (f')'$ and $(fg)' = fg' + f'g$.

3.11 Exercises

3.1 Two-dimensional incompressible flow can also be simulated using the vorticity ω and streamfunction ψ, instead of using the velocity and pressure variables. The vorticity field can be related to the velocity variables through the curl operation

$$\omega = \frac{\partial v}{\partial x} - \frac{\partial u}{\partial y}$$

and the velocity field (u, v) can be related to the streamfunction with

$$u = \frac{\partial \psi}{\partial y}, \quad v = -\frac{\partial \psi}{\partial x}. \tag{3.252}$$

1. For two-dimensional incompressible flow, derive the vorticity transport equation

$$\frac{\partial \omega}{\partial t} + u \frac{\partial \omega}{\partial x} + v \frac{\partial \omega}{\partial y} = \nu \left(\frac{\partial^2 \omega}{\partial x^2} + \frac{\partial^2 \omega}{\partial y^2} \right)$$

from the momentum equations with constant density and viscosity under no external forcing (also see Problem 1.2).

2. Using the continuity equation, show that

$$\frac{\partial^2 \psi}{\partial x^2} + \frac{\partial^2 \psi}{\partial y^2} = -\omega. \tag{3.253}$$

3. Show that the pressure field can be recovered from the streamfunction using the pressure Poisson equation.

4. Next, consider simulating flow inside a square lid-driven cavity as shown in Fig. 3.24a. Here, the flow is driven by the top wall translating at a constant U velocity with all other walls being stationary satisfying the no-slip boundary condition. For each side of the cavity, find the boundary conditions for vorticity and streamfunction. Note that the top left and right corners are singularity points; but they do not need to be explicitly handled in the numerical computation.

5. Let us discretize the spatial domain with a uniform grid ($\Delta x = \Delta y$) and place the variables according to Fig. 3.24b. Applying the second-order central difference to Eq. (3.252), show that second-order approximation of the continuity equation

$$\frac{-u_{i-1,j} + u_{i,j}}{\Delta x} + \frac{-v_{i,j-1} + v_{i,j}}{\Delta y} = 0$$

is satisfied.

6. Use the second-order finite-difference scheme to discretize Eq. (3.253).

7. Compare and contrast the computational techniques that use the vorticity -streamfunction formulation and velocity-pressure approach. Discuss the benefits in using the vorticity-streamfunction formulation.

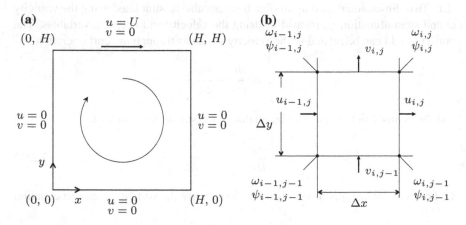

Fig. 3.24 Cavity flow setup using the vorticity-streamfunction formulation. **a** Domain setup. **b** Spatial discretization

3.2 It is possible to remove the pressure variable from the fractional step formulation using the *nullspace* method [3]. In this approach, we construct a discrete *curl* operator C, analogous to the continuous curl operator $(\nabla \times)$, that lies in the nullspace of the divergence operator D:

$$\nabla \cdot \nabla \times \equiv 0 \quad \rightarrow \quad DC \equiv 0$$

Using the curl operator, we can establish the velocity field as

$$u = Cs$$

where s is the discrete streamfunction for two-dimensional flow settings.[9]

1. Verify that the velocity field expressed with $u = Cs$ always satisfies the discrete continuity equation.
2. Show that Eq. (3.42)

$$\begin{bmatrix} R & \Delta t G \\ D & 0 \end{bmatrix} \begin{bmatrix} u^{n+1} \\ p^{n+1} \end{bmatrix} = \begin{bmatrix} Su^n + \frac{\Delta t}{2}(3A^n - A^{n-1}) \\ 0 \end{bmatrix}.$$

can be transformed to a simpler problem of

$$C^T R C s^{n+1} = C^T \left[Su^n + \frac{\Delta t}{2}(3A^n - A^{n-1}) \right]$$

with $u^{n+1} = Cs^{n+1}$. Assume the use of a staggered grid, in which case the discrete divergence and gradient operators can be formed such that $D = -G^T$.
3. Comment on the relationship between the nullspace approach and the formulation discussed in Problem 3.1.

3.3 The coefficient matrix of Eq. (3.40)

$$R = I - \frac{\Delta t}{2}\nu L$$

premultiplied to the intermediate velocity vector u^F or u^P results from the implicit treatment of the viscous term in incompressible flow solvers (see Sect. 3.3.5). Show that matrix R can be inverted analytically using series expansion

$$R^{-1} = I + \frac{\Delta t}{2}\nu L + \left(\frac{\Delta t}{2}\nu L \right)^2 + \dots$$

3.4 Specification of a laminar boundary layer profile as an inflow condition is often used in incompressible flow simulations. Let us consider the Blasius profile, which is a similarity solution for the steady laminar boundary layer on a flat plate.

---

[9]for three-dimensional flows, s is the discrete streamfunction vector (or vector potential); also see [3].

Fig. 3.25 Flat-plate laminar
boundary layer

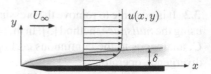

1. Assume two-dimensional, steady flow with no imposed pressure gradient as
   shown in Fig. 3.25. Show that the Navier–Stokes equations inside the boundary layer can be reduced to

$$u\frac{\partial u}{\partial x} + v\frac{\partial u}{\partial y} = \nu\frac{\partial^2 u}{\partial y^2}, \quad \frac{\partial p}{\partial y} = 0, \quad \text{and} \quad \frac{\partial u}{\partial x} + \frac{\partial v}{\partial y}.$$

2. Consider the streamfunction ψ (with $u = \frac{\partial\psi}{\partial y}$ and $v = -\frac{\partial\psi}{\partial x}$) and a coordinate
   transform from (x, y) to (x, η), where

$$\eta \equiv \frac{y}{\sqrt{\nu x/U_\infty}} = Re_x\frac{y}{x}.$$

Using the above and a streamfunction in the form $\psi = \sqrt{\nu x U_\infty}f(\eta)$, show that
the governing equations can be reduced to

$$ff'' + 2f''' = 0$$

with boundary conditions of $f(0) = f'(0) = 0$ and $\lim_{\eta\to\infty} f(\eta) = 1$. This
equation is referred to as the Blasius equation and its solution is known as the
Blasius profile.
3. Solve the Blasius equation numerically. Note that this is a boundary-value problem. You can instead solve an initial-value problem with $f''(0) = \alpha$ and iterate
   over the values of α such that the far-field slope of f reaches 1. The shooting
   method can be used with the Newton–Raphson solver or the bisection method
   [26] to find the correct α. Compare your solution to the one reported in Fig. 3.19.

References

1. Amsden, A.A., Harlow, F.H.: A simplified MAC technique for incompressible fluid flow calculations. J. Comput. Phys. **6**, 322–325 (1970)
2. Cain, A.B., Ferziger, J.H., Reynolds, W.C.: Discrete orthogonal function expansions for nonuniform grids using the fast Fourier transform. J. Comput. Phys. **56**, 272–286 (1984)
3. Chang, W., Giraldo, F., Perot, B.: Analysis of an exact fractional step method. J. Comput. Phys.
   180, 183–199 (2002)
4. Chorin, A.J.: Numerical solution of the Navier-Stokes equations. Math. Comput. **22**, 745–762
   (1968)

5. Doering, C.R., Gibbon, J.D.: Applied Analysis of the Navier-Stokes Equations. Cambridge Univ. Press (1995)
6. Dukowicz, J.K., Dvinsky, A.S.: Approximate factorization as a high order splitting for the implicit incompressible flow equations. J. Comput. Phys. **102**(2), 336–347 (1992)
7. Golub, G.H., Loan, C.F.V.: Matrix Computations, 3rd edn. Johns Hopkins Univ. Press (1996)
8. Gresho, P.M.: Incompressible fluid dynamics: some fundamental formulation issues. Annu. Rev. Fluid Mech. **23**, 413–453 (1991)
9. Gresho, P.M.: Some current CFD issures related to the incompressible Navier-Stokes equations. Comput. Methods Appl. Mech. Engrg. **87**, 201–252 (1991)
10. Harlow, F.H., Welch, J.E.: Numerical calculation of time-dependent viscous incompressibel flow of fluid with free surface. Phys. Fluids **8**(12), 2182–2189 (1965)
11. Hirt, C.W., Cook, J.L.: Calculating three-dimensional flows around structures and over rough terrain. J. Comput. Phys. **10**(2), 324–340 (1972)
12. Kajishima, T.: Conservation properties of finite difference method for convection. Trans. Japan Soc. Mech. Eng. B **60**(574), 2058–2063 (1994)
13. Kajishima, T.: Upstream-shifted interpolation method for numerical simulation of incompressible flows. Trans. Japan Soc. Mech. Eng. B **60**(578), 3319–3326 (1994)
14. Kajishima, T.: Finite-difference method for convective terms using non-uniform grid. Trans. Japan Soc. Mech. Eng. B **65**(633), 1607–1612 (1999)
15. Kawamura, T., Kuwahara, K.: Computation of high Reynolds number flow around circular cylinder with surface roughness. AIAA Paper 84-0340 (1984)
16. Kim, J., Moin, P.: Application of a fractional-step method to incompressible Navier-Stokes equations. J. Comput. Phys. **59**(2), 308–323 (1985)
17. Lambert, J.D.: Computational Methods in Ordinary Differential Equations. Wiley, London (1973)
18. Leonard, B.P.: A stable and accurate convective modelling procedure based on quadratic upstream interpolation. Comput. Methods Appl. Mech. Eng. **19**(1), 59–98 (1979)
19. LeVeque, R.J.: Finite Difference Methods for Ordinary and Partial Differential Equations: Steady-State and Time-Dependent Problems. SIAM (2007)
20. Miyauchi, T., Tanahashi, M., Suzuki, M.: Inflow and outflow boundary conditions for direct numerical simulations. JSME Int. J. B **39**(2), 305–314 (1996)
21. Morinishi, Y., Lund, T.S., Vasilyev, O.V., Moin, P.: Fully conservative higher order finite difference schemes for incompressible flow. J. Comput. Phys. **143**, 90–124 (1998)
22. Patankar, S.: Numerical Heat Transfer and Fluid Flow. Hemisphere, Washington (1980)
23. Patankar, S.V., Spalding, D.B.: A calculation procedure for heat, mass and momentum transfer in three-dimensional parabolic flows. Int. J. Heat Mass Trans. **15**(10), 1787–1806 (1972)
24. Perot, J.B.: An analysis of the fractional step method. J. Comput. Phys. **108**, 51–58 (1993)
25. Piacsek, S.A., Williams, G.P.: Conservation properties of convection difference schemes. J. Comput. Phys. **6**(3), 392–405 (1970)
26. Quarteroni, A., Sacco, R., Saleri, F.: Numerical Mathematics, 2nd edn. Springer (2006)
27. Rai, M.M., Moin, P.: Direct simulations of turbulent flow using finite-difference schemes. J. Comput. Phys. **91**(1), 15–53 (1991)
28. Saad, Y.: Iterative Methods for Sparse Linear Systems, 2nd edn. SIAM (2003)
29. Schlichting, H.: Boundary-Layer Theory. McGraw-Hill (1979)
30. Tamura, A., Kikuchi, K., Takahashi, T.: Residual cutting method for elliptic boundary value problems: application to Poisson's equation. J. Comput. Phys. **137**, 247–264 (1997)
31. Tomiyama, A., Hirano, M.: An improvement of the computational efficiency of the SOLA method. JSME Int. J. B **37**(4), 821–826 (1994)
32. van der Vorst, H.A.: Bi-CGSTAB: a fast and smoothly converging variant Bi-CG for the solution of nonsymmetric linear systems. SIAM. J. Sci. Stat. Comput. **13**(2), 631–644 (1992)
33. Williamson, J.H.: Low-storage Runge-Kutta schemes. J. Comput. Phys. **35**(1), 48–56 (1980)
34. Yoshida, T., Watanabe, T., Nakamura, I.: Numerical study of outflow boundary conditions for time-dependent incompressible flows. Trans. Japan Soc. Mech. Eng. B **61**(588), 123–131 (1995)

Chapter 4
Incompressible Flow Solvers for Generalized Coordinate System

4.1 Introduction

For simulating flows around bodies, it is possible to use the Cartesian grid formulation from Chap. 3. However, the Cartesian grid representation of a body with complex surface geometry can be a challenge. The body can be approximated in a staircase (pixelated, rasterized) manner as shown in Fig. 4.1 (top) but one can imagine that the numerical solution can be affected greatly from such staircase representation, especially for high-Reynolds number flows in which the boundary layer can be very thin. If we need to accurately represent the boundary geometry, the location of the boundary and its orientation with respect to the cells need to be captured with very fine mesh size if a Cartesian grid is used. This approach however inefficiently refines the grid for the sole purpose of representing the smooth body surface accurately on a grid that is not aligned with the boundary geometry. In recent years, adaptive mesh refinement [1] and immersed boundary methods [5, 8, 10] (see Chap. 5) have been developed to utilize Cartesian grids for flows over bodies with complex geometries. The use of Cartesian grid methods remain a challenge for accurately simulating turbulent flows that require highly refined grid near the wall boundary.

Finite volume and finite element methods can simulate flows over bodies with complex surface geometry by employing unstructured grids. For finite-difference methods, structured grids are in general generated to be aligned along the boundary surface as shown in Fig. 4.1 (bottom). This type of grid is called the *boundary-fitted grid* and is representative of the *generalized coordinate system*. The axes for the generalized coordinate system can be curved and need not intersect orthogonally. For this coordinate system, the computational method presented in Chap. 3 for a Cartesian staggered grid requires some modification. The finite-difference formulation and the discussion on the compatibility between the continuity equation and the pressure field discussed in the previous chapter are in principle the same here.

In this chapter, we present finite-difference techniques to solve the governing equations for flows on a generalized coordinate system and discuss how the velocity components should be chosen for discretizing the governing equations. We refer to

T. Kajishima and K. Taira, *Computational Fluid Dynamics*,
DOI 10.1007/978-3-319-45304-0_4

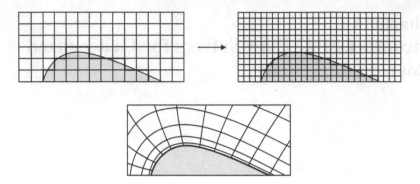

Fig. 4.1 Representation of the body geometry with (*top*) refinement of a Cartesian grid and (*bottom*) with body-fitted grid

the unknown variables (including the appropriate velocity components) selected for deriving the appropriate discretization schemes as the *basic variables*. For details on differentiation of vectors and tensors on generalized coordinate systems, we ask the readers to refer to Appendix A if needed. In general, the coordinate system can be time-varying if there is a body in motion in the flow field of if the geometry of the computational domain changes in time (e.g., free surface flow). Here, we limit our discussions to time-invariant coordinate systems.

4.2   Selection of Basic Variables

Let us represent a generalized coordinate system with components

$$\xi^1(=\xi), \quad \xi^2(=\eta), \quad \xi^3(=\zeta) \tag{4.1}$$

and their corresponding contravariant velocity components with

$$U^1(=U), \quad U^2(=V), \quad U^3(=W). \tag{4.2}$$

Since there is no difference between the contravariant and covariant components on a Cartesian grid, we denote the coordinate with

$$x_1 = x, \quad x_2 = y, \quad x_3 = z \tag{4.3}$$

and the corresponding velocity with

$$u_1 = u, \quad u_2 = v, \quad u_3 = w. \tag{4.4}$$

The *Jacobian* of the coordinate transformation is defined as[1]

$$J = \begin{vmatrix} x_\xi & x_\eta & x_\zeta \\ y_\xi & y_\eta & y_\zeta \\ z_\xi & z_\eta & z_\zeta \end{vmatrix}, \quad \frac{1}{J} = \begin{vmatrix} \xi_x & \xi_y & \xi_z \\ \eta_x & \eta_y & \eta_z \\ \zeta_x & \zeta_y & \zeta_z \end{vmatrix}. \tag{4.5}$$

with $J = \sqrt{\bar{g}/g}$. Note that $J = \sqrt{\bar{g}}$ for a Cartesian grid since $\sqrt{g} = 1$.

The transformation between the velocity components can be expressed in the following manner. The transformation from the Cartesian components to the generalized coordinate contravariant components is

$$U^j = \frac{\partial \xi^j}{\partial x_i} u_i = A_i^j u_i \quad \Rightarrow \quad \begin{bmatrix} U \\ V \\ W \end{bmatrix} = \begin{bmatrix} \xi_x u + \xi_y v + \xi_z w \\ \eta_x u + \eta_y v + \eta_z w \\ \zeta_x u + \zeta_y v + \zeta_z w \end{bmatrix} \tag{4.6}$$

and its inverse transformation is

$$u_i = \frac{\partial x_i}{\partial \xi^j} U^j = \overline{A}_j^i U^j \quad \Rightarrow \quad \begin{bmatrix} u \\ v \\ w \end{bmatrix} = \begin{bmatrix} x_\xi U + x_\eta V + x_\zeta W \\ y_\xi U + y_\eta V + y_\zeta W \\ z_\xi U + z_\eta V + z_\zeta W \end{bmatrix}, \tag{4.7}$$

where we have defined

$$A_i^j \equiv \frac{\partial \xi^j}{\partial x_i} \quad \text{and} \quad \overline{A}_j^i \equiv \frac{\partial x_i}{\partial \xi^j} \tag{4.8}$$

to denote the elements of the coordinate transformations.

To represent the conservation laws, Eq. (1.6), on generalized coordinates, one can choose various representations of the velocity components. It may seem that we can select the Cartesian components $u_j(u, v, w)$ and consider the spatial derivatives

$$\frac{\partial}{\partial x_j} = \frac{\partial \xi^i}{\partial x_j} \frac{\partial}{\partial \xi^i} \tag{4.9}$$

to be transformed to the coordinates $\xi^i(\xi, \eta, \zeta)$ to perform the computation. However, it may be difficult to derive a compatible finite-difference scheme because the velocity components $u_j(u, v, w)$ are not aligned with the curvilinear coordinates ξ^j. Therefore, we consider the formulation for the generalized coordinates $\xi^i(\xi, \eta, \zeta)$ for the following two cases only:

- Represent the velocity vector only with the contravariant components $U^i(U, U, W)$ (see Appendix A.2).

---

[1] Some may refer to $|\partial \xi^j / \partial x_i|$ as the Jacobian. The notation in Eq. (4.5) is used throughout this book.

- Represent the velocity vector with the Cartesian coordinate components $u_j(u, v, w)$ and use the contravariant components $U^i(U, V, W)$ for the advective velocity.

Both of these approaches are suitable for the generalized coordinate systems.

4.3   Strong Conservation Form of the Governing Equations

One can take a momentum conservation equation for the Cartesian coordinates and derive the same equation for the generalized coordinates, which is called the strong conservation form, as discussed below. In this form, the velocity field is represented using both the Cartesian and generalized coordinate components.

4.3.1   Strong Conservation Form

The conservation equation (without sink or source) for Cartesian coordinates $x_j(x, y, z)$ can be expressed as

$$\frac{\partial f}{\partial t} + \frac{\partial F^j}{\partial x_j} = 0, \tag{4.10}$$

where f is the conserved quantity and F^j is the flux. Let us represent the divergence of flux in the above equation with the coordinate transform

$$\frac{\partial F^j}{\partial x_j} = \frac{\partial \xi^k}{\partial x_j} \frac{\partial F^j}{\partial \xi^k}. \tag{4.11}$$

Multiplying both sides of Eq. (4.10) by the Jacobian of the coordinate transform, J ($=\sqrt{\hat{g}}$), we obtain

$$J\frac{\partial f}{\partial t} + \frac{\partial}{\partial \xi^k}\left(J\frac{\partial \xi^k}{\partial x_j}F^j\right) = F^j\frac{\partial}{\partial \xi^k}\left(J\frac{\partial \xi^k}{\partial x_j}\right). \tag{4.12}$$

Because the Jacobian is
$$J = |\overline{A}^i_j| = e^{lmn}\overline{A}^1_l\overline{A}^2_m\overline{A}^3_n, \tag{4.13}$$

we have
$$J\frac{\partial \xi^k}{\partial x_j} = e^{lmn}\overline{A}^1_l\overline{A}^2_m\overline{A}^3_n A^k_j = e^{lmn}\frac{\partial x_1}{\partial \xi^l}\frac{\partial x_2}{\partial \xi^m}\frac{\partial x_3}{\partial \xi^n}\frac{\partial \xi^k}{\partial x_j}, \tag{4.14}$$

which tells us that the right-hand side of Eq. (4.12) is always zero (also see Problem 4.1).

Thus, it follows that the conservation equation, Eq. (4.10), for the generalized coordinates is

$$\frac{\partial f}{\partial t} + \frac{1}{J}\frac{\partial}{\partial \xi^k}(J A_j^k F^j) = 0. \tag{4.15}$$

In this equation, the Christoffel symbol does not appear, unlike the governing equation using the contravariant components shown in Appendix A.2. This form of the equation on the ξ^j coordinates is called the *strong conservation form*.

For a time-invariant grid, we have $\partial J/\partial t = 0$ which makes Eq. (4.15)

$$\frac{\partial (Jf)}{\partial t} + \frac{\partial}{\partial \xi^k}(J A_j^k F^j) = 0, \tag{4.16}$$

where f is the conserved quantity per unit volume and J corresponds to the volume of the grid cell. The strong form of the equation represents the balance of the quantity f in terms of the integral quantity Jf.

4.3.2 Mass Conservation

Substituting ρ for f and ρu_j for F^j and using the relationship $A_j^k u_j = U^k$ in Eq. (4.15), we find the mass conservation equation

$$\frac{\partial \rho}{\partial t} + \frac{1}{J}\frac{\partial (J\rho U^k)}{\partial \xi^k} = 0. \tag{4.17}$$

For incompressible flows ($D\rho/Dt = 0$), the above equation becomes

$$\frac{1}{J}\frac{\partial (JU^k)}{\partial \xi^k} = 0. \tag{4.18}$$

For a scalar conservation equation, only the contravariant components of the velocity are included in the flux. Equations (A.60) and (A.62) correspond to Eqs. (4.17) and (4.18), respectively that are expressed solely by the contravariant components, yielding the strong conservation forms.

4.3.3 Momentum Conservation

Equation of Motion for Compressible Flow

Substituting ρu_i for f and $(\rho u_i u_j - T_{ij})$ for F^j in Eq. (4.15), we obtain the momentum conservation equation

$$\frac{\partial (\rho u_i)}{\partial t} + \frac{1}{J}\frac{\partial}{\partial \xi^k}\left[J A_j^k (\rho u_i u_j - T_{ij})\right] = 0, \tag{4.19}$$

which can be reexpressed as

$$\frac{\partial(\rho u_i)}{\partial t} + \frac{1}{J}\frac{\partial}{\partial \xi^k}\left(J\rho U^k u_i - J A_j^k T_{ij}\right) = 0. \tag{4.20}$$

Here, the stress tensor term is

$$A_j^k T_{ij} = -A_i^k\left(p + \frac{2}{3}\mu D_{mm}\right) + 2\mu A_j^k D_{ij}, \tag{4.21}$$

where T_{ij} is the Cartesian coordinate representation of the stress tensor

$$T_{ij} = -\left(p + \frac{2}{3}\mu D_{mm}\right)\delta_{ij} + 2\mu D_{ij} \tag{4.22}$$

and

$$D_{ij} = \frac{1}{2}\left(\frac{\partial u_i}{\partial x_j} + \frac{\partial u_j}{\partial x_i}\right) \tag{4.23}$$

is the rate of strain tensor.

Using the representation based on Eqs. (4.1)–(4.4), the equation of motion can be expressed as

$$\frac{\partial}{\partial t}\begin{bmatrix} \rho u \\ \rho v \\ \rho w \end{bmatrix} + \frac{1}{J}\frac{\partial}{\partial \xi}J\begin{bmatrix} \rho u U + \xi_x p - (\xi_x \tau_{xx} + \xi_y \tau_{xy} + \xi_z \tau_{xz}) \\ \rho v U + \xi_y p - (\xi_x \tau_{yx} + \xi_y \tau_{yy} + \xi_z \tau_{yz}) \\ \rho w U + \xi_z p - (\xi_x \tau_{zx} + \xi_y \tau_{zy} + \xi_z \tau_{zz}) \end{bmatrix}$$

$$+ \frac{1}{J}\frac{\partial}{\partial \eta}J\begin{bmatrix} \rho u V + \eta_x p - (\eta_x \tau_{xx} + \eta_y \tau_{xy} + \eta_z \tau_{xz}) \\ \rho v V + \eta_y p - (\eta_x \tau_{yx} + \eta_y \tau_{yy} + \eta_z \tau_{yz}) \\ \rho w V + \eta_z p - (\eta_x \tau_{zx} + \eta_y \tau_{zy} + \eta_z \tau_{zz}) \end{bmatrix} \tag{4.24}$$

$$+ \frac{1}{J}\frac{\partial}{\partial \zeta}J\begin{bmatrix} \rho u W + \zeta_x p - (\zeta_x \tau_{xx} + \zeta_y \tau_{xy} + \zeta_z \tau_{xz}) \\ \rho v W + \zeta_y p - (\zeta_x \tau_{yx} + \zeta_y \tau_{yy} + \zeta_z \tau_{yz}) \\ \rho w W + \zeta_z p - (\zeta_x \tau_{zx} + \zeta_y \tau_{zy} + \zeta_z \tau_{zz}) \end{bmatrix} = 0,$$

where $\tau_{ij} = 2\mu D_{ij}^a = 2\mu(D_{ij} - \frac{1}{3}\delta_{ij}D_{kk})$.

Equation of Motion for Incompressible Flow

By setting $D\rho/Dt = 0$ in Eq. (4.20), we find

$$\frac{\partial u_i}{\partial t} + \frac{1}{J}\frac{\partial(JU^k u_i)}{\partial \xi^k} - \frac{1}{\rho J}\frac{\partial}{\partial \xi^k}\left(-J A_i^k p + 2\mu J A_j^k D_{ij}\right) = 0, \tag{4.25}$$

which is the strong form of the momentum conservation equation for incompressible flow.

Next, we derive the gradient form. Using the continuity equation, Eq. (4.18), the advective term can be expressed as

$$\frac{\partial(JU^k u_i)}{\partial \xi^k} = JU^k \frac{\partial u_i}{\partial \xi^k}. \tag{4.26}$$

Also from the relationship of $\partial(JA_j^k)/\partial \xi^k = 0$, we have

$$\frac{\partial(JA_j^k h_{ij})}{\partial \xi^k} = JA_j^k \frac{\partial h_{ij}}{\partial \xi^k}, \tag{4.27}$$

which leads to the equation of motion be written as

$$\frac{\partial u_i}{\partial t} + U^k \frac{\partial u_i}{\partial \xi^k} + \frac{A_i^k}{\rho} \frac{\partial p}{\partial \xi^k} - 2 \frac{A_j^k}{\rho} \frac{\partial(\mu D_{ij})}{\partial \xi^k} = 0. \tag{4.28}$$

When the dynamic viscosity coefficient μ can be treated as constant, Eq. (4.28) can be rewritten as

$$\frac{\partial u_i}{\partial t} + U^k \frac{\partial u_i}{\partial \xi^k} + \frac{A_i^k}{\rho} \frac{\partial p}{\partial \xi^k} - \nu \frac{1}{J} \frac{\partial}{\partial \xi^k} \left(\gamma^{kl} \frac{\partial u_i}{\partial \xi^l} \right) = 0, \tag{4.29}$$

where

$$\gamma^{kl} = J \frac{\partial \xi^k}{\partial x_m} \frac{\partial \xi^l}{\partial x_m} \tag{4.30}$$

is a symmetric tensor. The momentum equation is expressed in the divergence (strong conservation) form with Eq. (4.25) and in the gradient (advective) form with Eqs. (4.28) and (4.29).

4.4   Basic Variables and Coordinate System

To construct the finite-difference formula for a generalized coordinate system, choices can be made for the velocity components of the basic variables and their locations with respect to the grid. Here, the velocity vector can be expressed in a number of ways

$$\mathbf{u} = u_i \mathbf{e}_i = U^j \mathbf{g}_j = U_k \mathbf{g}^k, \tag{4.31}$$

using the following velocity components

- Cartesian coordinate physical components: u_i (u, v, w)
- contravariant components: U^j (U, V, W)
- physical contravariant components: $U^{(j)}$

Fig. 4.2 Fluxes through cell
(i, j, k)

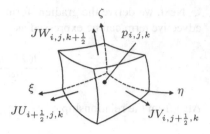

- covariant components: U_k
- physical covariant components: $U_{(k)}$

as appropriate basic variables.

Between the contravariant and covariant components, the former representation is used more widely. The mass conservation, Eq. (4.18), states that JU^j corresponds to the flux along the generalized coordinate axis through each cell face, as shown in Fig. 4.2. Hence, the use of contravariant components is appropriate for the generalized coordinate system. The governing equation however becomes complicated using the physical contravariant or covariant components. For this reason, it is rare to see the physical components being used with generalized coordinates, except for the limited cases of cylindrical and spherical coordinate systems. The common two choices for the basic variables are the Cartesian coordinate physical components u_i or the generalized coordinate contravariant components U^j.

Representations with Contravariant and Cartesian Components

First, let us write the equation of motion only with the contravariant velocity components. The momentum equation in divergence form becomes

$$\frac{\partial U^i}{\partial t} + (U^i U^j)|_j + \frac{1}{\rho} \left[\widehat{g}^{ij} p - \mu(\widehat{g}^{kj} U^i|_k + \widehat{g}^{ki} U^j|_k) \right]\Big|_j = 0 \tag{4.32}$$

and in gradient form is expressed as

$$\frac{\partial U^i}{\partial t} + (U^i U^j)|_j + \frac{\widehat{g}^{ij}}{\rho} \frac{\partial p}{\partial \xi^j} - \frac{1}{\rho} \left[\mu(\widehat{g}^{kj} U^i|_k + \widehat{g}^{ki} U^j|_k) \right]\Big| j = 0. \tag{4.33}$$

Discussions on these two equations are provided in Appendix A.2.3 (see Eqs. (A.78)–(A.80)).

Note that issues surface with the covariant derivative terms in Eqs. (4.32) and (4.33)

$$U^i|_j = \frac{\partial U^i}{\partial \xi^j} + \left\{ {}^i_{jk} \right\} U^k \tag{4.34}$$

$$H^{ij}|_j = \frac{1}{J} \frac{\partial}{\partial \xi^j} \left(J H^{ij} \right) + \left\{ {}^i_{jk} \right\} H^{jk}. \tag{4.35}$$

The first term in each of the above two equations is in conservative form representing the flux across the cell face. The second term with the Christoffel symbol[2] $\left\{ {i \atop j\,k} \right\}$ makes the computation clumsy. This term appears as a source term, which makes it difficult to conserve kinetic energy through the finite-difference equation constructed only with contravariant components. The need to save the Christoffel symbol term at each cell increases memory usage and is unattractive. Even with symmetry ($\left\{ {i \atop j\,k} \right\} = \left\{ {i \atop k\,j} \right\}$) taken into consideration, storing all six independent components for the two-dimensional case and eighteen components for the three-dimensional case at all grid points can be a computational burden.

The momentum equation that uses the physical Cartesian components u_i as the basic variables is expressed in the strong conservation form, Eq. (4.25), or in the gradient (advection) form, Eq. (4.28). Both forms do not contain the Christoffel symbol, which makes the momentum flux balance on the ξ^k grid system straightforward. For the advection velocity, the contravariant velocity U^k is used. The mass conservation equation, Eq. (4.18), also utilizes U^k. Thus, transformation from physical Cartesian components to contravariant components is necessary if we choose u_i as the basic variables for discretizing the governing equations. Nonetheless, it is advantageous to have a strong conservation form of the equation from the point of view of discretization.

Staggered and Collocated Grids

In addition to the selection of the basic variables, there are some choices available in the placement of their components with respect to the cells. We have already discussed in Chap. 3 that the regular grid, which positions all variables at the same point in space, can trigger the checkerboard instability. The regular grid is not suitable either for the generalized coordinate system due to the same reason.

Let us consider extending the staggered grid formulation for incompressible flow presented in Chap. 3 to the generalized coordinate formulation. Figure 4.3 shows the staggered arrangement of variables with Cartesian coordinate components. The staggered grid is beneficial when the velocity components are aligned in the directions created by connecting the adjacent pressure positions. In the case of Fig. 4.3, some of the velocity vectors are aligned almost tangential to the cell faces, reducing the benefit of the staggered grid formulation. One approach to remedy this issue is to place all velocity components in a staggered fashion. Such formulation, however results in increased computational load of double and triple for the two and three-dimensional momentum equations and is not efficient.

What is appropriate is to use the velocity components that are aligned with the underlying coordinate system. The mass flux (JU^j) corresponds to the flux across each cell face in Eq. (4.18). Therefore, it is rational to place the contravariant components at the cell face to couple the velocity and pressure fields through mass conservation. In other words, we can select U^j or JU^j as the basic variables, and use the variable placement shown in Fig. 4.4.

---

[2]See Sect. A.1.2 for details on the Christoffel symbol.

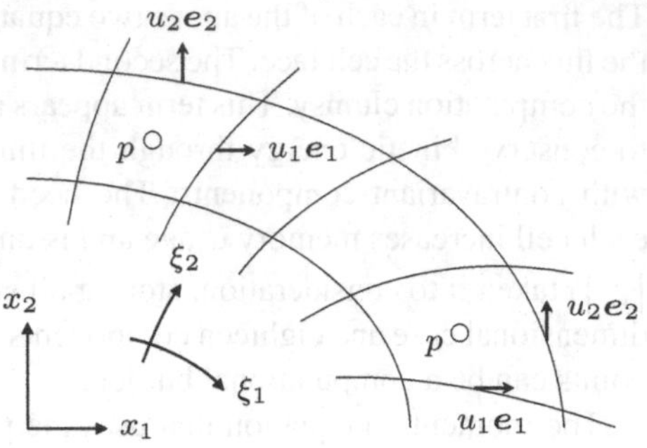

Fig. 4.3 Staggered grid arrangement with Cartesian velocity components for generalized coordinate system

Fig. 4.4 Staggered grid arrangement for generalized coordinate system

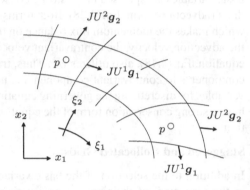

Fig. 4.5 Collocated grid arrangement for generalized coordinate system

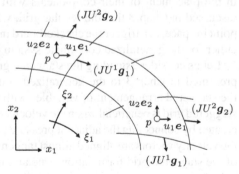

If the physical Cartesian grid components are chosen as the basic variables, it is convenient to place all variables at the same location in a similar manner to the regular grid, as shown in Fig. 4.5. The advection flux is given as (JU^j) in Eqs. (4.25) and (4.26) and is placed in a staggered fashion as an auxiliary variable. This grid arrangement is called the *collocated grid* and is distinguished from the *regular grid*. The collocated grid has been introduced by Rhie and Chow [9]. The mass conservation is enforced on (JU^j), which is positioned in a staggered fashion.

Combination of Basic Variables and Coordinate System

We can select a formulation that uses the contravariant components on a staggered grid or the Cartesian components on a collocated grid as the basic variables. If we rely only on the contravariant components, constructing conservative finite-difference equations becomes difficult due to the increased complexity. Furthermore, the required memory becomes large due to the presence of the Christoffel symbol. In contrast, if we utilize the collocated grid, we require the transformation from the Cartesian components to the contravariant components. However, this approach is advantageous from the view point of conservative discretization. For these reasons, we select the collocated grid for the generalized coordinate system hereafter.

4.5 Incompressible Flow Solvers Using Collocated Grids

In this section, we present the numerical procedure for the SMAC method on a collocated grid. Using the second-order accurate central-difference and interpolation, we can write for the ξ-direction,

$$[\delta_\xi f]_{i,j,k} = -f_{i-\frac{1}{2},j,k} + f_{i+\frac{1}{2},j,k}, \tag{4.36}$$

$$[\overline{f}^\xi]_{i,j,k} = \frac{f_{i-\frac{1}{2},j,k} + f_{i+\frac{1}{2},j,k}}{2}, \tag{4.37}$$

$$[\delta'_\xi f]_{i,j,k} = \frac{-f_{i-1,j,k} + f_{i+1,j,k}}{2}, \tag{4.38}$$

where we have chosen the grid spacing in the computational space to be $\Delta \xi = 1$ as it is often performed. Similarly, we can set $\Delta \eta = 1$ and $\Delta \zeta = 1$ for the other directions. In what follows, we simply denote the differences $\delta_{\xi^k} f$, $\delta'_{\xi^k} f$, and the interpolation $\overline{f}^{\xi^k}$ as $\delta_k f$, $\delta'_k f$, and $\overline{f}^k$, respectively. Following the indicial notation used in the previous chapter, we relax the treatment of the summation convention used for interpolation (e.g., $-^k$). We relax the summation convention here again, as we discussed on p. 114.

We take the contravariant components on a staggered grid to be constrained by the continuity equation

$$\frac{1}{J} \delta_j \left(J U^j \right)^{n+1} = 0. \tag{4.39}$$

Thus, we should solve for ϕ such that the sum of the predicted contravariant components and the gradient of ϕ satisfies the continuity equation at the next time step.

Explicit Scheme

Below we show the explicit time stepping procedure based on the second-order Adams–Bashforth method for both the advective and viscous terms. As in Chap. 3, we let the density ρ to be constant and set $P = p/\rho$. Explicit time stepping is performed with the following four steps:

1. Compute the advective, viscous, and pressure gradient terms at the cell center. Predict the velocity field u_i^P with

$$u_i^P = u_i^n - \Delta t \frac{\partial \xi^k}{\partial x_i} \overline{\delta_k P^n}^k + \Delta t \frac{3F_i^n - F_i^{n-1}}{2}, \qquad (4.40)$$

where F_i includes the advective and viscous terms, as well as the external force if present.

2. Perform transformation to obtain the contravariant velocity components and multiply them by the Jacobian J. Interpolate this product on the staggered grid to find $(JU^j)^P$:

$$\left(JU^j\right)^P = J \overline{\frac{\partial \xi^j}{\partial x_i} u_i^P}^j . \qquad (4.41)$$

The value of (JU^j) should be provided for the staggered grid instead of U^j. Note that we interpolate after J has been premultiplied at the position where u_i is defined.[3]

3. Solve the pressure Poisson equation (e.g., with SOR or BiCGSTAB methods)

$$\delta_j \left(\overline{\gamma^{jk}}^j \delta_k \phi\right) = \frac{1}{\Delta t} \delta_j \left(JU^j\right)^P \qquad (4.42)$$

to determine the pressure increment ϕ.

4. Correct (JU^j) with the gradient of ϕ. Add the gradient of ϕ to the Cartesian coordinate component u_i, and update P:

$$\left(JU^j\right)^{n+1} = \left(JU^j\right)^P - \Delta t \, \overline{\gamma^{jk}}^j \delta_k \phi \qquad (4.43)$$

$$u_i^{n+1} = u_i^P - \Delta t \frac{\partial \xi^k}{\partial x_i} \overline{\delta_k \phi}^k \qquad (4.44)$$

$$P^{n+1} = P^n + \phi, \qquad (4.45)$$

where the finite-difference formulas with δ are expanded about $\pm\frac{1}{2}, \frac{3}{2}, \ldots$.

Following the above algorithm, we obtain the variables u_i^{n+1}, $(JU^j)^{n+1}$, and P^{n+1} at the next time level t^{n+1}.

Implicit Treatment of the Viscous Term

We can treat the advection term explicitly with the second-order Adams–Bashforth method and integrate the viscous term implicitly with the Crank–Nicolson method.

---

[3]Inagaki and Abe [3] suggest increasing the order of accuracy of this particular interpolation operation and altering the differencing scheme in the pressure Poisson equation in a consistent manner. They report that these modifications enable the collocated grid-based solvers to achieve the lower level of error comparable to what is achieved by the staggered grid formulation.

Below, we present the numerical procedure taken to implement these time stepping schemes for solving for the time evolution of incompressible flow.

1. Compute the advection, viscous, and pressure gradient terms at the cell center and predict the velocity field u_i^P

$$u_i^P - \frac{\Delta t}{2} \frac{\nu}{J} \delta_k \left(\overline{\gamma^{kl}}^k \delta_l \overline{u}_i^P \right) = u_i^n + \frac{\Delta t}{2} \frac{\nu}{J} \delta_k \left(\overline{\gamma^{kl}}^k \delta_l \overline{u}_i^n \right)$$
$$- \Delta t \frac{\partial \xi^k}{\partial x_i} \overline{\delta_k P^n}^k + \Delta t \frac{3A_i^n - A_i^{n-1}}{2}, \quad (4.46)$$

where A^n is the advection term, which can include the volume force.[4]

2. Perform interpolation using Eq. (4.41) to evaluate $(JU^j)^P$ at the staggered location in a manner similar to the aforementioned explicit method.

3. Solve the Poisson equation, Eq. (4.42), for ϕ in the same manner as the explicit method.

4. Correct (JU^j) with the gradient of ϕ through Eq. (4.43) and u_i with Eq. (4.44) the same way as the above explicit method. Pressure P is updated in the following fashion

$$P^{n+1} = P^n + \phi - \frac{\Delta t}{2} \frac{\nu}{J} \delta_k \left(\overline{\gamma^{kl}}^k \delta_l \phi \right). \quad (4.47)$$

Let us point out some differences between this scheme and that for the Cartesian staggered grid presented in Chap. 3. The difference between the above steps and Eqs. (3.187)–(3.190) is the addition of Eq. (4.41) to interpolate the predicted velocity onto the staggered grid after transforming the variables from the contravariant representation. Another difference is the need for adding the gradient of ϕ to both the contravariant and Cartesian components, in a similar manner to Eqs. (4.43) and (4.44). For these reasons, we must take the following points (i)–(iii) into consideration:

(i) Continuity Equation

If the discretized Poisson equation, Eq. (4.42), is solved accurately, then $(JU^j)^{n+1}$ obtained from Eq. (4.43) satisfies the difference equation, Eq. (4.39), for the continuity equation, Eq. (4.18). However there is no constraint directly imposed on the velocity u^{n+1} in Eq. (4.44), since the discrete continuity equation is not provided for u^{n+1}. That is, the discrete version of $\partial u_i / \partial x_i = 0$, which is

$$\frac{\partial \xi^j}{\partial x_i} \delta_j' u_i^{n+1} = 0 \quad (4.48)$$

does not necessarily hold numerically. The influence of the error introduced from the discretization of the continuity equation is discussed by Kajishima et al. [4] and Morinishi [7].

---

[4]It is possible to treat only the $k = l$ term in the viscous term implicitly.

(ii) Arbitrariness in the Correction Step

It may appear that we can use an equation other than Eq. (4.44) to determine the Cartesian velocity components u_i^{n+1}, since they are not constrained by the continuity equation. We can also consider inverse transforming the updated contravariant velocity components from Eq. (4.44)

$$u_i^{n+1} = \frac{1}{J} \frac{\partial x_i}{\partial \xi^j} \overline{(JU^j)^{n+1}}^j. \tag{4.49}$$

However, this approach in general should not be adopted. The reason can be understood by substituting Eq. (4.43) in $(JU^j)^{n+1}$ above and subsequently Eqs. (4.41) and (4.40) in $(JU^j)^P$. It can be noticed that instead of F_i acting on u_i, we have $\overline{\overline{F_i}^j}^j$ (no summation implied) acting on u_i. Here, the term $\overline{\overline{F_i}^j}^j$ has been smoothed through the double interpolation. If we inverse transform the updated contravariant components that were interpolated onto the cell centers, the terms in the momentum equation become regularized (diffused). Spatial resolution may be lost with Eq. (4.49) compared to Eq. (4.44). Although one may potentially benefit in terms of numerical stability from the induced numerical diffisivity, we however emphasize that this type of stabilization is not a fundamental solution to address numerical instability.

(iii) Placement of Coordinate Transformation Coefficients

For the schemes described by Eqs. (4.40)–(4.45), the Jacobian J and the elements of the coordinate transformation matrix ($\partial \xi^j / \partial x_i$ and $\partial x_i / \partial \xi^j$) are all positioned at the cell center (same as the locations for P and u_i). These values are interpolated to other locations as needed. For instance, instead of performing the interpolation after multiplying J as in Eq. (4.41), J can be multiplied to the interpolated contravariant component at the staggered position

$$\left(JU^j\right)^P = J \overline{\frac{\partial \xi^j}{\partial x_i} u_i^P}^j. \tag{4.50}$$

This requires J to be precomputed on the staggered grid or necessitates J to be interpolated from the cell centers, both of which would not be advisable. Furthermore, there is also arbitrariness in how we can add the gradient of pressure or the gradient of the pressure difference ϕ to the velocity field. Further discussion on this matter is offered in Sect. 4.6.

4.6 Spatial Discretization of Pressure Gradient Term

4.6.1 Pressure Gradient Term

The time stepping scheme for collocated grids discussed in Sect. 4.5 adds the gradient of pressure or pressure correction ϕ to the physical Cartesian components in

Fig. 4.6 Finite-difference stencil of the pressure gradient with the physical components

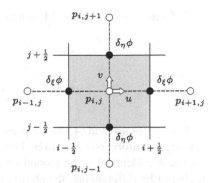

Eqs. (4.40) and (4.44) as well as to the contravariant components in Eq. (4.43). Below we consider the implications of the finite-difference approximation of the pressure gradient term for the two-dimensional momentum equation.

Pressure Gradient with Cartesian (Physical) Components

Let us examine the finite-difference stencil for the pressure gradient term using Cartesian components in Eq. (4.40) and for the pressure correction (ϕ) gradient term in Eq. (4.44). The variables u, v, and p (as well as ϕ) are defined at the cell centers as shown in Fig. 4.6, which prevents us from utilizing a finite-difference stencil with $\pm\frac{1}{2}$ indices.

Here, three approaches presented below can be suggested. The correction step, Eq. (4.44), can be

- Differenced over the $\pm\frac{1}{2}$ stencil and then interpolated

$$
\left.
\begin{aligned}
u^{n+1} &= u^P - \Delta t \left(\frac{\partial \xi}{\partial x} \overline{\delta_\xi \phi}^\xi + \frac{\partial \eta}{\partial x} \overline{\delta_\eta \phi}^\eta \right) \\
v^{n+1} &= v^P - \Delta t \left(\frac{\partial \xi}{\partial y} \overline{\delta_\xi \phi}^\xi + \frac{\partial \eta}{\partial y} \overline{\delta_\eta \phi}^\eta \right)
\end{aligned}
\right\}
\tag{4.51}
$$

- Interpolated at $\pm\frac{1}{2}$ positions and then differenced

$$
\left.
\begin{aligned}
u^{n+1} &= u^P - \Delta t \left(\frac{\partial \xi}{\partial x} \delta_\xi \overline{\phi}^\xi + \frac{\partial \eta}{\partial x} \delta_\eta \overline{\phi}^\eta \right) \\
v^{n+1} &= v^P - \Delta t \left(\frac{\partial \xi}{\partial y} \delta_\xi \overline{\phi}^\xi + \frac{\partial \eta}{\partial y} \delta_\eta \overline{\phi}^\eta \right)
\end{aligned}
\right\}
\tag{4.52}
$$

- Differenced (δ') over the ± 1 stencil

$$
\left.\begin{aligned}
u^{n+1} &= u^P - \Delta t \left(\frac{\partial \xi}{\partial x} \delta'_\xi \phi + \frac{\partial \eta}{\partial x} \delta'_\eta \phi \right) \\
v^{n+1} &= v^P - \Delta t \left(\frac{\partial \xi}{\partial y} \delta'_\xi \phi + \frac{\partial \eta}{\partial y} \delta'_\eta \phi \right)
\end{aligned}\right\}
\tag{4.53}
$$

For Eqs. (4.40) and (4.44), we presented the form of Eq. (4.51). Because the grid spacing is uniform (set here to be 1) in the computational space, Eqs. (4.51)–(4.53) all become identical if the second-order central-difference scheme is utilized. For higher-order differencing, the above three approaches are not necessarily identical. While it is not clear which one of the three is optimal, the chosen method should at least have the same finite-difference scheme as in the predictive step, Eq. (4.40), and the correction (projection) step, Eq. (4.44), so that they are compatible with Eq. (4.45).

Pressure Gradient with Contravariant Components

Let us examine the finite-difference stencil for the pressure gradient term using contravariant components in Eq. (4.43). Figure 4.7 is shown for reference. For each point where $(JU)_{i+\frac{1}{2},j}$ or $(JV)_{i,j+\frac{1}{2}}$ is located, the pressure (○) is positioned $\pm 1/2$ cell width away in the ξ or η-directions, respectively. For points where (JU) or (JV) is positioned, there is no discrete pressure placed along the η and ξ-directions, respectively. Therefore, it becomes necessary to provide interpolated values of pressure at locations denoted by (●) in Fig. 4.7. Another approach is to determine the derivatives at (●), which then can be interpolated to the positions of (JU) and (JV) at ($\Rightarrow$) and ($\Uparrow$), respectively. Here, we select the former approach. The pressure correction term can be interpolated from the neighboring four points

$$
\overline{\phi}^{\xi\eta}_{i+\frac{1}{2},j+\frac{1}{2}} = \frac{\phi_{i,j} + \phi_{i+1,j} + \phi_{i,j+1} + \phi_{i+1,j+1}}{4},
\tag{4.54}
$$

Fig. 4.7 Finite-difference stencil of the pressure gradient for the contravariant components with the control volume highlighted in *gray*

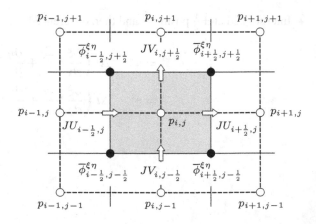

where we should recall that $\phi_{i,j}$ is positioned at the same location as $p_{i,j}$. Because the coefficient γ^{jk} in the pressure gradient term is not available at the staggered grid points, we must interpolate it from the adjacent cell centers. We then arrive at the spatial difference approximation

$$
\left.
\begin{aligned}
(JU)^{n+1} &= (JU)^P - \Delta t \left(\overline{\gamma^{\xi\xi}}^{\xi} \delta_\xi \phi + \overline{\gamma^{\xi\eta}}^{\xi} \delta_\eta \overline{\phi}^{\xi\eta} \right) \\
(JV)^{n+1} &= (JV)^P - \Delta t \left(\overline{\gamma^{\eta\xi}}^{\eta} \delta_\xi \overline{\phi}^{\xi\eta} + \overline{\gamma^{\eta\eta}}^{\eta} \delta_\eta \phi \right)
\end{aligned}
\right\}
\tag{4.55}
$$

Since $\gamma^{jk} \neq 0$ $(j \neq k)$ for non-orthogonal grids, the interpolated value $\overline{\phi}^{\xi\eta}$ is necessitated which leaves some arbitrariness. This makes it difficult to avoid the loss of numerical accuracy with any interpolation schemes.

4.6.2 Pressure Poisson Equation

As emphasized in Sect. 3.4.1, the pressure Poisson equation should be discretized such that its solution yields the pressure gradient term that enforces incompressibility on the velocity field at the next time step. That is, Eq. (4.42), should be discretized by substituting the finite-difference expression of Eq. (4.43) into Eq. (4.39). We can examine the formulation in further details with Fig. 4.7. For cell (i, j), the continuity equation, Eq. (4.39), becomes

$$
\frac{1}{J_{i,j}} \left[-(JU)^{n+1}_{i-\frac{1}{2},j} + (JU)^{n+1}_{i+\frac{1}{2},j} - (JV)^{n+1}_{i,j-\frac{1}{2}} + (JV)^{n+1}_{i,j+\frac{1}{2}} \right] = 0.
\tag{4.56}
$$

The terms in the above equation can be discretized using Eq. (4.55) to yield

$$
\left.
\begin{aligned}
(JU)^{n+1}_{i-\frac{1}{2},j} &= (JU)^P_{i-\frac{1}{2},j} - \Delta t \left[\overline{\gamma^{\xi\xi}}^{\xi}_{i-\frac{1}{2},j}(-\phi_{i-1,j} + \phi_{i,j}) \right. \\
&\left. \qquad + \overline{\gamma^{\xi\eta}}^{\xi}_{i-\frac{1}{2},j}(-\overline{\phi}^{\xi\eta}_{i-\frac{1}{2},j-\frac{1}{2}} + \overline{\phi}^{\xi\eta}_{i-\frac{1}{2},j+\frac{1}{2}}) \right] \\
(JU)^{n+1}_{i+\frac{1}{2},j} &= (JU)^P_{i+\frac{1}{2},j} - \Delta t \left[\overline{\gamma^{\xi\xi}}^{\xi}_{i+\frac{1}{2},j}(-\phi_{i,j} + \phi_{i+1,j}) \right. \\
&\left. \qquad + \overline{\gamma^{\xi\eta}}^{\xi}_{i+\frac{1}{2},j}(-\overline{\phi}^{\xi\eta}_{i+\frac{1}{2},j-\frac{1}{2}} + \overline{\phi}^{\xi\eta}_{i+\frac{1}{2},j+\frac{1}{2}}) \right] \\
(JV)^{n+1}_{i,j-\frac{1}{2}} &= (JV)^P_{i,j-\frac{1}{2}} - \Delta t \left[\overline{\gamma^{\eta\xi}}^{\eta}_{i,j-\frac{1}{2}}(-\overline{\phi}^{\xi\eta}_{i-\frac{1}{2},j-\frac{1}{2}} + \overline{\phi}^{\xi\eta}_{i+\frac{1}{2},j-\frac{1}{2}}) \right. \\
&\left. \qquad + \overline{\gamma^{\eta\eta}}^{\eta}_{i,j-\frac{1}{2}}(-\phi_{i,j-1} + \phi_{i,j}) \right] \\
(JV)^{n+1}_{i,j+\frac{1}{2}} &= (JV)^P_{i,j+\frac{1}{2}} - \Delta t \left[\overline{\gamma^{\eta\xi}}^{\eta}_{i,j+\frac{1}{2}}(-\overline{\phi}^{\xi\eta}_{i-\frac{1}{2},j+\frac{1}{2}} + \overline{\phi}^{\xi\eta}_{i+\frac{1}{2},j+\frac{1}{2}}) \right. \\
&\left. \qquad + \overline{\gamma^{\eta\eta}}^{\eta}_{i,j+\frac{1}{2}}(-\phi_{i,j} + \phi_{i,j+1}) \right]
\end{aligned}
\right\}
\tag{4.57}
$$

Substituting these expressions (corrective step for the contravariant components) into Eq. (4.56), we arrive at

$$
\overline{\gamma^{\eta\eta}}^{\eta}_{i,j-\frac{1}{2}}\phi_{i,j-1} + \overline{\gamma^{\xi\xi}}^{\xi}_{i-\frac{1}{2},j}\phi_{i-1,j}
$$

$$
- \left(\overline{\gamma^{\xi\xi}}^{\xi}_{i-\frac{1}{2},j} + \overline{\gamma^{\xi\xi}}^{\xi}_{i+\frac{1}{2},j} + \overline{\gamma^{\eta\eta}}^{\eta}_{i,j-\frac{1}{2}} + \overline{\gamma^{\eta\eta}}^{\eta}_{i,j+\frac{1}{2}} \right) \phi_{i,j}
$$

$$
+ \overline{\gamma^{\xi\xi}}^{\xi}_{i+\frac{1}{2},j}\phi_{i+1,j} + \overline{\gamma^{\eta\eta}}^{\eta}_{i,j+\frac{1}{2}}\phi_{i,j+1}
$$

$$
+ \left(\overline{\gamma^{\xi\eta}}^{\xi}_{i-\frac{1}{2},j} + \overline{\gamma^{\eta\xi}}^{\eta}_{i,j-\frac{1}{2}} \right) \overline{\phi}^{\xi\eta}_{i-\frac{1}{2},j-\frac{1}{2}} - \left(\overline{\gamma^{\xi\eta}}^{\xi}_{i+\frac{1}{2},j} + \overline{\gamma^{\eta\xi}}^{\eta}_{i,j-\frac{1}{2}} \right) \overline{\phi}^{\xi\eta}_{i+\frac{1}{2},j-\frac{1}{2}}
$$

$$
- \left(\overline{\gamma^{\xi\eta}}^{\xi}_{i-\frac{1}{2},j} + \overline{\gamma^{\eta\xi}}^{\eta}_{i,j+\frac{1}{2}} \right) \overline{\phi}^{\xi\eta}_{i-\frac{1}{2},j+\frac{1}{2}} + \left(\overline{\gamma^{\xi\eta}}^{\xi}_{i+\frac{1}{2},j} + \overline{\gamma^{\eta\xi}}^{\eta}_{i,j+\frac{1}{2}} \right) \overline{\phi}^{\xi\eta}_{i+\frac{1}{2},j+\frac{1}{2}}
$$

$$
= \frac{1}{\Delta t} \left[-(JU)^{P}_{i-\frac{1}{2},j} + (JU)^{P}_{i+\frac{1}{2},j} - (JV)^{P}_{i,j-\frac{1}{2}} + (JV)^{P}_{i,j+\frac{1}{2}} \right]. \tag{4.58}
$$

Comparing this result with that for the Cartesian coordinate system, lines 1 to 3 in the above equation share the same forms and lines 4 and 5 appear due to the non-orthogonality of the grid.

If we choose to use the four-point interpolation from Eq. (4.54), we then obtain

$$
\frac{\overline{\gamma^{\xi\eta}}^{\xi}_{i-\frac{1}{2},j} + \overline{\gamma^{\eta\xi}}^{\eta}_{i,j-\frac{1}{2}}}{4}\phi_{i-1,j-1} + \left(\overline{\gamma^{\eta\eta}}^{\eta}_{i,j-\frac{1}{2}} + \frac{\overline{\gamma^{\eta\xi}}^{\xi}_{i-\frac{1}{2},j} - \overline{\gamma^{\xi\eta}}^{\xi}_{i+\frac{1}{2},j}}{4} \right) \phi_{i,j-1}
$$

$$
- \frac{\overline{\gamma^{\xi\eta}}^{\xi}_{i+\frac{1}{2},j} + \overline{\gamma^{\eta\xi}}^{\eta}_{i,j-\frac{1}{2}}}{4}\phi_{i+1,j-1} + \left(\overline{\gamma^{\xi\xi}}^{\eta}_{i-\frac{1}{2},j} + \frac{\overline{\gamma^{\eta\xi}}^{\eta}_{i,j-\frac{1}{2}} - \overline{\gamma^{\eta\xi}}^{\eta}_{i,j+\frac{1}{2}}}{4} \right) \phi_{i-1,j}
$$

$$
- \left(\overline{\gamma^{\xi\xi}}^{\xi}_{i-\frac{1}{2},j} + \overline{\gamma^{\xi\xi}}^{\xi}_{i+\frac{1}{2},j} + \overline{\gamma^{\eta\eta}}^{\eta}_{i,j-\frac{1}{2}} + \overline{\gamma^{\eta\eta}}^{\eta}_{i,j+\frac{1}{2}} \right) \phi_{i,j}
$$

$$
+ \left(\overline{\gamma^{\xi\xi}}^{\xi}_{i+\frac{1}{2},j} - \frac{\overline{\gamma^{\eta\xi}}^{\eta}_{i,j-\frac{1}{2}} - \overline{\gamma^{\eta\xi}}^{\eta}_{i,j+\frac{1}{2}}}{4} \right) \phi_{i+1,j} - \frac{\overline{\gamma^{\xi\eta}}^{\xi}_{i-\frac{1}{2},j} + \overline{\gamma^{\eta\xi}}^{\eta}_{i,j+\frac{1}{2}}}{4}\phi_{i-1,j+1}
$$

$$
+ \left(\overline{\gamma^{\eta\eta}}^{\eta}_{i,j+\frac{1}{2}} - \frac{\overline{\gamma^{\xi\eta}}^{\xi}_{i-\frac{1}{2},j} - \overline{\gamma^{\xi\eta}}^{\xi}_{i+\frac{1}{2},j}}{4} \right) \phi_{i,j+1} + \frac{\overline{\gamma^{\xi\eta}}^{\xi}_{i+\frac{1}{2},j} + \overline{\gamma^{\eta\xi}}^{\eta}_{i,j+\frac{1}{2}}}{4}\phi_{i+1,j+1}
$$

$$
= \frac{1}{\Delta t} \left[-(JU)^{P}_{i-\frac{1}{2},j} + (JU)^{P}_{i+\frac{1}{2},j} - (JV)^{P}_{i,j-\frac{1}{2}} + (JV)^{P}_{i,j+\frac{1}{2}} \right]. \tag{4.59}
$$

The orthogonal two-dimensional stencil is comprised of five points (i, j), $(i \pm 1, j)$, $(i, j \pm 1)$. In the case of non-orthogonal grids, the stencil consists of nine points with the addition of points $(i \pm 1, j \pm 1)$. For the three-dimensional case, the non-orthogonal grid requires 19 points for the stencil, while the orthogonal grid requires 7 points.

4.6.3   Iterative Solver for the Pressure Poisson Equation

Let us present a basic iterative solution technique, following the discussion from Sect. 3.4.2. First, let us write Eq. (4.59) in the following simplified manner

$$
\begin{aligned}
& B_{i,j}^{-x,-y}\phi_{i-1,j-1} + B_{i,j}^{0,-y}\phi_{i,j-1} + B_{i,j}^{+x,-y}\phi_{i+1,j-1} \\
& + B_{i,j}^{-x,0}\phi_{i-1,j} - B_{i,j}^{0,0}\phi_{i,j} + B_{i,j}^{+x,0}\phi_{i+1,j} \\
& + B_{i,j}^{-x,+y}\phi_{i-1,j+1} + B_{i,j}^{0,+y}\phi_{i,j+1} + B_{i,j}^{+x,+y}\phi_{i+1,j+1} \\
& - \psi_{i,j} = 0.
\end{aligned}
\tag{4.60}
$$

The SOR method for the above equation can then be implemented as

$$
\begin{aligned}
\phi_{i,j}^{\langle m+1\rangle} = \; & \phi_{i,j}^{\langle m\rangle} \\
& + \frac{\beta}{B_{i,j}^{0,0}}\Big[B_{i,j}^{-x,-y}\phi_{i-1,j-1}^{\langle m+1\rangle} + B_{i,j}^{0,-y}\phi_{i,j-1}^{\langle m+1\rangle} + B_{i,j}^{+x,-y}\phi_{i+1,j-1}^{\langle m+1\rangle} \\
& + B_{i,j}^{-x,0}\phi_{i-1,j}^{\langle m+1\rangle} - B_{i,j}^{0,0}\phi_{i,j}^{\langle m\rangle} + B_{i,j}^{+x,0}\phi_{i+1,j}^{\langle m\rangle} \\
& + B_{i,j}^{-x,+y}\phi_{i-1,j+1}^{\langle m\rangle} + B_{i,j}^{0,+y}\phi_{i,j+1}^{\langle m\rangle} + B_{i,j}^{+x,+y}\phi_{i+1,j+1}^{\langle m\rangle} - \psi_{i,j}\Big].
\end{aligned}
\tag{4.61}
$$

As explained in Sect. 3.4.2, we replace $\phi^{\langle m\rangle}$ by $\phi^{\langle m+1\rangle}$ if they are evaluated by the time the DO loop reaches the ith and jth component. In the computer program, we do not need to treat $\langle m\rangle$ and $\langle m+1\rangle$ differently, as the new value simply can overwrite the old value during the iteration.

While we do not discuss in detail the special treatment used to accelerate computations for vectorized or parallel computers, we note in passing that to accelerate the calculation, the non-orthogonal components can be included in the right-hand side as

$$
\begin{aligned}
\psi_{i,j}^{\prime\langle m\rangle} = \; & \psi_{i,j} - \Big[B_{i,j}^{-x,-y}\phi_{i-1,j-1}^{\langle m\rangle} + B_{i,j}^{+x,-y}\phi_{i+1,j-1}^{\langle m\rangle} \\
& + B_{i,j}^{-x,+y}\phi_{i-1,j+1}^{\langle m\rangle} + B_{i,j}^{+x,+y}\phi_{i+1,j+1}^{\langle m\rangle}\Big]
\end{aligned}
\tag{4.62}
$$

so that the the equation can be formulated in the following manner

$$
\begin{aligned}
\phi_{i,j}^{\langle m+1\rangle} = \; & \phi_{i,j}^{\langle m\rangle} + \frac{\beta}{B_{i,j}^{0,0}}\Big[B_{i,j}^{0,-y}\phi_{i,j-1}^{\langle m+1\rangle} + B_{i,j}^{-x,0}\phi_{i-1,j}^{\langle m+1\rangle} - B_{i,j}^{0,0}\phi_{i,j}^{\langle m\rangle} \\
& + R_{i,j}^{+x,0}\phi_{i+1,j}^{\langle m\rangle} + B_{i,j}^{0,+y}\phi_{i,j+1}^{\langle m\rangle} - \psi_{i,j}^{\prime\langle m\rangle}\Big].
\end{aligned}
\tag{4.63}
$$

For flows solved on strongly non-orthogonal grids, high level of acceleration of the iterative scheme cannot be expected.

4.7 Spatial Discretization of Advection Term

4.7.1 Compatibility and Conservation

Through the use of the continuity equation, Eq. (4.18), the two forms of the momentum equation, Eqs. (4.25) and (4.28), can be shown to be equivalent differential equations. While we only deal directly with the continuity and momentum equations for incompressible flow, we must also implicitly satisfy the energy equation that is dependent on the continuity and momentum equations. As it was discussed in Sect. 3.5, it is crucial that the discretization process takes compatibility and conservation properties into account.[5] Here, we examine the compatibility and conservation properties of the numerical schemes used to simulate incompressible flows on generalized coordinate collocated grids.

The two discrete forms of the advection term, namely the divergence form, Eq. (4.25), and the gradient form, Eq. (4.28), should match. Finite differencing the advection term at the cell center (generalized coordinate) leads to the divergence form to be expressed in two ways. The first one is

$$\frac{1}{J}\frac{\partial(JU^k u_i)}{\partial \xi^k} = \frac{1}{J}\left[\delta_\xi(JU\overline{u_i}^\xi) + \delta_\eta(JV\overline{u_i}^\eta) + \delta_\zeta(JW\overline{u_i}^\eta)\right] \qquad (4.64)$$

that corresponds to Eqs. (3.119) and (3.120). The second expression is

$$\frac{1}{J}\frac{\partial(JU^k u_i)}{\partial \xi^k} = \frac{1}{J}\left[\delta'_\xi(\overline{JU^\xi}u_i) + \delta'_\eta(\overline{JV^\eta}u_i) + \delta'_\zeta(\overline{JW^\eta}u_i)\right] \qquad (4.65)$$

that corresponds to Eqs. (3.121) and (3.122). The advective interpolation scheme for the gradient form can be formulated as

$$U^k\frac{\partial u_i}{\partial \xi_k} = \overline{U\delta_\xi u_i}^\xi + \overline{V\delta_\eta u_i}^\eta + \overline{W\delta_\zeta u_i}^\zeta \qquad (4.66)$$

that corresponds to Eqs. (3.129) and (3.130), or be discretized as

$$U^k\frac{\partial u_i}{\partial \xi_k} = \overline{U}^\xi\delta'_\xi u_i + \overline{V}^\eta\delta'_\eta u_i + \overline{W}^\zeta\delta'_\zeta u_i \qquad (4.67)$$

that corresponds to Eqs. (3.125) and (3.126). Collocated grids do not have the issues of the interpolation schemes being directivity dependent as in the case for staggered grids. However, the use of finite-difference schemes based on Eqs. (4.64)–(4.67) can lead to different results.

---

[5] *Compatibility*: regardless of the forms of the equation, the same numerical results should be obtained. *Conservation*: not only mass and momentum should be conserved but kinetic energy should also be implicitly conserved.

Instead of using U^k, let us consider using $(JU)^k$ at the cell boundary and select the advective form of the interpolation to yield

$$\frac{1}{J}(JU)^k \frac{\partial u_i}{\partial \xi_k} = \frac{1}{J}\left[\overline{JU\delta_\xi u_i}^{\xi} + \overline{JV\delta_\eta u_i}^{\eta} + \overline{JW\delta_\zeta u_i}^{\zeta}\right]. \tag{4.68}$$

The difference between Eqs. (4.64) and (4.68) is

$$\frac{u_i}{J}\left[\delta_\xi(JU) + \delta_\eta(JV) + \delta_\zeta(JW)\right], \tag{4.69}$$

which tells us that if the difference equation

$$\frac{1}{J}\left[\delta_\xi(JU) + \delta_\eta(JV) + \delta_\zeta(JW)\right] = 0 \tag{4.70}$$

holds at the cell center, then the difference approximation of the continuity equation is zero. If the pressure equation, Eq. (4.42), compatible with the above equation is accurately solved, then Eqs. (4.64) and (4.68) match within numerical tolerance.

Next, let us examine the conservation of kinetic energy. Setting $i = 1$ and multiplying u to Eqs. (4.64) and (4.68), they become

$$\frac{1}{J}\left[\delta_\xi\left(JU\frac{\widetilde{u^2}^\xi}{2}\right) + \delta_\eta\left(JV\frac{\widetilde{u^2}^\eta}{2}\right) + \delta_\zeta\left(JW\frac{\widetilde{u^2}^\zeta}{2}\right)\right]$$
$$\pm \frac{u^2}{2J}[\delta_\xi(JU) + \delta_\eta(JV) + \delta_\zeta(JW)], \tag{4.71}$$

where

$$\widetilde{u^2}^\xi_{i+\frac{1}{2},j,k} = u_{i,j,k} \times u_{i+1,j,k}. \tag{4.72}$$

This quadratic variable is defined similarly for the other directions $(\widetilde{u^2}^\eta, \widetilde{u^2}^\zeta)$. The second term in Eq. (4.71) can be neglected if the continuity equation, Eq. (4.70), is satisfied. In such case, the kinetic energy balance can be expressed solely by the first term in Eq. (4.71). This tells us that the volume integral (summation over all cells) of the quadratic quantities over the computational domain is influenced only by the influx and efflux through the boundary. It also means that there should not be any unphysical increase or decrease in the quadratic variable within the computational domain. In that context, we can say that Eqs. (4.64) and (4.68) achieve the conservation of kinetic energy. Furthermore, we can consider what is called the *mixed form* (*skew-symmetric form*) by taking the average of Eqs. (4.64) and (4.68) to cancel the second term (due to the $\pm$). The mixed form always conserves kinetic energy due to the cancelation.

For collocated grids, finite-difference schemes for the advection term to satisfy compatibility and kinetic energy conservation can be summarized as follows. For the divergence form, Eq. (4.64) leads to

$$\frac{1}{J}\frac{\partial(JU^k u_i)}{\partial\xi^k} = \frac{1}{J}\delta_k[(JU^k)\overline{u}_i^k].$$ (4.73)

and for the gradient form, Eq. (4.68) gives

$$\frac{1}{J}(JU^k)\frac{\partial u_i}{\partial\xi^k} = \frac{1}{J}\overline{(JU^k)\delta_k u_i}^k.$$ (4.74)

By taking the average of the above two forms, we have

$$\frac{1}{2}\left[\frac{1}{J}\frac{\partial(JU^k u_i)}{\partial\xi^k} + \frac{1}{J}(JU^k)\frac{\partial u_i}{\partial\xi^k}\right]$$
$$= \frac{1}{2J}\left\{\delta_k[(JU^k)\overline{u}_i^k] + \overline{(JU^k)\delta_k u_i}^k\right\},$$ (4.75)

which is the quadratic conservation form.

4.7.2 Upwinding Schemes

Let us extend the upwinding schemes introduced in Sect. 3.5.3 for the Cartesian staggered grid to the generalized coordinate collocated grid. For the collocated grid, the transported quantity f is positioned at the cell center. In the case of the momentum equation, f is the Cartesian coordinate velocity components (u, v, w).

Upwinding the Divergence Form of the Advection Term

Computing the flux $(JU^k)\overline{f}^k$ at the cell center with two-point finite differencing based on Eq. (4.73) gives

$$\left[\frac{1}{J}\frac{\partial(JU^k f)}{\partial\xi^k}\right]_{i,j,k} = \frac{1}{J_{i,j,k}}\left\{-[(JU)\overline{f}^\xi]_{i-\frac{1}{2},j,k} + [(JU)\overline{f}^\xi]_{i+\frac{1}{2},j,k}\right.$$
$$- [(JV)\overline{f}^\eta]_{i,j-\frac{1}{2},k} + [(JV)\overline{f}^\eta]_{i,j+\frac{1}{2},k}$$ (4.76)
$$\left.- [(JW)\overline{f}^\zeta]_{i,j,k-\frac{1}{2}} + [(JW)\overline{f}^\zeta]_{i,j,k+\frac{1}{2}}\right\}.$$

For upwinding the above divergence form, the interpolation of the transported quantity $\overline{f}^k$ at the cell interface is performed by weighing the upstream values more than the downstream ones based on the direction of the advection velocity U^k.

Let us for example consider the evaluation of $[JU]_{i+\frac{1}{2},j,k}$. For simplicity, let us remove subscripts j and k. While the regular second-order central-difference scheme is

$$[(JU)\overline{f}^\xi]_{i+\frac{1}{2}} = (JU)_{i+\frac{1}{2}}\frac{f_i + f_{i+1}}{2},$$ (4.77)

the donor cell method yields

$$[(JU)\overline{f}^\xi]_{i+\frac{1}{2}} = \begin{cases} (JU)_{i+\frac{1}{2}} f_i, & U \geq 0 \\ (JU)_{i+\frac{1}{2}} f_{i+1}, & U < 0 \end{cases} \tag{4.78}$$

where we utilize the upstream value. Rewriting the above equation without the use of classifying the expression based on the sign of U, we obtain

$$[(JU)\overline{f}^\xi]_{i+\frac{1}{2}} = (JU)_{i+\frac{1}{2}} \frac{f_i + f_{i+1}}{2} - |JU|_{i+\frac{1}{2}} \frac{-f_i + f_{i+1}}{2}$$

$$= (JU)_{i+\frac{1}{2}} \overline{f}^\xi_{i+\frac{1}{2}} - \frac{\Delta\xi |JU|_{i+\frac{1}{2}}}{2} [\delta_\xi f]_{i+\frac{1}{2}} \tag{4.79}$$

(note that $\Delta\xi$ is set to 1). The QUICK method can be written as

$$[(JU)\overline{f}^\xi]_{i+\frac{1}{2}} = \begin{cases} (JU)_{i+\frac{1}{2}} \dfrac{-f_{i-1} + 6f_i + 3f_{i+1}}{8}, & U \geq 0 \\ (JU)_{i+\frac{1}{2}} \dfrac{3f_i + 6f_{i+1} - f_{i+2}}{8}, & U < 0 \end{cases} \tag{4.80}$$

and alternatively as

$$[(JU)\overline{f}^\xi]_{i+\frac{1}{2}} = (JU)_{i+\frac{1}{2}} \frac{-f_{i-1} + 9f_i + 9f_{i+1} - f_{i+2}}{16}$$

$$- |JU|_{i+\frac{1}{2}} \frac{-f_{i-1} + 3f_i - 3f_{i+1} + f_{i+2}}{16} \tag{4.81}$$

$$= (JU)_{i+\frac{1}{2}} \overline{f}^\xi_{i+\frac{1}{2}} + \frac{(\Delta\xi)^3 |JU|_{i+\frac{1}{2}}}{16} [\delta_\xi^3 f]_{i+\frac{1}{2}}.$$

Based on the above discussion, the upwinded flux, Eq. (3.163), can be extended to the generalized coordinate formulation as

$$(JU^k)f = (JU^k)\overline{f}^k + (-1)^{\frac{m}{2}} \alpha (\Delta\xi^k)^{m-1} |JU^k| \delta_k^{m-1} f, \tag{4.82}$$

where m is an even number. The first term on the right-hand side is the m-point mth order accurate interpolation and the second term is added from the $(m-1)$th derivative. Equation (4.76) utilizes the second-order scheme which limits the overall accuracy to second order even if $m \geq 2$ is selected. The added term introduces the mth order artificial viscosity to the momentum equations with additional differentiation performed to compute the advection term.

Upwinding the Gradient Form of the Advection Term

For the gradient form, we choose the corrected upwind finite differencing that adds artificial viscosity to the appropriate central differencing such as Eq. (4.74) presented in Sect. 3.5.3. The corrected upwind finite difference, Eq. (3.170), can be extended to the generalized coordinate system as

$$\frac{1}{J}(JU^k)\frac{\partial f}{\partial \xi^k} = \frac{1}{J}\overline{(JU^k)\delta_k f}^k + \frac{1}{J}(-1)^{\frac{m}{2}}\alpha(\Delta\xi^k)^{m-1}|JU^k|\delta'^m_k f, \qquad (4.83)$$

where again m is even. The first and second terms on the right-hand side are the mth order accurate central difference and the artificial viscosity due to the mth derivative, respectively. This expression provides the $(m-1)$th order accurate upwind finite-difference scheme. Higher order accurate central differencing is discussed later in Sect. 4.10.

4.8　Spatial Discretization of Viscous Term

The viscous term in the divergence form of the momentum equation, Eq. (4.25), is

$$\frac{1}{J}\frac{\partial}{\partial \xi^k}\left(\nu J\frac{\partial \xi^k}{\partial x_j}D_{ij}\right) = \frac{1}{J}\delta_k\left(\nu J\frac{\overline{\partial \xi^k}}{\partial x_j}^k D_{ij}\right), \qquad (4.84)$$

where D_{ij} needs to be evaluated at the cell interface where (JU^k) is defined. This term is discretized as

$$D_{ij} = \frac{\overline{\partial \xi^l}}{\partial x_j}^k \delta_l \overline{u}_i^{k*} + \frac{\overline{\partial \xi^l}}{\partial x_i}^k \delta_l \overline{u}_j^{k*}. \qquad (4.85)$$

where finite difference in the ξ^l direction at the (JU^k) point is computed with

$$\delta_l \overline{u}_i^{k*} = \begin{cases} \delta_l u_i & k = l \\ \delta_l \overline{u}_i^{kl} & k \neq l \end{cases} \qquad (4.86)$$

depending on whether u_l is defined along that direction. For a constant kinematic viscosity ν, the viscous term can be discretized as

$$\nu\frac{1}{J}\frac{\partial}{\partial \xi^k}\left(\gamma^{kl}\frac{\partial u_i}{\partial \xi^l}\right) = \nu\frac{1}{J}\delta_k(\overline{\gamma^{kl}}^k \delta_l \overline{u}_i^{k*}) \qquad (4.87)$$

following the formation given in Eq. (4.29).

4.9　Boundary Conditions

Let us consider, the treatment of the boundary condition for a two-dimensional example where the boundary is described by a constant η as illustrated in Fig. 4.8. The basic idea of the SMAC method for incompressible flow on collocated grids is to

Fig. 4.8 Cell adjacent to the
boundary surface

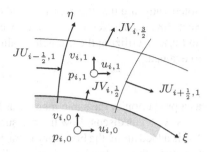

(i) predict the Cartesian components of velocity (u, v) at the cell center, (ii) transform
the velocity to the generalized coordinate contravariant components and interpolate
them at the staggered positions, and (iii) solve for the pressure field that enforces the
continuity equation based on (JU, JV).

Consider predicting the velocity $(u_{i,1}, v_{i,1})$ at the cells adjacent to the boundary
through the momentum equation, as illustrated in Fig. 4.8. To accomplish this, we
can utilize virtual cells outside of the computational domain and predetermine the
elements of the coordinate transformation matrix $(\partial \xi^k / \partial x_j)$ and the Jacobian $J_{i,0}$ at
the centers of the virtual cells. The momentum equation requires JV at the boundary
locations for $(JV)_{i,\frac{1}{2}}$ as well as

$$\begin{cases} \bar{u}^\eta, \ \bar{v}^\eta & \text{for divergence form of advection term} \\ \delta_\eta u, \ \delta_\eta v & \text{for gradient form of advection term and viscous term} \\ \delta_\xi \bar{u}^{\xi\eta}, \ \delta_\xi \bar{v}^{\xi\eta} & \text{for viscous term (when grid is non-orthogonal at boundary)} \end{cases}$$

The interpolation and gradient calculations in the η-direction follow those for the
Cartesian grid described in the previous chapter. With the Dirichlet boundary condition, the values of $\bar{u}^\eta$ and $\bar{v}^\eta$ are available along the boundary, which enables us
to compute $\delta_\eta u$ and $\delta_\eta v$. With the Neumann boundary condition, the wall-normal
gradient of the velocity $\delta_\eta u$ and $\delta_\eta v$ are prescribed, which can be used to extrapo-
late the values of $u_{i,0}$ and $v_{i,0}$. The contribution from quantities $\delta_\xi \bar{u}^{\xi\eta}$ and $\delta_\xi \bar{v}^{\xi\eta}$ are
important when the grid is non-orthogonal (i.e., $\gamma^{\xi\eta} \neq 0$). If the ξ and η axes intersect
orthogonally at the boundary surface, these terms are not needed. For these values,
we can compute the finite difference of $\delta'_\xi u$ and $\delta'_\xi v$ on a ± 1 stencil since we already
know $u_{i,\frac{1}{2}}$ and $v_{i,\frac{1}{2}}$ (or $u_{i,0}$ and $v_{i,0}$ at the virtual cells). With the aforementioned
boundary values evaluated, we can now predict $u_{i,1}$ and $v_{i,1}$.

Next, let us predict $(JV)_{i,\frac{1}{2}}$ at the boundary. Given the boundary velocity values
u_wall and v_wall as the Dirichlet boundary condition, we can determine

$$(JV)_{i,\frac{1}{2}} = [\overline{J\delta_\eta x}^\eta u_\text{wall} + \overline{J\delta_\eta y}^\eta v_\text{wall}]_{i,\frac{1}{2}}. \tag{4.88}$$

If the no-slip boundary condition $(u_\text{wall} = 0, v_\text{wall} = 0)$ is prescribed, $(JV)_{i,\frac{1}{2}} = 0$.
When the boundary condition is directly used as the predicted flow field, it should be

noted with care that the pressure field cannot be directly determined, as discussed in Sect. 3.8. For the Neumann boundary condition, we can extrapolate the values of $u_{i,0}$ and $v_{i,0}$ by utilizing the predicted values of $u_{i,1}$ and $v_{i,1}$ and the boundary condition to evaluate

$$(JV)_{i,\frac{1}{2}} = [\overline{J\delta_\eta x u}^\eta + \overline{J\delta_\eta y v}^\eta]_{i,\frac{1}{2}} \tag{4.89}$$

as a prediction in the same manner as how JV is computed for the inner field.

Next, let us discuss how the pressure gradient ∇P or pressure correction gradient $\nabla\phi$ is added along the boundary in the prediction step or correction step, respectively, to satisfy the continuity equation. When the normal velocity v_n is specified at the boundary on a Cartesian staggered grid, the pressure gradient $\delta_n p$ is not needed at the boundary and $\delta_n\phi = 0$ can be used because ϕ should not alter v_n. However, in the case of collocated grids, the value of pressure $p_{i,0}$ outside of the computational domain becomes necessary for computing the pressure gradient term in Eq. (4.40). There is flexibility in choosing how to determine $p_{i,0}$. One possible approach is to use extrapolation based on $\delta_n\phi = 0$. Furthermore, we should enforce

$$\overline{\gamma^{jk}}^j \delta_k\phi = 0 \tag{4.90}$$

as the boundary condition for ϕ such that the prescribed velocity boundary condition in the correction step, Eq. (4.43), is not altered. In this case, we construct relations for the three points of $\phi_{i-1,0}$, $\phi_{i,0}$, and $\phi_{i+1,0}$ (nine points for three-dimensional problems), instead of an extrapolation formula to compute the value at a point outside of the domain. This set of equations involving all outside points adjacent to the boundary should be solved. If the coordinate axes are orthogonal along the boundary, Eq. (4.90) reduces to $\delta_n\phi = 0$ because $\gamma^{\xi\eta} = 0$, resulting in a simple extrapolation relation.

For facilitating the evaluation of the viscous term along the boundary and the implementation of the pressure boundary condition, having the grid orthogonal in the vicinity of the boundary is beneficial. The extent of how far the grid should remain orthogonal from the boundary is determined by the stencil size which is dependent on the order of accuracy of the chosen finite-difference scheme. Even if it is impossible to have the grid orthogonal along the entire boundary, partial coverage of the boundary with orthogonal grid is still beneficial.

4.10 High-Order Accurate Spatial Discretization

For the generalized coordinate collocated grid, we can construct high-order accurate finite-difference schemes by replacing the interpolation and differencing calculations with fourth, sixth, or higher-order accurate schemes, similar to how it is performed for the Cartesian staggered grid as presented in Sect. 3.9.1. For the advection term, it is necessary to adopt methods similar to the ones discussed in Sect. 3.9.2 for satisfying local compatibility and conservation.

In this section, we extend the method presented by Morinishi [6] to the generalized coordinate collocated grid. For the divergence form of Eq. (4.73), we have

$$\frac{1}{J}\frac{\partial (JU)^k u_i}{\partial \xi^k} = \frac{9}{8J}\delta_{1k}[(JU)^k \overline{u_i}^{1k}] - \frac{1}{8J}\delta_{3k}[(JU)^k \overline{u_i}^{3k}] \qquad (4.91)$$

for fourth-order accuracy and

$$\frac{1}{J}\frac{\partial (JU^k u_i)}{\partial \xi^k} = \frac{150}{128J}\delta_{1k}[(JU)^k \overline{u_i}^{1k}$$
$$- \frac{25}{128J}\delta_{3k}[(JU)^k \overline{u_i}^{3k}] + \frac{3}{128J}\delta_{5k}[(JU)^k \overline{u_i}^{5k}] \qquad (4.92)$$

for sixth-order accuracy. For the gradient form of Eq. (4.74), we have

$$\frac{1}{J}(JU)^k \frac{\partial u_i}{\partial \xi^k} = \frac{9}{8J}\overline{(JU)^k \delta_{1k} u_i}^{1k} - \frac{1}{8J}\overline{(JU)^k \delta_{3k} u_i}^{3k} \qquad (4.93)$$

for fourth-order accuracy and

$$\frac{1}{J}(JU)^k \frac{\partial u_i}{\partial \xi^k} = \frac{150}{128J}\overline{(JU)^k \delta_{1k} u_i}^{1k}$$
$$- \frac{25}{128J}\overline{(JU)^k \delta_{3k} u_i}^{3k} + \frac{3}{128J}\overline{(JU)^k \delta_{5k} u_i}^{5k} \qquad (4.94)$$

for sixth-order accuracy.

With increase in the order of accuracy, the difference stencil becomes wider. This can lead to the stencil extending outside of the computational domain. Approaches to address such issue can follow those discussed in Sect. 3.9.3.

4.11 Evaluation of Coordinate Transform Coefficients

Here, we briefly discuss some details on computing the coefficients for the coordinate transform between the physical and computational space. For details on grid generation, readers should refer to [2, 11].

Preservation of Uniform Flow

Let us consider the conservation of mass, Eq. (4.18), for incompressible flow from a different point of view. We assume that the background flow is time-invariant uniform flow u^∞. Transforming such velocity field to the contravariant velocity $U_\infty^k = (\partial \xi^k / \partial x_j) u_j^\infty$ and substituting it into the continuity equation, Eq. (4.18), we have

$$\frac{u_j^\infty}{J}\frac{\partial}{\partial \xi^k}\left(J\frac{\partial \xi^k}{\partial x_j}\right) = 0. \qquad (4.95)$$

For arbitrary uniform flow, we must satisfy

$$\frac{\partial}{\partial \xi^k} \left(J \frac{\partial \xi^k}{\partial x_j} \right) = 0, \tag{4.96}$$

as we can see from Eq. (4.14). It is necessary that Eq. (4.96) be satisfied numerically also. If it is not satisfied numerically, the numerical solution becomes distorted even if uniform flow is prescribed.

Taking the condition to preserve uniform flow in the x-direction, we have

$$(J\xi_x)_\xi + (J\eta_x)_\eta + (J\zeta_x)_\zeta = 0, \tag{4.97}$$

where $J\xi_x$, $J\eta_x$, and $J\zeta_x$ can be determined from Eq. (A.10). This equation can be rewritten as

$$(y_\eta z_\zeta - y_\zeta z_\eta)_\xi + (y_\zeta z_\xi - y_\xi z_\zeta)_\eta + (y_\xi z_\eta - y_\eta z_\xi)_\zeta = 0. \tag{4.98}$$

For the two-dimensional case, we have $z_\xi = 0$, $z_\eta = 0$, and $z_\zeta = 1$, which lead to

$$(y_\eta)_\xi - (y_\xi)_\eta = 0. \tag{4.99}$$

Equations (4.97)–(4.99) always hold in a continuous sense. The question is whether they are satisfied discretely also.

Let us further examine the two-dimensional case. Discretizing Eq. (4.97) with the second-order central-difference scheme at the cell center (i, j) of the collocated grid, we have

$$-[\overline{J\xi_x}^\xi]_{i-\frac{1}{2},j} + [\overline{J\xi_x}^\xi]_{i+\frac{1}{2},j} - [\overline{J\eta_x}^\eta]_{i,j-\frac{1}{2}} + [\overline{J\eta_x}^\eta]_{i,j+\frac{1}{2}} = 0. \tag{4.100}$$

If we use the second-order interpolation (averaging) for the cell interface, the above equation becomes

$$-[J\xi_x]_{i-1,j} + [J\xi_x]_{i+1,j} - [J\eta_x]_{i,j-1} + [J\eta_x]_{i,j+1} = 0, \tag{4.101}$$

where $J\xi_x = y_\eta$ and $J\eta_x = -y_\xi$ at the cell center. Hence we arrive at

$$-[y_\eta]_{i-1,j} + [y_\eta]_{i+1,j} - [-y_\xi]_{i,j-1} + [-y_\xi]_{i,j+1} = 0. \tag{4.102}$$

Employing the second-order central differencing for the evaluation of y_ξ and y_η,

$$\left. \begin{array}{l} [y_\xi]_{i,j} = -\overline{y}^\xi_{i-\frac{1}{2},j} + \overline{y}^\xi_{i+\frac{1}{2},j} = \dfrac{-y_{i-1,j} + y_{i+1,j}}{2} \\[3mm] [y_\eta]_{i,j} = -\overline{y}^\eta_{i,j-\frac{1}{2}} + \overline{y}^\eta_{i,j+\frac{1}{2}} = \dfrac{-y_{i,j-1} + y_{i,j+1}}{2} \end{array} \right\} \tag{4.103}$$

Substituting these expression into Eq. (4.102), we indeed make the left-hand side equal to zero. Equation (4.99) always hold for higher-order accurate finite-difference schemes. This concept is related to the condition, Eq. (2.3), and is tied back to compatibility, discussed in Sect. 2.1. As long as central differencing is employed, the order of differentiation is irrelevant in the discrete equations as well, resulting in preserving uniform flow in two dimensions.

In the case of three-dimensional flows, Eq. (4.98) must also hold discretely. This requires the conditions expressed by Eqs. (2.2) and (2.3) be satisfied simultaneously. When the coordinate transform coefficients are computed, not only should we use central differencing but also the scheme should have compatibility to enforce $(fg)' = fg' + f'g$ as discussed in Sect. 3.5.1. For higher order differencing, further care such as the ones mentioned in Sect. 3.9.2 must be taken into consideration.

Accuracy of Transform Coefficients

The Jacobian corresponds to the cell volume ratio between the physical and computational space. If we set the grid size to be unity uniformly in the computational space, the sum of the Jacobian must match the total volume of the computational domain. We have noted that the Jacobian $J = |\partial x_j / \partial \xi^k|$ should be evaluated with central-difference schemes to preserve compatibility. The computed Jacobian may not necessarily be strictly equal to the cell volume in the physical space. Such error should not directly cause the simulation to fail. That is because the mass conservation is satisfied as long as (JU^k) is properly balanced even if J contains some level of error.

Nonetheless, we still would like to achieve sufficient accuracy for both J and U^k. For this purpose, it is desirable to use a grid that is close to orthogonal with the its spacing changing gradually over the computational domain. The neighboring cells should have sizes with a ratio no larger than 1:3, as suggested in Sect. 2.3.1, to reliably compute the coordinate transformation coefficient, even if high-order accurate differencing is not utilized. The ratio is recommended to be less than 1:1.2 to obtain data for fundamental research and 1:1.5 for practical flow problems, based on the authors' experience.

4.12 Remarks

In this chapter, the finite-difference schemes for the Cartesian coordinate systems presented in Chap. 3 have been extended to the generalized coordinate systems. With the extension, we can accurately analyze a wide variety of incompressible flow problems for boundary-fitted grids. While we have used the staggered grid for discretizing the governing equations for the Cartesian coordinate system, we have instead adopted the collocated grid for the generalized coordinate system. We of course can employ a formulation based on the collocated grid for the Cartesian coordinate system, as well. However, computations based on a collocated grid often results in lower numerical resolution in comparison to results based on a staggered grid. Hence, a collocated grid for the Cartesian coordinate systems generally require finer grid resolution or higher order accurate discretization to match the results from a staggered grid. One of the reasons for the difference can be attributed to the frequent use of interpolation in the collocated formulation, which may regularize the flow profiles. It has also been pointed out that the cell-centered velocity (Cartesian velocity components) chosen as the basic variable need not satisfy continuity in a strict sense and may introduce associated errors, which can act in a dissipative manner. In general, collocated grids can be easily extended to the generalized coordinate systems but may not achieve the same level of numerical resolution as the staggered grid formulation. Considering these pros and cons, it would be advisable to use the staggered grid formation for the Cartesian grid system and the collocated grid formulation for the generalized coordinate system.

4.13 Exercises

4.1 Show that right-hand side of Eq. (4.12) is identically zero by proving

$$\frac{\partial}{\partial \xi^k}\left(J\frac{\partial \xi^k}{\partial x_1}\right) \equiv 0.$$

4.2 Evaluate the difference between the fourth-order gradient form and the fourth-order divergence form, Eqs. (4.91) and (4.93), respectively. Discuss when this difference is zero.

4.3 As an example of governing equations on the generalized coordinate system, let us consider, the cylindrical coordinate system (although it is a special example of an orthogonal coordinate system). We denote the cylindrical coordinates with ξ^i, where $\xi^1 = r, \xi^2 = \theta, \xi^3 = z$, and the Cartesian coordinates with x^i, where $x^1 = x$, $x^2 = y, x^3 = z$. The coordinate transforms between the two systems are

$$x^1 = \xi^1 \cos \xi^2, \quad x^2 = \xi^1 \sin \xi^2, \quad x^3 = \xi^3$$

and

$$\xi^1 = \sqrt{(x^1)^2 + (x^2)^2}, \quad \xi^2 = \tan^{-1}(x^2/x^1), \quad \xi^3 = x^3.$$

Additional details are offered in Appendix A.

1. Find all elements of the metric tensor $\hat{g}_{ij}$ and the metric tensor for the reciprocal basis $\hat{g}^{ij}$.
2. Determine all elements of the Christoffel symbol $\{_j{}^i{}_k\}$.
3. Evaluate the Jacobian $J = \sqrt{\hat{g}/g}$ for the coordinate transform.
4. Show that the continuity equation can be expressed as

$$\frac{1}{\xi^1}\frac{\partial(\xi^1 U^1)}{\partial \xi^1} + \frac{\partial U^2}{\partial \xi^2} + \frac{\partial U^3}{\partial \xi^3} = 0.$$

5. Find all components of the covariant derivative

$$U^i|_j = \partial U^i/\partial \xi^j + \{{}_k{}^i{}_j\}U^k.$$

6. For the momentum equation in the form

$$\frac{\partial U^i}{\partial t} + H^{ij}|_j = 0,$$

find all components of the flux

$$H^{ij} = U^i U^j + \hat{g}^{ij}\frac{p}{\rho} - \nu\left(\hat{g}^{kj}U^i|_k + \hat{g}^{ki}U^j|_k\right).$$

7. The contravariant components U^i, H^{ij} and the physical contravariant components $U^{(i)}$, $H^{(ij)}$ are related by

$$U^{(i)} = \sqrt{\hat{g}_{ii}}U^i, \quad H^{(ij)} = \sqrt{\hat{g}_{ii}}\sqrt{\hat{g}_{jj}}H^{ij}.$$

From these relationships, show that the continuity equation using the physical components for the cylindrical coordinate system is

$$\frac{1}{r}\frac{\partial(ru_r)}{\partial r} + \frac{1}{r}\frac{\partial u_\theta}{\partial \theta} + \frac{\partial u_z}{\partial z} = 0$$

and the momentum equation becomes

$$\frac{\partial u_r}{\partial t} + \frac{1}{r}\frac{\partial r h_{rr}}{\partial r} + \frac{1}{r}\frac{\partial h_{r\theta}}{\partial \theta} + \frac{\partial h_{rz}}{\partial z} - \frac{h_{\theta\theta}}{r} = 0,$$

$$\frac{\partial u_\theta}{\partial t} + \frac{1}{r}\frac{\partial r h_{\theta r}}{\partial r} + \frac{1}{r}\frac{\partial h_{\theta\theta}}{\partial \theta} + \frac{\partial h_{\theta z}}{\partial z} + \frac{h_{\theta r}}{r} = 0,$$

$$\frac{\partial u_z}{\partial t} + \frac{1}{r}\frac{\partial r h_{zr}}{\partial r} + \frac{1}{r}\frac{\partial h_{z\theta}}{\partial \theta} + \frac{\partial h_{zz}}{\partial z} = 0.$$

Also determine h_{rr}, $h_{\theta\theta}$, h_{zz}, $h_{r\theta}$ (= $h_{\theta r}$), $h_{z\theta}$ (= $h_{\theta z}$), and h_{rz} (= h_{zr}).
8. Find the gradient form of the momentum equation.

References

1. Berger, M.J., Colella, P.: Local adaptive mesh refinement for shock hydrodynamics. J. Comput. Phys. **83**, 64–84 (1989)
2. Fujii, K.: Numerical Methods for Computational Fluid Dynamics. University Tokyo Press, Tokyo (1994)

3. Inagaki, M., Abe, K.: An improvement of prediction accuracy of large eddy simulation on colocated grids. Trans. Jpn. Soc. Mech. Eng. B **64**(623), 1981–1988 (1998)
4. Kajishima, T., Ohta, T., Okazaki, K., Miyake, Y.: High-order finite-difference method for incompressible flows using collocated grid system. JSME Int. J. B **41**(4), 830–839 (1998)
5. Mittal, R., Iaccarino, G.: Immersed boundary methods. Annu. Rev. Fluid Mech. **37**, 239–261 (2005)
6. Morinishi, Y., Lund, T.S., Vasilyev, O.V., Moin, P.: Fully conservative higher order finite difference schemes for incompressible flow. J. Comput. Phys. **143**, 90–124 (1998)
7. Morinishi, Y.: An improvement of collocated finite differene scheme with regard to kinetic energy conservation. Trans. Japan Soc. Mech. Eng. B **65**(630), 505–512 (1999)
8. Peskin, C.S.: The immersed boundary method. Acta Numer. **11**, 479–517 (2002)
9. Rhie, C.M., Chow, W.L.: Numerical study of the turbulent flow past an airfoil with trailing edge separation. AIAA J. **21**(11), 1525–1532 (1983)
10. Roma, A.M., Peskin, C.S., Berger, M.J.: An adaptive version of the immersed boundary method. J. Comput. Phys. **153**, 509–534 (1999)
11. Thompson, J.F., Warsi, Z.U.A., Mastin, C.W.: Numerical Grid Generation: Foundations and Applications. North-Holland (1985)

Chapter 5
Immersed Boundary Methods

5.1 Introduction

Analysis of flow over bodies with complex surface geometry can pose a challenge in terms of spatial discretization. Creating a high quality boundary fitted mesh can not only be difficult but also time consuming especially when there are complex flow structures over intricate boundary geometry that need to be resolved in the simulations. This is especially true for bodies encountered in engineering applications such as fluid flow around an automotive engine and undercarriage as well as aircraft landing gears. Furthermore, if we have problems involving fluid-structure interaction, the location of the moving or deforming interface needs to be determined numerically. In such cases, the need to re-mesh the flow field around the body at every time step can introduce an added computational burden. Similar situation arises when one attempts to simulate particle-laden flows in which the interface between different phases needs to be tracked and resolved accurately.

Regardless of the position of the body boundary or interface, it would be attractive to perform the fluid flow simulation on a predefined time-invariant mesh. The *immersed boundary method* is a technique that simulates flows around bodies of arbitrary geometry by solving for the flow field on a fixed *Eulerian grid* and represent the immersed boundary with a *Lagrangian grid* that can be time-varying. The effect of the immersed boundary would need to be represented in some manner either by introducing a momentum source or by altering the spatial discretization near the surface to satisfy the no-slip boundary condition. For immersed boundaries in motion or deformation, only the Lagrangian grid moves over time, hence eliminating the need to re-mesh the underlying spatial discretization for the flow field. The key to accurately representing the flow around the immersed objects is to discretize the governing equations to conserve physical variables between the Eulerian and Lagrangian grids.

Peskin [27] first introduced the method by describing the flow field with an Eulerian discretization and representing the immersed surface with a set of Lagrangian points. The Eulerian grid is not required to conform to the body geometry as the

© Springer International Publishing AG 2017   179
T. Kajishima and K. Taira, *Computational Fluid Dynamics*,
DOI 10.1007/978-3-319-45304-0_5

no-slip boundary condition is enforced at the Lagrangian points by adding appropriate boundary forces. The boundary forces that exist as singular functions along the surface in the continuous equations are described by discrete delta functions that smear (regularize) the forcing effect over the neighboring Eulerian cells. Peskin originally used the immersed boundary method to examine blood flow inside the beating heart with the forcing function being computed by Hooke's law [27, 28]. The overall scheme that incorporates fluid–structure interaction into the immersed boundary formulation is discussed in this chapter for both flexible and rigid bodies.

Immersed boundary methods can be classified broadly into two methods [23]. The first method introduces a boundary force along the immersed surface to counteract the surrounding flow to satisfy the no-slip boundary condition. The boundary force is added to the continuous Navier–Stokes equations and is discretized afterwards to ensure the interaction (communication) between the Eulerian flow field and the Lagrangian representation of the boundary surface. This approach relies on the use of a discrete delta function and was used by the original formulation of the immersed boundary method of Peskin [27]. This approach is referred to as the *continuous forcing approach* [23].

The second approach alters the Eulerian spatial discretization only near the immersed surface to embed the no-slip boundary condition into the discrete spatial operators. This approach is called the *discrete forcing approach* since the discretization scheme (operator) is modified to account for the immersed object in the flow field [23]. This latter approach requires derivation of differentiation schemes fitted to the boundary which can be somewhat strenuous compared to the continuous forcing approach. However, the discrete forcing approach generally yields better spatial accuracy by embedding the appropriate differencing schemes into the spatial operators that appear in the discretization of the Navier–Stokes equations.

In what follows, we present both approaches and highlight the characteristic features from each of the methods. The discussions in this chapter on immersed boundary methods are built upon the incompressible flow solvers from Chap. 3 and utilize the same notations. There are also immersed boundary methods for compressible flows [11, 16] but they will not be covered in this book. For a comprehensive overview of the immersed boundary methods and additional details, readers should also consult with the review articles by Peskin [29] and Mittal and Iacarrino [23] and the citations therein.

5.2   Continuous Forcing Approach

The original immersed boundary method was developed to study hemodynamics inside the heart by Peskin [27, 28]. The flow field is represented by a Cartesian grid and the boundary of the immersed body (heart) is represented by a collection of Lagrangian points where boundary forces are introduced to satisfy the no-slip boundary condition, as illustrated in Fig. 5.1. We represent the spatial vector on the

Fig. 5.1 The continuous forcing approach of the immersed boundary method. The immersed boundary S$_B$ in *solid line* is represented in the method with a set of Lagrangian points ξ, where boundary force F is applied to satisfy the no-slip boundary condition. The *shaded boxes* around the Lagrangian points depict the support of the discrete delta function over which the boundary force is regularized

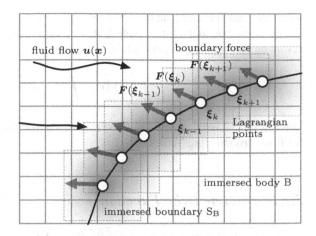

Cartesian grid domain D as x and the spatial vector on the body surface S$_B$ as ξ, which has a surface (boundary) velocity $u_b(\xi)$.

The continuous representation of the original immersed boundary method [27] to analyze fluid-structure interaction around an elastic immersed body B is based on the Navier–Stokes equations, Eqs. (1.39) and (1.36), with boundary forcing f

$$\frac{\partial u}{\partial t} + \nabla \cdot (uu) = -\nabla P + \nu \nabla^2 u + f \qquad (5.1)$$

$$\nabla \cdot u = 0 \qquad (5.2)$$

where $P = p/\rho$. Here, the immersed boundary force f acts on the Cartesian grid to enforce the no-slip condition along the body surface $\xi \in S_B$

$$\frac{d\xi}{dt} = u_b(\xi) = \int_{x \in D} u(x)\delta(x - \xi)dx. \qquad (5.3)$$

The force that is applied to the flow field from the elastic body is expressed as an external body force by

$$f(x) = \int_{\xi(s) \in S_B} F(s)\delta(\xi(s) - x)ds, \qquad (5.4)$$

where s is used to parametrize the surface and the boundary force F defined on the immersed surface is given from the boundary configuration and the elastic property. In short-hand notation, we write the constitutive relation for the boundary force F with

$$F = K(\xi) \qquad (5.5)$$

where K is an operator (can be nonlinear) that relates the position of the Lagrangian points to the boundary force. Note that f is acting on the Eulerian grid and F is acting on the Lagrangian grid, while satisfying

$$\int_{x \in D} f(x) dx = \int_{\xi(s) \in S_B} F(s) ds. \tag{5.6}$$

Later in this chapter, we extend the above discussion to simulation methodologies for flows around rigid bodies.

5.2.1 Discrete Delta Functions

Let us consider how we can discretize the Dirac delta function in Eqs. (5.3) and (5.4) of the immersed boundary method. Here, we describe the conditions that we require the discrete delta functions to satisfy so that the resulting immersed boundary method conserves desired physical quantities, such as mass and momentum. We will also discuss different types of delta functions introduced in past literature and examine how the choice of the discrete delta function affects the overall scheme. Detailed discussions on deriving discrete delta functions are provided in Peskin [29] and Roma et al. [30].

The Dirac delta function δ is modeled with the discrete delta function $\widetilde{\delta}$ in the computation. We construct the three-dimensional delta function from its one-dimensional version $d(x/\Delta s)/\Delta s$, where Δs is the Lagrangian grid spacing. That is

$$\delta(x) = \delta(x)\delta(y)\delta(z) \approx \widetilde{\delta}(x) = \frac{1}{\Delta s^3} d\left(\frac{x}{\Delta s}\right) d\left(\frac{y}{\Delta s}\right) d\left(\frac{z}{\Delta s}\right), \tag{5.7}$$

where the scalings for the argument $x/\Delta s$ and the function $d/\Delta s$ are chosen to make $\widetilde{\delta} \to \delta$ as $\Delta s \to 0$. In delta function based immersed boundary methods, Δs is chosen to be approximately the size of the Cartesian grid Δx. Here, we only note that this choice is made to ensure that the flow does not penetrate through the immersed surface. However, further discussions on this selection are given at the end of this subsection.

To ensure physical quantities are conserved as information from the Eulerian discretization is passed over to the Lagrangian grid and vice versa, care must be taken to discretize the Dirac delta function (convolution). Here, we require the following properties to be satisfied discretely by the discrete version of the delta function:

$$\int \delta(x - \xi) dx = 1 \quad \rightarrow \quad \sum_{i,j,k} \widetilde{\delta}(x_{ijk} - \xi_n) \Delta s^3 = 1 \tag{5.8}$$

$$\int (x - \xi)\delta(x - \xi) dx = 0 \quad \rightarrow \quad \sum_{i,j,k} (x_{ijk} - \xi_n)\widetilde{\delta}(x_{ijk} - \xi_n) \Delta s^3 = 0 \tag{5.9}$$

for all ξ_n, where indices i, j, and k are for the x, y, and z directions. The conservation of mass and force is satisfied with the use of Eq. (5.8) and similarly for torque with Eq. (5.9).

To satisfy the above properties of the delta function, we enforce the following discrete properties on the one-dimensional discrete delta function d. First, the solution must be continuous throughout the domain, which requires the delta function to be at least continuous (C^0) also

$$d(x/\Delta s) \in C^0 (\text{continuous}) \quad \text{for all } x. \tag{5.10}$$

We also require Eqs. (5.8) and (5.9) to be satisfied discretely

$$\sum_i d(x/\Delta s - i) = 1 \quad \text{for all } x \tag{5.11}$$

$$\sum_i (x/\Delta s - i) d(x/\Delta s - i) = 0 \quad \text{for all } x. \tag{5.12}$$

These again are used to conserve mass, force, and torque between the Eulerian and Lagrangian discretizations.[1]

Another property that we impose on the discrete delta function is for computational practicality. The discrete delta function can have infinite support over the entire spatial domain, such as the exponential function which has been used by Briscolini and Santangelo in spectral method calculations [4].[2] However, it is not desirable to have a wide stencil from a computational standpoint. Hence, we design the discrete delta function to have minimal finite support (over m grid cells)

$$d(x/\Delta s) = 0 \quad \text{for } |x/\Delta s| \geq m/2. \tag{5.13}$$

The last property we impose on the delta function is

$$\sum_i [d(x/\Delta s - i)]^2 = C \quad \text{for all } x, \tag{5.14}$$

where C is a constant independent of x. This condition is a weaker statement that is needed to guarantee translational invariance of the immersed boundary method. Since the exact invariance cannot be specified, we enforce this weaker condition.[3] For further details on this condition, refer to the discussion offered by Peskin [29].

---

[1]For a collocated spatial discretization, a stronger requirement of $\sum_{i=\text{odd}} d(x/\Delta s - i) = \sum_{i=\text{even}} d(x/\Delta s - i) = 1/2$ is imposed to avoid checkerboard oscillation from naively transmitting the boundary force onto the Eulerian grid.

[2]The exponential form of the discrete delta function is infinitely continuous which is useful for spectral methods to avoid the introduction of spatial oscillations.

[3]The translational invariance would need a stronger statement of $\sum_i d(x_1/\Delta s - i) d(x_2/\Delta s - i) = \text{fnc}(x_1/\Delta s - x_2/\Delta s)$ for all x.

We list some discrete delta functions that have been used in past studies and comment on the characteristics of the delta functions.

Linear/Triangular Delta Function

In an earlier study by Saiki and Biringen [32], a linear interpolation (triangular function)

$$d_{\text{linear}}(x/\Delta s) = \begin{cases} 1 - \frac{|x|}{\Delta s} & \text{for } |x| \leq \Delta s \\ 0 & \text{otherwise} \end{cases} \tag{5.15}$$

was used in place of the Dirac delta function with support over two grid cells. While this is one of the simplest delta functions that one can use in the immersed boundary formulation, a naive application does not lead to conservation of variables.

Exponential Delta Function

A delta function that is based on the exponential function has been utilized by Briscolini and Santangelo [4] in their Fourier spectral method. Since spectral methods require functions used in the formulation to be infinitely continuous to avoid Gibbs phenomena to appear in the solution or its higher order derivatives, the exponential function of the form of

$$d_{\text{exp}}(x/\Delta s) = \frac{1}{\sqrt{2\pi}} \exp\left(-\frac{x^2}{2\Delta s^2}\right). \tag{5.16}$$

is considered in their application. While this function is continuous for all derivatives, this delta function has an infinitely wide support (although it is possible to truncate the function), which makes it unattractive as it leads to full matrices. Sparse matrices are preferred from a computational standpoint.

Cosine Delta Function

This delta function has been used by Peskin in his earlier work [28] and approximates the four-point delta function (see below) very well as shown by its overlap in Fig. 5.2.

$$d_{\cos}(x/\Delta s) = \begin{cases} \frac{1}{4}[1 + \cos(\frac{\pi x}{2\Delta s})] & \text{for } |x| \leq 2\Delta s \\ 0 & \text{otherwise} \end{cases} \tag{5.17}$$

Four-Grid Cell Delta Function

This particular discrete delta function has been designed for its use on collocated grids. In addition to the imposed properties discussed above, it has a balanced distribution over the even and odd numbered grid points briefly described in an earlier footnote to suppress the onset of checkerboard instability. For the staggered grid formulation, this additional requirement is not necessary.

$$
d_4(x/\Delta s) = \begin{cases} \frac{1}{8}\left[3 - 2\frac{|x|}{\Delta s} + \sqrt{1 + 4\frac{|x|}{\Delta s} - 4(\frac{|x|}{\Delta s})^2}\right] & \text{for } |x| \leq \Delta s, \\[2ex] \frac{1}{8}\left[5 - 2\frac{|x|}{\Delta s} - \sqrt{-7 + 12\frac{|x|}{\Delta s} - 4(\frac{|x|}{\Delta s})^2}\right] & \text{for } \Delta s \leq |x| \leq 2\Delta s, \\[2ex] 0 & \text{otherwise} \end{cases}
$$

$$(5.18)$$

Three-Grid Cell Delta Function

Roma et al. [30] derived a discrete delta function that is appropriate for its use on staggered grids, which we will also use in the immersed boundary projection method discussed later in Sect. 5.2.3. This delta function satisfies all of the aforementioned properties, Eqs. (5.11)–(5.14) with the support of only three cell grids.

$$
d_3(x) = \begin{cases} \frac{1}{3}\left[1 + \sqrt{-3\left(\frac{x}{\Delta s}\right)^2 + 1}\right] & \text{for } |x| \leq 0.5\Delta s, \\[2ex] \frac{1}{6}\left[5 - 3\frac{|x|}{\Delta s} - \sqrt{-3\left(1 - \frac{|x|}{\Delta x}\right)^2 + 1}\right] & \text{for } 0.5\Delta s \leq |x| \leq 1.5\Delta s, \\[2ex] 0 & \text{otherwise} \end{cases}
$$

$$(5.19)$$

Discussions on Delta Functions

We show the discrete delta functions in Fig. 5.2 and summarize the characteristics of these delta functions in Table 5.1. It can be seen that the three and four-grid cell delta functions (d_3 and d_4, respectively) satisfy all or most of the desired properties. For finite-difference or volume methods, we recommend the use of these functions. The four and three-grid cell delta functions are suitable for their use on collocated and staggered grids, respectively. In the delta function-based formulation of the immersed boundary method described later, we use a staggered grid formulation to avoid the checkerboard instability and use the narrowly supported three-point delta

Fig. 5.2 Comparison of discrete delta functions, Eqs. (5.15)–(5.19)

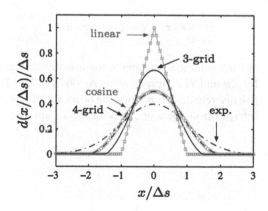

Table 5.1 Comparison of discrete delta functions and their properties

$d(r), r = x/\Delta s$	Linear Eq. (5.15)	Exponential Eq. (5.16)	Cosine Eq. (5.17)	4-grid δ Eq. (5.18)	3-grid δ Eq. (5.19)
Support ($m \Delta s$)	$2\Delta s$	∞	$4\Delta s$	$4\Delta s$	$3\Delta s$
$d(r) \in C^0$	o	o	o	o	o
$d(r) = 0$ for $r > m/2$	o		o	o	o
$\sum_i d(r-i) = 1$	o	o	o	o	o
$\sum_i (r-i)d(r-i) = 0$	o	o		o	o
$\sum_i [d(r-i)]^2 = C$		o	o		o
Notes	Linear interpolation	Used for spectral method	Approximates 4-pt δ well	Suitable for collocated grids	Suitable for staggered grids

The symbol o denotes the condition being satisfied

function by Roma et al. [30]. While the linear delta function has a narrow support, the discontinuity in its first derivative often leads to spatial oscillations in the solution.

There are other possible choices of discrete delta functions that one could use. For certain applications, it may be desirable to have higher moments conserved with the delta function. By increasing the support width, these additional constraints can be accommodated [34]. Also, one can extend the delta function to be formulated as one-sided [3]. It is possible to construct a delta function to suit one's own need in a chosen formulation. Detailed discussions on the derivation of discrete delta functions can be found in [29, 36].

Interpolation and Regularization Operators

In immersed boundary methods, the influence of the boundary force needs to be transmitted to the Cartesian grid and the velocity field on the Cartesian grid must be interpolated onto the Lagrangian points on the immersed surface. Such communication between the two spatial grids can be expressed as operators. Here, we describe how to construct these two operators based on the discrete delta functions.

The convolutions of variables with the delta function appear in the continuous immersed boundary formulation as

$$\int_{x \in D} u(x)\delta(x - \xi(s))dx \quad \text{and} \quad \int_{\xi(s) \in S_B} F(s)\delta(\xi(s) - x)ds. \quad (5.20)$$

We will refer to these operations as *interpolation* and *regularization* and denote them by Eu and HF, respectively. As observed by Peskin [27] and Beyer and LeVeque [3], discrete delta functions can be used both for interpolation and regularization. The interpolation operator can be derived from discretizing the convolution of u and δ,

$$u(\xi) = \int_x u(x)\delta(x - \xi)\mathrm{d}x \tag{5.21}$$

yielding

$$u(\xi_k) = \Delta x \Delta y \sum_i u(x_i)d(x_i - \xi_k)d(y_i - \eta_k) \tag{5.22}$$

for the two-dimensional case, where $u(x_i)$ is the discrete velocity vector defined on the staggered grid $x_i = (x_i, y_i)$ and $u(\xi_k)$ is the discrete boundary velocity at the k-th Lagrangian point $\xi_k = (\xi_k, \eta_k)$. For the three-dimensional case an extra factor of $\Delta z d(z_i - \zeta_k)$ is needed. Accordingly, the no-slip boundary condition, Eq. (5.3), on the Lagrangian points can be expressed as

$$E_{ki}u(x_i) = u_b(\xi_k), \tag{5.23}$$

where

$$E_{ki} = \alpha d(x_i - \xi_k)d(y_i - \eta_k) \tag{5.24}$$

and α is the scaling factor. Note here that E is a rectangular matrix.

The regularization operator is a discrete version of the convolution operator that is used to transmit the boundary force to the neighboring Cartesian cells, similar to the interpolation operator. The continuous expression, Eq. (5.4), can be discretized as

$$f(x_i) = \Delta s \sum_k F(\xi_k)d(\xi_k - x_i)d(\eta_k - y_i). \tag{5.25}$$

We denote this regularization as

$$f(x_i) = H_{ik}F(\xi_k), \tag{5.26}$$

where

$$H_{ik} = \beta d(\xi_k - x_i)d(\eta_k - y_i) = \frac{\beta}{\alpha}E_{ki}^T. \tag{5.27}$$

with β being the numerical integration factor proportional to ds. By absorbing the offset in the scaling ratio into the unknown boundary force, we find that

$$H = E^T, \tag{5.28}$$

which is a useful property for solving the overall set of equations. Note that this symmetry between E and H is not necessary for discretization, but it allows us to solve the overall system in an efficient manner, especially for the immersed boundary projection method [35] presented later. There are unexplored possibilities using different discrete delta functions for interpolation and regularization operators. Beyer and LeVeque [3] consider such cases in a one-dimensional model problem.

In discretizing the immersed boundary in the Lagrangian framework, no repeating Lagrangian points are allowed to avoid discrete matrices from becoming rank-deficient. That is because duplicate Lagrangian points give emergence to duplicate column vectors in the regularization operator (and duplicate row vectors in interpolation operator). Also, to achieve a reasonable condition number of the matrices to be inverted in immersed boundary methods and to prevent penetration of streamlines through the boundary caused by a lack of Lagrangian points, the distance between adjacent Lagrangian points (Δs) should be set approximately to the Cartesian grid spacing (Δx or Δy).

5.2.2   Original Immersed Boundary Method

Since the introduction of the immersed boundary method by Peskin [27, 28] in the 1970s to analyze unsteady blood flow in the heart, there has been numerous extensions to improve the temporal and spatial accuracy by modifying the time integration schemes and discrete delta functions [29]. We will refer to the family of the immersed boundary methods sharing the original spirit of Peskin [27, 28] as the original immersed boundary method in a broad sense. In this section, we focus our discussion on how the original immersed boundary method handles fluid–structure interaction and extends the approach to rigid body problems.

For fluid–structure interaction problems, a formally second-order accurate immersed boundary method has been developed by Lai and Peskin [18]. They call this method formally second-order since it is actually first-order accurate in time with the presence of the delta function. When the discrete delta function is absent from the spatial domain, the method achieves second-order temporal accuracy. The extended method by Lai and Peskin relies on a two-step approach, where a preliminary step is performed to estimate the location of the Lagrangian points at time $\Delta t/2$ into the future based on the current position of the Lagrangian points. That estimate is then used to correct the interpolation and regularization operations. The Navier–Stokes equations are discretized in a manner similar to the fractional-step formulation.

First, the *preliminary step* takes the current immersed surface location $\boldsymbol{\xi}^n$ and computes the boundary force using the constitutive relation $\boldsymbol{K}$. The velocity field on the Eulerian grid is integrated in time to estimate how the Lagrangian points should move over a half step of $\Delta t/2$.

$$\boldsymbol{F}^n = \boldsymbol{K}(\boldsymbol{\xi}^n) \tag{5.29}$$

$$\frac{\boldsymbol{u}^{n+\frac{1}{2}} - \boldsymbol{u}^n}{\Delta t/2} = \boldsymbol{A}^n - \boldsymbol{G}\widetilde{P}^{n+\frac{1}{2}} + \nu \boldsymbol{L}\boldsymbol{u}^{n+\frac{1}{2}} + \boldsymbol{H}(\boldsymbol{\xi}^n)\boldsymbol{F}^n \tag{5.30}$$

$$\boldsymbol{D}\boldsymbol{u}^{n+\frac{1}{2}} = 0 \tag{5.31}$$

$$\frac{\boldsymbol{\xi}^{n+\frac{1}{2}} - \boldsymbol{\xi}^n}{\Delta t/2} = \boldsymbol{E}(\boldsymbol{\xi}^n)\boldsymbol{u}^{n+\frac{1}{2}} \tag{5.32}$$

where $\widetilde{P}^{n+\frac{1}{2}}$ is the intermediate pressure field. Based on the above estimate of the Lagrangian points $\xi^{n+\frac{1}{2}}$, the correction is performed in the *main step* of the time integration shown below.

$$F^{n+\frac{1}{2}} = K(\xi^{n+\frac{1}{2}}) \tag{5.33}$$

$$\frac{u^{n+1} - u^n}{\Delta t} = A^{n+\frac{1}{2}} - GP^{n+\frac{1}{2}}$$

$$+ \tfrac{1}{2}\nu L(u^n + u^{n+1}) + H(\xi^{n+\frac{1}{2}})F^{n+\frac{1}{2}} \tag{5.34}$$

$$Du^{n+1} = 0 \tag{5.35}$$

$$\frac{\xi^{n+1} - \xi^n}{\Delta t} = E(\xi^{n+\frac{1}{2}})\tfrac{1}{2}(u^n + u^{n+1}) \tag{5.36}$$

In the main step, the regularization and interpolation operations are performed using $\xi^{n+\frac{1}{2}}$. The viscous term and the last time advancement equation are integrated in time with the Crank–Nicolson method to stabilize the overall scheme at a reasonable computational cost.

In the work of Lai and Peskin, they have used a regular grid and the four-grid cell supported delta function given by Eq. (5.18). Because the velocity profile is in general discontinuous across the immersed surface in its derivative, the delta function leads to first-order convergence in the L_∞ sense for this method. The spatial accuracy can be improved with some of the discrete forcing approaches, discussed later in Sect. 5.3.

This immersed boundary formulation allows the use of matrix solvers from the traditional fractional-step method because the implicit treatment of the viscous and pressure terms are not altered. Hence the implementation of the above immersed boundary method into a preexisting fractional-step based incompressible flow solver requires minimal effort.

Extension to Rigid Bodies

The above immersed boundary formulation can be extended to simulate flows over rigid bodies. The boundary force has been estimated for rigid surfaces with the use of constitutive relations and feedback law. Lai and Peskin [18] considered the use of Hooke's law

$$F = \kappa(\xi^e - \xi), \tag{5.37}$$

where κ is the spring constant and ξ^e is the equilibrium position for the boundary surface, to compute the boundary force from the Lagrangian point positions. For rigid bodies, $\kappa \gg 1$ can be chosen to prevent the Lagrangian points from moving away from their equilibrium positions.

Instead of the Hooke's law, a feedback controller

$$F = -\kappa_1 \int_0^t [u(\xi, \tau) - u_b]\,d\tau - \kappa_2 [u(\xi, t) - u_b] \tag{5.38}$$

with large gains ($\kappa_1 \gg 1$ and $\kappa_2 \gg 1$) has also been considered to compute the boundary force [12], which results in a similar formulation to the one based on the Hooke's law [35].

However, large gains used in such constitutive relations add stiffness to the governing system, thus prohibiting the use of high CFL numbers. For instance, CFL numbers used in [12, 18] are $\mathcal{O}(10^{-3})$ to $\mathcal{O}(10^{-1})$ for simulations of flow over a rigid circular cylinder. Discussions on the stability margins are offered in [12, 19]. It is possible to use higher CFL numbers by lowering the gains at the expense of relaxing the no-slip condition. We show below that flow over rigid bodies can be simulated without introducing the aforementioned gain parameters (e.g., κ, κ_1, and κ_2), removing the issue of stiffness from the flow solver. For additional discussions, see Problem 5.3.

5.2.3   Immersed Boundary Projection Method

If the rigid body position and its surface velocity are known, the availability of the Navier–Stokes equations itself should be sufficient to determine the velocity and pressure fields as well as the boundary force. The use of constitutive relations discussed above should not be needed if the forces are solved for implicitly. Below, we present an immersed boundary method that satisfies the boundary condition through projection (see Sect. 3.3.1).

Given the discretized Navier–Stokes equations, we can view the pressure variable P as a Lagrange multiplier needed to satisfy the continuity constraint.[4] We can further consider appending additional algebraic constraints by increasing the number of Lagrange multipliers. Based on this observation, Taira and Colonius [35] incorporated into the fractional-step framework the no-slip condition for the immersed surface as an additional kinematic constraint with boundary force as an additional Lagrange multiplier. They discretized the Navier–Stokes equations, Eqs. (5.1)–(5.3), as

$$\frac{u^{n+1} - u^n}{\Delta t} = \tfrac{3}{2}A^n - \tfrac{1}{2}A^{n-1} - GP^{n+1} + \tfrac{1}{2}\nu L(u^{n+1} + u^n) + HF^{n+1} \qquad (5.39)$$

$$Du^{n+1} = 0 \qquad (5.40)$$

$$Eu^{n+1} = u_b^{n+1}, \qquad (5.41)$$

where the last equation describes the no-slip boundary condition for the immersed object. The advective and viscous terms are integrated in time with the second-order Adams–Bashforth and Crank–Nicolson methods, respectively, following Eq. (3.39). Pressure and boundary forces are treated similarly in an implicit fashion. For this

---

[4]The discrete Navier–Stokes equations can be formulated as a Karush–Kuhn–Tucker (KKT) system [25, 35] (see: Eqs. (3.26) and (5.45) with $D = -G^T$).

method, we consider the use of a staggered grid such that the operators G and D are related by $D = -G^T$ (see [5, 26] for details).

Let us express the above set of equations in matrix form following [26]:

$$
\begin{bmatrix} R & \Delta t G & -\Delta t H \\ D & 0 & 0 \\ E & 0 & 0 \end{bmatrix} \begin{bmatrix} u^{n+1} \\ P^{n+1} \\ F^{n+1} \end{bmatrix} = \begin{bmatrix} r^n \\ 0 \\ u_b^{n+1} \end{bmatrix}. \tag{5.42}
$$

We should recall that the interpolation and regularization operators can be expressed as transpose of each other (i.e., $E = H^T$). The staggered grid formulation allows us to use the three-cell delta function d_3 from Eq. (5.19). While the use of the same discrete delta function for E and H is not absolutely necessary, it leads to a desirable symmetric operator in the final form of the algebraic equations.

In the above form of the equation, both the pressure and boundary forcing functions act as Lagrange multipliers to enforce constraints (incompressibility and no-slip) on the flow field. Algebraically speaking, it is no longer necessary to make a distinction between the two. By organizing the sub-matrices and vectors in Eq. (5.42) in the following fashion:

$$
Q \equiv [G, \, E^T], \quad \lambda \equiv \Delta t \begin{bmatrix} P \\ -F \end{bmatrix},
$$

$$
r_1 \equiv S u^n + \Delta t \left(\frac{3}{2} A^n - \frac{1}{2} A^{n-1} \right), \quad r_2 \equiv \begin{bmatrix} 0 \\ u_b^{n+1} \end{bmatrix}, \tag{5.43}
$$

Eq. (5.42) can be simplified to

$$
\begin{bmatrix} R & Q \\ Q^T & 0 \end{bmatrix} \begin{bmatrix} u^{n+1} \\ \lambda^{n+1} \end{bmatrix} = \begin{bmatrix} r_1 \\ r_2 \end{bmatrix}, \tag{5.44}
$$

which is now in a form of an equation identical to how the Navier–Stokes equations without forcing can be discretized, Eq. (3.42). This hence provides motivation to apply the same fractional-step technique discussed in Chap. 3 to solve the above matrix equations. Performing an LU decomposition of Eq. (5.44),

$$
\begin{bmatrix} R & 0 \\ Q^T & -Q^T R^{-1} Q \end{bmatrix} \begin{bmatrix} I & R^{-1} Q \\ 0 & I \end{bmatrix} \begin{bmatrix} u^{n+1} \\ \lambda^{n+1} \end{bmatrix} = \begin{bmatrix} r_1 \\ r_2 \end{bmatrix}. \tag{5.45}
$$

As in the original fractional-step method, there is an N-th order splitting error. In this method, the three-term expansion for $R^{-1} \approx C^N = I + \frac{\Delta t}{2} \nu L + (\frac{\Delta t}{2} \nu)^2 L^2$ is recommended, as discussed in [26, 35] (see Eq. (3.53)). This expansion leads to second order temporal accuracy regardless of the existence of the immersed boundary.

Fig. 5.3 Illustration of the projection approach in the immersed boundary projection method

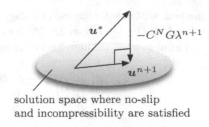

solution space where no-slip and incompressibility are satisfied

The above LU decomposition provides the following three-step fractional-step algorithm that is algebraically identical to Eqs. (3.44)–(3.46) but with λ^{n+1} replacing P^{n+1} and Q replacing G

$$Ru^* = r_1, \tag{5.46}$$

$$Q^T R^{-1} Q \lambda^{n+1} = Q^T u^* - r_2, \tag{5.47}$$

$$u^{n+1} = u^* - R^{-1} Q \lambda^{n+1}, \tag{5.48}$$

where again R^{-1} is approximated by C^N. The main differences between the above and the traditional fractional-step methods are in the Poisson equation and the projection step. Here, the pressure and boundary force are simultaneously determined implicitly from the modified Poisson equation. The projection step removes the divergent and slip components of the velocity from the intermediate velocity field in a single step as illustrated in Fig. 5.3. Hence, this method is called the *immersed boundary projection method*.

The numerical constraint of no-slip is applied only at the Lagrangian points, hence making the dimensions of H and F considerably smaller than those of G and P, respectively. It is encouraging that there is no significant increase in size of $Q^T C^N Q$ in the modified Poisson equation from $G^T C^N G$ in the classical fractional-step method.

For rigid bodies, the immersed boundary projection method solves for the boundary force fully implicitly with no constitutive relations and behaves similarly to the traditional fractional-step method in terms of temporal stability. Hence simulations can be performed with a Courant number as high as 1. We can still solve Eqs. (5.46) and (5.47) with the conjugate gradient method as both left-hand side operators are symmetric and positive-definite. The immersed boundary projection method satisfies the continuity equation and the no-slip condition exactly to machine precision or, if desired, to a prescribed tolerance.

This technique with the three-cell delta function d_3, Eq. (5.14), yields a spatial accuracy of order ≈ 1.5 in the L_2 sense due to the smearing of the delta function. We note that while this projection approach used a discrete delta function in the spatial discretization, any form of E and H can be used in general [3]. Higher-order spatial representation of these operators may possibly lead to increase in spatial accuracy.

In the case of moving immersed bodies, the location of the Lagrangian points must be updated at each time and so must E, i.e.,

$$E_{ki} = E_{ki}^{n+1} = E(\xi_k(t^{n+1}), \mathbf{x}_i) \qquad (5.49)$$

and similarly for H. These operators can be pre-computed at each time step by knowing the location of the Lagrangian points a priori. The current technique is not limited to rigid bodies and can model flexible moving bodies if we are provided with the location of S_B at time level $n + 1$. The immersed boundary projection method has also been extended to solve fluid–structure interaction problems [17, 20].

5.3 Discrete Forcing Approach

5.3.1 Direct Forcing Method

To enforce the no-slip boundary condition at an immersed surface, the boundary force can be estimated using the momentum equation. The direct forcing method [9, 24] relies on the observation that the boundary force can be estimated using the momentum equation and the boundary velocity. Let us discretize the momentum equation with explicit Euler integration for the purpose of this discussion[5]

$$\frac{u^{n+1} - u^n}{\Delta t} = \text{RHS}^n + H F^n, \qquad (5.50)$$

where $\text{RHS} = A - GP + \nu Lu$ represents the right-hand side of the momentum equation containing the advective, pressure gradient, and viscous terms. The boundary force that is needed to enforce the desired boundary velocity u_b^{n+1} at the next time step can be estimated as

$$F^n = \frac{u_b^{n+1} - Eu^n}{\Delta t} - E(\text{RHS}^n). \qquad (5.51)$$

For Lagrangian points not aligned with the Cartesian grids, linear interpolation can be used for constructing the interpolation operator E. In actual implementation, $H F^n$ need not be computed but can be directly implemented into the spatial discretization scheme. See [9] for full details.

The direct forcing method is attractive due to its simplicity to create complex immersed objects in a straightforward manner. For the actual incompressible flow solution (e.g., fractional-step method), u^{n+1} in Eq. (5.50) is replaced by u^*. This means that the direct forcing method prescribes no-slip on u^* but not on u^{n+1}. With the fractional-step scheme, the projection step is applied after the above steps to project the intermediate velocity, u^*, onto the divergence-free (solenoidal) solution space, which can offset the no-slip boundary condition. To ensure that u^{n+1} does indeed satisfy the no-slip condition, iterations over the entire fractional-step algorithm are

---

[5]In Fadlun et al. [9], the third-order Runge–Kutta method is used.

suggested at each time level [9]. In space, this direct forcing method achieves second-order accuracy if the flow is linear near the immersed boundary.

While the velocity field compensates for the boundary condition on the immersed surface, the pressure field is not treated in a special manner. The linear assumption of velocity profile near the immersed boundary leads to the wall-normal pressure gradient to be $\partial P/\partial n = 0$, which does not generally hold. While the specification of this Neumann condition on pressure may be small for laminar boundary layers, it can cause discrepancy for turbulent flows. Below a method that extends the direct forcing method is presented.

5.3.2 Consistent Direct Forcing Method

To address the temporal inconsistency in enforcing the no-slip condition on the immersed boundary and the continuity equation on the flow field in the original direct forcing method, Ikeno and Kajishima [14] utilized consistent differencing schemes for both the velocity and pressure fields.

In this immersed boundary formulation, a two-point forcing approach is used to convey the effect of the immersed boundary to the neighboring Cartesian grids. Denoting the velocity at the location where boundary forcing is applied with $u^{f(k)}$, we have

$$u^{f(k)} = Wu^{(k)}, \tag{5.52}$$

where the operator W incorporates the no-slip condition through linear interpolation, as depicted in Fig. 5.4. Here, we take the immersed body to be stationary. The superscript (k) represents the component in the kth direction. For the two-dimensional case, this operator W takes the form of

$$
\begin{aligned}
[u^{f(k)}]_{i,j} =& W[u^{(k)}]_{i,j} \\
=& w_{u^{(k)}}^{-1x}[u^{(k)}]_{i-1,j} + w_{u^{(k)}}^{+1x}[u^{(k)}]_{i+1,j} \\
&+ w_{u^{(k)}}^{-1y}[u^{(k)}]_{i,j-1} + w_{u^{(k)}}^{+1y}[u^{(k)}]_{i,j+1} \\
&+ w_{u^{(k)}}^{-2x}[u^{(k)}]_{i-2,j} + w_{u^{(k)}}^{+2x}[u^{(k)}]_{i+2,j} \\
&+ w_{u^{(k)}}^{-2y}[u^{(k)}]_{i,j-2} + w_{u^{(k)}}^{+2y}[u^{(k)}]_{i,j+2},
\end{aligned}
\tag{5.53}
$$

where $w_{u^{(k)}}^{1j}$ performs interpolation to enforce the no-slip boundary condition and $w_{u^{(k)}}^{2j}$ applies extrapolation to make the velocity field zero within the body. These operations take the form of

$$w_{u^{(k)}}^{1j} = \alpha \frac{d_{u^{(k)}}^j - \Delta x^{(k)}}{d_{u^{(k)}}^j + \Delta x^{(k)}} \quad \text{and} \quad w_{u^{(k)}}^{2j} = \alpha \frac{d_{u^{(k)}}^j - \Delta x^{(k)}}{d_{u^{(k)}}^j + \Delta x^{(k)}}, \tag{5.54}$$

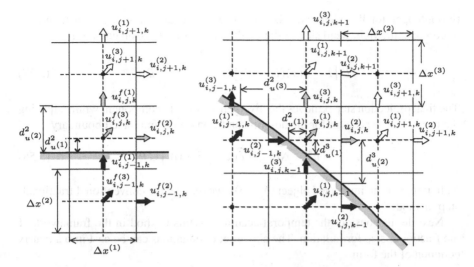

Fig. 5.4 Illustration of cells affected by boundary forcing in the method of Ikeno and Kajishima for (*left*) one and (*right*) two-dimensional settings [14] (reproduced with permission from Elsevier)

where $\Delta x^{(k)}$ is the grid size in the k-th direction and $\alpha = 1/n$ (n being the number of dimensions for the flow of interest). The interpolation scheme is one-dimensional in each direction. The location of the immersed boundary with respect to the Cartesian grid is expressed with $d^j_{u^{(k)}}$, as illustrated in Fig. 5.4. If the boundary velocity $u^{(k)}_b$ is nonzero, the forced velocity becomes

$$u^{f(k)} = W u^{(k)} + (1 - |W|) u^{(k)}_b, \tag{5.55}$$

where $|W|$ denotes the summation of $w^{1j}_{u^{(k)}}$. With this spatial discretization scheme, this immersed boundary method achieves second-order spatial accuracy. Extension of the above formula to three-dimensions is fairly straightforward. Readers can refer to [14] for additional details.

Ikeno and Kajishima incorporated the above immersed boundary formulation into the fractional-step method. First, the momentum equation is solved with modification due to the existence of the immersed boundary

$$\left[I - \frac{\Delta t}{2} \nu L - \widehat{\delta}(\boldsymbol{x}) \left(E - \frac{\Delta t}{2} \nu L \right) \right] \boldsymbol{u}^*$$
$$= [1 - \widehat{\delta}(\boldsymbol{x})] \left[\left(I + \frac{\Delta t}{2} \nu L \right) \boldsymbol{u}^n + \Delta t \left(\frac{3}{2} A^n - \frac{1}{2} A^{n-1} \right) \right] \tag{5.56}$$
$$+ \widehat{\delta}(\boldsymbol{x})(1 - |W|) \boldsymbol{u}^n_b,$$

where we have dropped the superscript (k) for clarity. The delta function $\widehat{\delta}(\boldsymbol{x})$ takes the value of unity when $\boldsymbol{x}$ lies on a location where boundary forcing is introduced

through operator W; otherwise $\widehat{\delta}$ is zero. The intermediate velocity u^* is then used
to compute the right-hand side of the modified pressure Poisson equation

$$D\left[I - \widehat{\delta}(x)(I - W)\right]GP^{n+1} = \frac{1}{\Delta t}Du^*. \tag{5.57}$$

The final projection step removes the divergence velocity component while adjusting
the projection to account for the no-slip condition on the immersed boundary.

$$u^{n+1} = u^* - \Delta t\left[1 - \widehat{\delta}(x) + \widehat{\delta}(x)W\right]GP^{n+1}. \tag{5.58}$$

Note if there is no immersed object ($\widehat{\delta} \to 0$), we recover the conventional fractional-
step method.

Next, let us consider the temporal accuracy of this method in the framework of
the fractional-step formulation. The above set of equations can be cast into a matrix
equation of the form

$$\begin{bmatrix} \widetilde{R} & [1 - \widehat{\delta}(x_i)]G \\ D & 0 \end{bmatrix}\begin{bmatrix} u^{n+1} \\ P^{n+1} \end{bmatrix} = \begin{bmatrix} \widetilde{r}^n \\ 0 \end{bmatrix}, \tag{5.59}$$

where

$$\widetilde{R} = \frac{1}{\Delta t}\left\{I - \frac{\Delta t}{2}\left[1 - \widehat{\delta}(x_i)\right]\nu L - \widehat{\delta}(x_i)E\right\} \tag{5.60}$$

and

$$\widetilde{r}^n = [1 - \widehat{\delta}(x)]\left[\left(I + \frac{\Delta t}{2}\nu L\right)u^n + \Delta t\left(\frac{3}{2}A^n - \frac{1}{2}A^{n-1}\right)\right]$$
$$+ \widehat{\delta}(x)(1 - |W|)u_b^n. \tag{5.61}$$

Ikeno and Kajishima approximate $[1 - \widehat{\delta}(x)]G$ in the above matrix equation with an
operator $(\widetilde{R}\widetilde{B})$ to yield

$$\begin{bmatrix} \widetilde{R} & \widetilde{R}\widetilde{B}G \\ D & 0 \end{bmatrix}\begin{bmatrix} u^{n+1} \\ P^{n+1} \end{bmatrix} = \begin{bmatrix} \widetilde{r}^n \\ 0 \end{bmatrix}. \tag{5.62}$$

Here, the approximation is chosen such that $\widetilde{R}\widetilde{B}G \approx [1 - \widehat{\delta}(x)]G$ with operator B
being

$$\widetilde{B} = \Delta t\left\{[1 - \widehat{\delta}(x)]I + \widehat{\delta}(x)W\right\} \tag{5.63}$$

The approximation that Eq. (5.63) takes in place of Eq. (5.59) has an error from the
use of B, which is

$$\left\{ [1 - \widehat{\delta}(\boldsymbol{x})]G - (\widetilde{R}\widetilde{B})G \right\} P^{n+1}$$

$$= \left\{ \widehat{\delta}(\boldsymbol{x})(W - W^2) - \frac{\Delta t}{2}[1 - \widehat{\delta}(\boldsymbol{x})]\nu L \right\} G P^{n+1}. \tag{5.64}$$

The above expression tells us that the temporal error is $\mathcal{O}(\Delta t^0)$ if $W \neq W^2$. For this reason, the two-point forcing (interpolation) operator in this method is designed to satisfy $W = W^2$ for the error to be $\mathcal{O}(\Delta t)$.

The above temporal error is generated from the incompatibility of the finite-difference schemes between the momentum equation and the pressure Poisson equation in the vicinity of the immersed boundary. To address this issue, Sato et al. [33] utilize appropriate differencing schemes to evaluate the pressure gradient near the immersed boundary such that it is compatible with the discrete velocity variables. We only mention the work by Sato et al. briefly since it is founded on a collocated grid, while this chapter focuses on the staggered grid formulation. However it is noteworthy that their approach benefits from the compatible formulation with sizable reduction in the spatial error, especially in the neighborhood of the immersed surface. Moreover, this approach has been extended to handle not only moving rigid bodies but also to simulate complex fluid–structure interaction problems [10].

5.3.3 Cut-Cell Immersed Boundary Method

It is also possible to relax the Cartesian grid discretization to adapt the finite volume cells in the vicinity of the immersed boundary. Let us briefly present another spatially second-order immersed boundary formulation by Ye et al. [39] and Udaykumar et al. [38]. This method employs a *cut-cell* approach with trapezoidal cells that conforms to the immersed boundary to satisfy the governing equations in a finite volume formulation. Mass and momentum balance can be performed using the fluxes shown in Fig. 5.5 (left).

To balance the fluxes for the highlighted trapezoidal cell, we need to sum all of the fluxes shown by the large arrows. These fluxes further require accurate evaluation of fluxes represented by the smaller arrows in the middle figure. The fluxes corresponding to these smaller arrows can be determined through interpolation with second-order spatial accuracy with a linear-quadratic interpolating function

$$g(x, y) = C_1 x y^2 + C_2 y^2 + C_3 x y + C_4 y + C_5 x + C_6, \tag{5.65}$$

where coefficients C_i $(i = 1, \ldots, 6)$ are determined locally for variables or their gradients near the immersed surface. For example, the flux shown in red can be evaluated using the values (stars) on the trapezoid shown in Fig. 5.5 (right). A linear-quadratic function is chosen to ensure second-order spatial convergence while only requiring six coefficients. Although one can choose a bi-quadratic formulation, it would require a total of nine coefficients imposing additional computational cost.

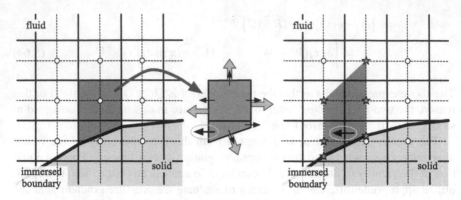

Fig. 5.5 Modification of flux balance performed for the shaded trapezoidal control volume adjacent to the immersed boundary (*left*). Fluxes leaving the cell faces denoted by the *smaller arrows* are needed to evaluate the combined fluxes depicted by the *wider arrows* (*middle*). Computation of each flux (e.g., the *circled arrow*) is performed using six points around the trapezoid shown by *stars* (*right*). Figures based on Ye et al. [39]

Cut cells (trapezoidal cells) are used to determine the coefficients and evaluate physical variables or their derivatives. Figure 5.5 (right) is shown to represent the cell structure used to determine the interpolation function. Once the trapezoidal cell is given, the following set of equations

$$
\begin{bmatrix}
x_1 y_1^2 & y_1^2 & x_1 y_1 & y_1 & x_1 & 1 \\
x_2 y_2^2 & y_2^2 & x_2 y_2 & y_2 & x_2 & 1 \\
\vdots & \vdots & \vdots & \vdots & \vdots & \vdots \\
x_6 y_6^2 & y_6^2 & x_6 y_6 & y_6 & x_6 & 1
\end{bmatrix}
\begin{bmatrix}
C_1 \\
C_2 \\
\vdots \\
C_6
\end{bmatrix}
=
\begin{bmatrix}
g_1 \\
g_2 \\
\vdots \\
g_6
\end{bmatrix}
\tag{5.66}
$$

can be solved to find the coefficients. The matrix to be inverted is known as the Vandermonde matrix. Once these coefficients are determined, the terms from the Navier–Stokes equations using the cut-cell-based discretization can be evaluated within the formulation of the fractional-step method.

The resulting spatial accuracy of this cut-cell method is both locally and globally second order. Furthermore, this approach is able to conserve both mass and momentum since the spatial discretization is handled with care to evaluate the fluxes in a consistent manner. Details on the implementation of the cut-cell method can be found in Ye et al. [39] and Udaykumar et al. [38].

5.4 Applications of Immersed Boundary Methods

To demonstrate the capabilities of the immersed boundary methods, we present some computational results of flow over bodies with non-grid conforming and complex geometries. In some cases, the bodies are in motion with fluid–structure interaction.

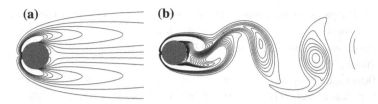

Fig. 5.6 Vorticity fields over a circular cylinder at **a** $Re = 40$ and **b** $Re = 200$ simulated with the immersed boundary projection method [35]

5.4.1 Flow Around a Circular Cylinder

Two-dimensional separated flow over a circular cylinder has been simulated with the immersed boundary projection method on a Cartesian grid [35]. The vorticity fields around the cylinder of diameter d in cross flow are shown in Fig. 5.6 for $Re = U_\infty d/\nu = 40$ and 200. The Cartesian grid was set up with a resolution of $\Delta x/d = 1/30$. The computed flow fields and the characteristic properties of the wake behind the cylinder are found to be in agreement with those reported in past studies [7, 8], including the ones obtained using other immersed boundary methods [18, 21, 39]. The comparison of wake characteristics are summarized in Table 5.2. Note that the immersed boundary methods are able to predict the behavior of steady and unsteady separated flows as well as the forces exerted on the cylinder in an accurate manner.

These cases test flows over rigid bodies. An interesting use of the immersed boundary projection method has been considered by Bagheri et al. [1]. In their study, they analyzed flow over a rigid circular cylinder with an elastic flapping filament attached at the rear end. Both the rigid and flexible parts of the bodies are generated with the immersed boundary formulation. The motion of the filament was computed using a predictor–corrector scheme. An interesting symmetry breaking flow creating lift and torque on the cylinder was observed for certain filament length (and rigidity) as shown in Fig. 5.7. Such results provide insights into how symmetry can be broken by some of the biological swimmers or flyers and highlight the strengths of immersed boundary formulations.

5.4.2 Turbulent Flow Through a Nuclear Rod Bundle

To demonstrate the capability of immersed boundary methods to predict flows around complex bodies, let us consider as an example turbulent flow through a nuclear rod bundle (see Chap. 6 for turbulent flows). The geometry of the rod bundle is inherently complex due to the cooling requirement. To enhance heat exchange, spacers and mixing vanes are placed between rods as illustrated in Fig. 5.8a. Ikeno and Kajishima [14] have performed large-eddy simulation based on the immersed boundary method introduced in Sect. 5.3.2 with a one-equation dynamic sub-grid scale model [15].

Table 5.2 Comparison of the computed characteristic properties of the wake behind the cylinder

Re	References	l/d	a/d	b/d	θ	C_D	C_L	St
20	Coutanceau and Bouard [7]*	0.93	0.33	0.46	45.0°	–	–	–
	Tritton [37]*	–	–	–	–	2.09	–	–
	Dennis and Chang [8]	0.94	–	–	43.7°	2.05	–	–
	Linnick and Fasel [21]†	0.93	0.36	0.43	43.5°	2.06	–	–
	Taira and Colonius [35]†	0.94	0.37	0.43	43.3°	2.06	–	–
	Ye et al. [39]†	0.92	–	–	–	2.03	0	0
40	Coutanceau and Bouard [7]*	2.13	0.76	0.59	53.8°	–	–	–
	Tritton [37]*	–	–	–	–	1.59	–	–
	Dennis and Chang [8]	2.35	–	–	53.8°	1.52	–	–
	Linnick and Fasel [21]†	2.28	0.72	0.60	53.6°	1.54	–	–
	Taira and Colonius [35]†	2.30	0.73	0.60	53.7°	1.54	–	–
	Ye et al. [39]†	1.52	–	–	–	2.27	0	0
200	Belov et al. [2]	–	–	–	–	1.19 ± 0.042	± 0.64	0.193
	Liu et al. [22]	–	–	–	–	1.31 ± 0.049	± 0.69	0.192
	Lai and Peskin [18]†	–	–	–	–	–	–	0.190
	Roshko [31]*	–	–	–	–	–	–	0.19
	Linnick and Fasel [21]†	–	–	–	–	1.34 ± 0.044	± 0.69	0.197
	Taira and Colonius [35]†	–	–	–	–	1.36 ± 0.043	± 0.69	0.197

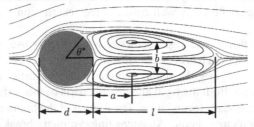

The inserted figure defines the characteristic wake dimensions. References with asterisk (*) denote experimental studies and those with dagger (†) represent computational studies with immersed boundary methods

The flow field predicted by the immersed boundary method is shown in Fig. 5.8b for a Reynolds number of $\approx 4 \times 10^3$ based on the rod pitch P and the bulk velocity. Observe that the unsteady flow structures created by the mixing vane are captured. The resolved turbulent flow field and associated properties can assess the cooling performance of a rod-bundle design. The strength of the immersed boundary method lies in its ability to examine a wide range of design parameters used in designs of complex surface geometries without having to construct a grid for each and every new design. This particular capability is crucial in many industrial designs where complex geometry is ubiquitous.

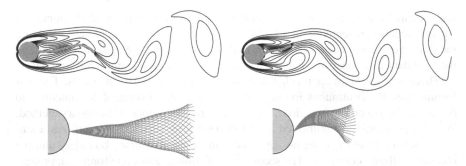

Fig. 5.7 Flow over a circular cylinder with a flexible membrane (*gray*) attached to its rear end having different lengths with the same flexibility. Shown on the *top* are the instantaneous vorticity fields at $Re = 100$. The membranes are shown on the *bottom* at different times. The *right* and *left* cases correspond to symmetric and symmetry breaking flows, respectively, caused by the length of the membrane and its flexibility. Flow field data were generously shared by S. Bagheri and visualizations were adapted with permission from [1], copyrighted by the American Physical Society

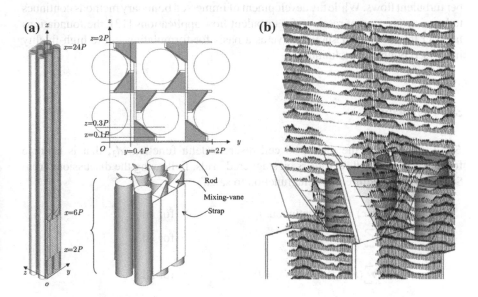

Fig. 5.8 **a** A representative nuclear rod bundle with spacers and mixing vanes are placed to enhance heat transfer. **b** Turbulent velocity field inside the nuclear rod bundle simulated by the consistent immersed boundary method [14] (reproduced with permission from Elsevier)

5.5 Remarks

To simulate flows over bodies with complex geometries or those that undergo deformation, spatial discretization of the flow field around these bodies can pose computational challenges. In this chapter, we presented immersed boundary methods that are capable of simulating flows over bodies on a Cartesian grid with the immersed surface being represented with a Lagrangian grid. The key in accurately resolving the flow over an immersed boundary is to discretize the governing equations around the boundary in a consistent manner to ensure mass and momentum are conserved.

There are two main approaches used in immersed boundary methods. The first formulation, the continuous forcing approach, uses the discrete delta functions to discretize the governing equations, such as the original immersed boundary method. We also presented the immersed boundary projection method that utilizes a single projection to enforce the no-slip condition on the immersed boundary and the incompressibility constraint. The second formulation, the discrete forcing approach, modifies the spatial discretization operators, such as the direct forcing method and its extensions that ensure consistency not only in the momentum equation but also in the pressure Poisson equation to satisfy mass and momentum conservation. The cut-cell immersed boundary method that utilizes second-order discretization based on trapezoidal control volumes was also mentioned. The conservation of mass and momentum near the boundary becomes especially important for high-Reynolds number turbulent flows. While the development of immersed boundary methods continues to be an active area of research for turbulent flow applications [13], the foundations laid out in this chapter should serve as a basis for formulating future high-fidelity immersed boundary techniques.

5.6   Exercises

5.1 Let us derive the three-grid cell discrete delta function, d_3, that is suitable for incompressible flow solvers on staggered grids. Following the discussions from Sect. 5.2.1, we require this delta function to satisfy

$$d_3(r) \in C^0 \text{(continuous)} \qquad \text{for all } r,$$

$$\sum_i d_3(r - i) = 1 \qquad \text{for all } r,$$

$$\sum_i (r - i) d_3(r - i) = 0 \qquad \text{for all } r,$$

$$\sum_i [d(r - i)]^2 = C \qquad \text{for all } r,$$

$$d_3(x/\Delta s) = 0 \qquad \text{for } |r| \geq 3/2,$$

where $r = x/\Delta s$ and $C = 1/2$ (see [30] for details).

1. Based on the above conditions, show that the following three equations need to be satisfied for $|r| < 1/2$:

$$d_3(r-1) + d_3(r) \qquad + d_3(r+1) = 1$$
$$(r-1)d_3(r-1) + rd_3(r) + (r+1)d_3(r+1) = 0$$
$$d_3^2(r-1) + d_3^2(r) + \qquad d_3^2(r+1) = 1/2$$

2. Solving the set of equations from Part 1, show that

$$d_3(r) = \frac{1}{3}\left(1 + \sqrt{-3r^2 + 1}\right).$$

3. Repeat the above steps for $1/2 \leq |r| \leq 3/2$ and verify Eq. (5.19).

5.2 For the immersed boundary projection method, an LU decomposition was applied to the system of equations to yield, Eq. (5.45),

$$\begin{bmatrix} R & 0 \\ Q^T & -Q^T R^{-1} Q \end{bmatrix} \begin{bmatrix} I & R^{-1}Q \\ 0 & I \end{bmatrix} \begin{bmatrix} u^{n+1} \\ \lambda^{n+1} \end{bmatrix} = \begin{bmatrix} r_1 \\ r_2 \end{bmatrix}.$$

Demonstrate that using an N-term expansion of $R^{-1} \approx C^N = I + \frac{\Delta t}{2}\nu L + (\frac{\Delta t}{2}\nu)^2 L^2 + \cdots + (\frac{\Delta t}{2}\nu)^N L^N$, this approach has an N-th order temporal error in the velocity variable, i.e., $\mathcal{O}(\Delta t^N)$. Discuss how high of an expansion order, N, should be chosen.

5.3 Simple extension of the original immersed boundary method to model rigid bodies using constitutive relations with large parameters, such as the spring constant κ in the Hooke's law (Eq. (5.37)), can lead to stiff set of equations.

This stiffness issue has also been observed for an incompressible flow solver called the *artificial compressibility method* [6] (even without an immersed body) that approximately satisfies the continuity equation with

$$\frac{1}{a^2}\frac{\partial p}{\partial t} + \nabla \cdot u = 0, \tag{5.67}$$

where a is an artificial speed of sound that is usually taken to be a large number (i.e., $a \gg 1$).

1. Show for the artificial compressibility method (without an immersed body) using the implicit Euler time-stepping method for the above modified continuity equation, we can have

$$\begin{bmatrix} R & \Delta t G \\ D & \alpha I \end{bmatrix} \begin{bmatrix} u^{n+1} \\ P^{n+1} \end{bmatrix} = \begin{bmatrix} r_1 \\ r_2 \end{bmatrix},$$

where r_1 and r_2 are the explicit terms. Use the Crank–Nicolson and the second-order Adams–Bashforth methods for the viscous and advective terms, respectively. Also, determine α.

2. Next, let us consider the use of a large spring constant in the Hooke's law to simulate flow over an immersed body. For this analysis, consider the time-derivative of the Hooke's law

$$\frac{\partial f}{\partial t} = \kappa \frac{\partial}{\partial t} (\xi_e - \xi)$$

and demonstrate that a null sub-matrix (bottom right) in Eq. (5.42) is no longer zero, analogous to the artificial compressibility method.

3. Repeat the analysis for the immersed boundary formulation when the feedback controller approach, Eq. (5.38), is used to generate a rigid body.
4. Comment on why the methods discussed in this problem lead to stiff set of equations, restricting the time step Δt.

5.4 Given an immersed boundary method code for viscous flow, discuss how the computer program can be used to find potential flow over a body (this can be accomplished without any changes to the program but by setting up the simulation in an appropriate manner).

References

1. Bagheri, S., Mazzino, A., Bottaro, A.: Spontaneous symmetry breaking of a hinged flapping filament generates lift. Phys. Rev. Let. **109**(15), 154,502 (2012)
2. Belov, A., Martinelli, L., Jameson, A.: A new implicit algorithm with multigrid for unsteady incompressible flow calculations. AIAA Paper 95-0049 (1995)
3. Beyer, R.P., LeVeque, R.J.: Analysis of a one-dimensional model for the immersed boundary method. SIAM J. Numer. Anal. **29**(2), 332–364 (1992)
4. Briscolini, M., Santangelo, P.: Development of the mast method for incompressible unsteady flows. J. Comput. Phys. **84**, 57–75 (1989)
5. Chang, W., Giraldo, F., Perot, B.: Analysis of an exact fractional step method. J. Comput. Phys. **180**, 183–199 (2002)
6. Chorin, A.J.: A numerical method for solving incompressible viscous flow problems. J. Comput. Phys. **2**(1), 12–26 (1969)
7. Coutanceau, M., Bouard, R.: Experimental determination of the main features of the viscous flow in the wake of a circular cylinder in uniform translation. Part 1. Steady flow. J. Fluid Mech. **79**(2), 231–256 (1977)
8. Dennis, S.C.R., Chang, G.: Numerical solutions for steady flow past a circular cylinder at Reynolds number up to 100. J. Fluid Mech. **42**(3), 471–489 (1970)
9. Fadlun, E.A., Verzicco, R., Orlandi, P., Mohd-Yusof, J.: Combined immersed-boundary finite-difference methods for three-dimensional complex flow simulations. J. Comput. Phys. **161**, 35–60 (2000)
10. Fukuoka, H., Takeuchi, S., Kajishima, T.: Interaction between fluid and flexible membrane structures by a new fixed-grid direct forcing method. In: AIP conference proceedings, vol. 1702, p. 190014 (2015)
11. Ghias, R., Mittal, R., Dong, H.: A sharp interface immersed boundary method for compressible viscous flows. J. Comput. Phys. **225**, 528–553 (2007)
12. Goldstein, D., Handler, R., Sirovich, L.: Modeling a no-slip flow boundary with an external force field. J. Comput. Phys. **105**, 354 (1993)
13. Iaccarino, G., Verzicco, R.: Immersed boundary technique for turbulent flow simulations. App. Mech. Rev. **56**(3), 331–347 (2003)
14. Ikeno, T., Kajishima, T.: Finite-difference immersed boundary method consistent with wall conditions for incompressible turbulent flow simulations. J. Comput. Phys. **226**, 1485–1508 (2007)

15. Kajishima, T., Nomachi, T.: One-equation subgrid scale model using dynamic procedure for the energy production. J. Appl. Mech. **73**(3), 368–373 (2006)
16. Karagiozis, K., Kamakoti, R., Pantano, C.: A low numerical dissipation immersed interface method for the compressible Navier-Stokes equations. J. Comput. Phys. **229**, 701–727 (2010)
17. Lācis, U., Taira, K., Bagheri, S.: A stable fluid-structure-interaction solver for low-density rigid bodies using the immersed boundary projection method. J. Comput. Phys. **305**, 300–318 (2016)
18. Lai, M.C., Peskin, C.S.: An immersed boundary method with formal second-order accuracy and reduced numerial viscosity. J. Comput. Phys. **160**, 705–719 (2000)
19. Lee, C.: Stability characteristics of the virtual boundary method in three-dimensional applications. J. Comput. Phys. **183**, 559–591 (2003)
20. Li, X., Hunt, M.L., Colonius, T.: A contact model for normal immersed collisions between a particle and a wall. J. Fluid Mech. **691**, 123–145 (2012)
21. Linnick, M.N., Fasel, H.F.: A high-order immersed interface method for simulating unsteady incompressible flows on irregular domains. J. Comput. Phys. **204**, 157–192 (2005)
22. Liu, C., Zheng, X., Sung, C.H.: Preconditioned multigrid methods for unsteady incompressible flows. J. Comput. Phys. **139**, 35–57 (1998)
23. Mittal, R., Iaccarino, G.: Immersed boundary methods. Annu. Rev. Fluid Mech. **37**, 239–261 (2005)
24. Mohd-Yusof, J.: Development of immersed boundary methods for complex geometries. pp. 325–336. Annual Research Brief, Center for Turbulence Research, Stanford, CA (1998)
25. Nocedal, J., Wright, S.J.: Numerical Optimization. Springer (1999)
26. Perot, J.B.: An analysis of the fractional step method. J. Comput. Phys. **108**, 51–58 (1993)
27. Peskin, C.S.: Flow patterns around heart valves: a numerical method. J. Comput. Phys. **10**, 252–271 (1972)
28. Peskin, C.S.: Numerical analysis of blood flow in the heart. J. Comput. Phys. **25**, 220–252 (1977)
29. Peskin, C.S.: The immersed boundary method. Acta Numerica **11**, 479–517 (2002)
30. Roma, A.M., Peskin, C.S., Berger, M.J.: An adaptive version of the immersed boundary method. J. Comput. Phys. **153**, 509–534 (1999)
31. Roshko, A.: On the development of turbulent wakes from vortex streets. Tech. Rep. 1191, NACA (1954)
32. Saiki, E.M., Biringen, S.: Numerical simulation of a cylinder in uniform flow: application of a virtual boundary method. J. Comput. Phys. **123**, 450–465 (1996)
33. Sato, N., Takeuchi, S., Kajishima, T., Inagaki, M., Horinouchi, N.: A consistent direct discretization scheme on Cartesian grids for convective and conjugate heat transfer. J. Comput. Phys. **321**, 76–104 (2016)
34. Stockie, J.M.: Analysis and computation of immersed boundaries, with application to pulp fibres. Ph.D. thesis, Univ. British Columbia (1997)
35. Taira, K., Colonius, T.: The immersed boundary method: a projection approach. J. Comput. Phys. **225**, 2118–2137 (2007)
36. Tornberg, A.K., Engquist, B.: Numerical approximation of singular source terms in differential equations. J. Comput. Phys. **200**, 462–488 (2004)
37. Tritton, D.J.: Experiments on the flow past a circular cylinder at low Reynolds number. J. Fluid Mech. **6**, 547–567 (1959)
38. Udaykumar, H.S., Mittal, R., Rampunggoon, P., Khanna, A.: A sharp interface cartesian grid method for simulating flows with complex moving boundaries. J. Comput. Phys. **174**, 345–380 (2001)
39. Ye, T, Mittal, R., Udaykumar, H.S., Shyy, W.: An accurate Cartesian grid method for viscous incompressible flows with complex immersed boundaries. J. Comput. Phys. **156**, 209–240 (1999)

Chapter 6
Numerical Simulation of Turbulent Flows

6.1 Introduction

Numerical simulations of turbulent flows can be performed to capture (1) the temporal fluctuations and (2) the time-averaged features in the flow field. For example, let us consider a simulated flow through an asymmetric diffuser. The three-dimensional instantaneous flow field obtained from numerically solving the Navier–Stokes equations [36] is shown in Fig. 6.1. The computation captures flow separation and the existence of the recirculation zone in the diffuser. The streamwise velocity profiles from the computation are compared to experimental measurements in Fig. 6.2.

Although many engineering problems may only seek the time-averaged solution that does not mean that turbulent fluctuations can be ignored. Since the averaged flow and turbulent fluctuations are coupled due to the nonlinear nature of turbulence, the effect of turbulence appears in the averaged governing equations as *turbulent stress*. The model for such stress is referred to as a *turbulence model*. As it has been shown in the past literature, the numerical results can be strongly dependent on the choice of turbulence model, such as in the example of flow through diffusers.

Turbulent flow is comprised of vortices with a range of various sizes. Except in limited cases, directly simulating all scales of vortices is not possible due to the restrictions imposed by the current availability of computational resources. Furthermore, we have not been able to find a universal turbulence model that works for all turbulent flows. As a remedy to these challenges, we can solve for the large-scale vortices and provide a model for the small-scale vortices. This approach to solve turbulent flow is called *large-eddy simulation*. In what follows, we discuss different methods for simulating turbulent flows with focus on large-eddy simulation toward the end of this book.

© Springer International Publishing AG 2017
T. Kajishima and K. Taira, *Computational Fluid Dynamics*,
DOI 10.1007/978-3-319-45304-0_6

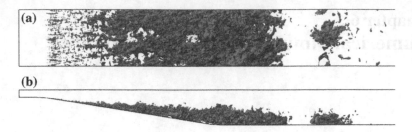

Fig. 6.1 Visualization of the instantaneous turbulent flow through an asymmetric diffuser captured by DNS [36] (Reprinted with permission from the Japan Society of Mechanical Engineers). **a** Top view. **b** Side view

Fig. 6.2 Streamwise velocity profiles from Fig. 6.1 compared to experimental measurements [36] (reprinted with permission from the Japan Society of Mechanical Engineers)

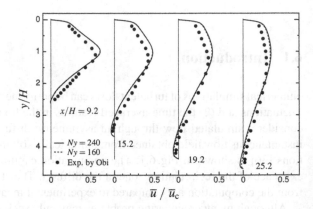

6.2 Direct Numerical Simulation of Turbulent Flows

6.2.1 Reynolds Number

For incompressible flows, the Navier–Stokes equation is

$$\frac{\partial u_i}{\partial t} + \frac{\partial (u_i u_j)}{\partial x_j} = -\frac{1}{\rho}\frac{\partial p}{\partial x_i} + \frac{\partial}{\partial x_j}(2\nu D_{ij}), \tag{6.1}$$

where the second term on the left-hand side is the advective (inertial) term and the second term on the right-hand side is the diffusive (viscous) term. The tensor

$$D_{ij} = \frac{1}{2}\left(\frac{\partial u_i}{\partial x_j} + \frac{\partial u_j}{\partial x_i}\right) \tag{6.2}$$

represents the rate-of-strain tensor. We can take the representative flow velocity and the size of the body to be the characteristic velocity U and length L, respectively. Based on these scales, the ratio between the inertial and viscous terms become

$$Re = \frac{U^2/L}{\nu U/L^2} = \frac{UL}{\nu}, \tag{6.3}$$

which is known as the *Reynolds number*. This is the most important non-dimensional parameter that characterizes viscous flows. If we were to describe turbulence in simple terms, it would be the three-dimensional high Reynolds number flow that exhibits irregular fluctuation.

Let us consider an alternative view of the Reynolds number by rewriting Eq. (6.3) as

$$Re = \frac{L^2/\nu}{L/U}. \tag{6.4}$$

The numerator and the denominator have units of time and represent the characteristic time it takes for a fluid element to travel over a distance L. The characteristic timescales for viscous diffusion and advection to convey information are L^2/ν and L/U, respectively. This equation tells us that for high Reynolds number flows, the time it takes for information to propagate by advection is much shorter than that by viscous diffusion.

The relative magnitude of inertial term with respect to that of viscous term is large for high Reynolds number flows as it appears in Eq. (6.3). This, however, does not imply that viscous effects are unimportant in turbulent flows. For flows with turbulence being prevalent, turbulent energy must be dissipated through heat by viscous effects. The stronger the turbulence is, the larger amount of turbulent energy should be lost through dissipation to balance the energy budget. The ratio in Eq. (6.3) only represents that the magnitude of the viscous term is small compared to the scales defined by U and L for high Reynolds number flows.

Now, let us consider the relationship between the Reynolds number Re and the length scales present in turbulent flow. Assuming that the flow is in a state that turbulence fluctuation is sustained, the supply of turbulent energy must be balanced by the dissipation of turbulent energy. For the supply side of turbulent energy, let us relate the turbulent energy supply per unit time ε to the wake scales represented by U and L (e.g., bluff body wake). The production of turbulent energy has the dimension of U^2 divided by a unit of time L/U:

$$\varepsilon = \mathcal{O}\left(\frac{U^3}{L}\right). \tag{6.5}$$

Next, consider the energy being dissipated with viscosity ν at a length scale of η. Based on these two parameters, dimensional analysis[1] leads to energy dissipation rate being

$$\varepsilon = \mathcal{O}\left(\frac{\nu^3}{\eta^4}\right). \tag{6.6}$$

---

[1] Additional discussion on dimensional analysis can be found in Problem 6.1.

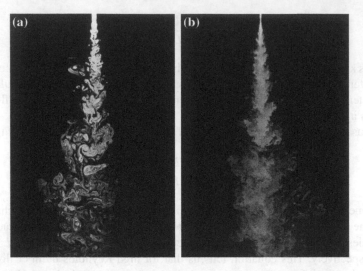

Fig. 6.3 Depiction of the influence of Reynolds number on jets. Wider range of length scales is observed for higher Reynolds number [9] (Reproduced with permission from Paul E. Dimotakis and Cambridge University Press). **a** Low Reynolds number. **b** High Reynolds number

Balancing the supply and dissipation of turbulent energy, we find that

$$\frac{L}{\eta} = \mathcal{O}\left(Re^{3/4}\right). \tag{6.7}$$

This tells us that the length scale at which dissipation takes place is much smaller than the characteristic length scale L of the flow with their ratio L/η being proportional to $Re^{3/4}$. As the Reynolds number of the flow is increased, the range of length scales observed in the flow becomes wider.

Illustrations based on laser-induced fluorescence imaging of turbulence jet flow for low and high Reynolds numbers are shown in Fig. 6.3 [9]. While the outer view of the jet may appear similar, the low Reynolds number jet lacks the small-scale structures that are present in high Reynolds number flows.

6.2.2   Full Turbulence Simulation

One of the features of turbulent flow is the wide range of scales in the irregularities we observe in space and time. Larger vortices generate the turbulent kinetic energy with scales of (L, T), which are passed on to smaller vortices with scales of (η, τ) until the energy is dissipated through heat. This flow of energy over the range of scales is called *energy cascade*. As shown in Eq. (6.7), the ratio of these scales L/η is related to the Reynolds number and takes a large value for turbulent flows.

If the governing equations for turbulent flow can be discretized with sufficient spatial resolution and high-order numerical accuracy along with appropriate initial and boundary conditions, one should be able to solve turbulent flow deterministically. Such perfect turbulent flow simulation is called *full turbulence simulation* (FTS). In order to simulate all scales in turbulent flows, the computational domain must be sufficiently larger than the largest characteristic length scale of the flow L and the grid size must be smaller than the finest turbulence scale η. Since turbulent flows are fundamentally three-dimensional phenomena, we would at least require $(L/\eta)^3$ grid points, which is proportional to $Re^{9/4}$.

Even for a relatively low Reynolds number of 10^4, FTS requires grid points at least on the order of $1000^3 = 10^9$. Some of the large-scale computations has been performed with $6.87 \times 10^{10} (=4096^3)$ points to simulate homogeneous turbulence in a periodic box [23] and 1.74×10^{11} points to analyze turbulent channel flow [1]. Most turbulent flows of engineering interest possess much higher Reynolds numbers with complex boundary surface geometries. The required computational resource becomes enormous also from the point of view of temporal resolution since the solution necessitates the number of time steps to be on the order of the number of grid points for numerical stability. For these reasons, FTS in most simulation is not a realistic approach for solving turbulent flows even with the use of supercomputers.

Suppose we have been able to secure computational resources with necessary speed and capacity to fully simulate turbulent flows. Even in that situation, predicting turbulence would be a challenge. One of the main reasons is the difficulty in preparing an appropriate initial condition that spans over all scales of vortices in the flow. While we can perform a large-scale simulation of the atmosphere for weather prediction, it would be impossible to collect weather data for setting up an exact initial condition. Even if an exact initial condition is available, the numerical calculation would have discretization and round-off errors. Since turbulent flows are chaotic due to the nonlinearity in the governing equations, even the smallest perturbation in the initial condition can grow large over time making the solution diverge from the exact solution.

A model that captures the essence of chaos in atmospheric flow called the *Lorenz system* [11, 28] and illustrates such behavior, as shown in Fig. 6.4. The Lorenz system can be described for example by the following set of equations:

Fig. 6.4 Numerical simulation of the Lorenz system [11, 28]. Even the smallest perturbation in the initial condition can generate solutions that are completely different from the unperturbed case

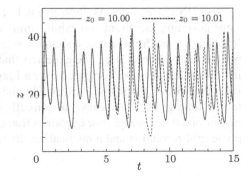

$$\frac{\mathrm{d}x}{\mathrm{d}t} = -10(x - y), \quad \frac{\mathrm{d}y}{\mathrm{d}t} = -xz + 28x - y, \quad \frac{\mathrm{d}z}{\mathrm{d}t} = xy - \frac{8}{3}z. \quad (6.8)$$

We can compare the numerical solution of these equations obtained with the Runge–Kutta method using initial conditions with $x_0 = y_0 = z_0 = 10.00$ and a similar initial condition with only z_0 slightly changed to 10.01. The comparison of $z(t)$ is shown in the figure. Even though the initial condition is different only by a small magnitude, the trajectories from the two initial conditions become strikingly different over time. Such phenomenon is called the Butterfly effect. While the overall pattern in the trajectory of the Lorenz system can be analyzed, it becomes rather impractical to accurately predict the solution at a particular time. This is especially true since numerical discretization errors can trigger changes in the trajectory in an unpredictable manner. The same observation can be made for turbulent flow simulation. While we can capture the qualitative evolution of turbulent flow, correctly predicting the precise state of turbulence using FTS for a particular location and time would be impossible and impractical for high Reynolds number flows.

6.2.3 Direct Numerical Simulation of Turbulence

Numerically solving the Navier–Stokes equations without any alterations is referred to as *direct numerical simulation* (DNS). For high Reynolds number flows, it is rare to fully solve the governing equations with grid points of $\mathcal{O}(Re^{9/4})$, based on Eq. (6.7), and is often not equivalent to FTS strictly speaking. Most DNS performed does not take into account the vortices with size smaller than the grid size. The reason why such simulations are able to obtain acceptable results is due to the majority of energy dissipation occurring at length scales about an order of magnitude larger than the smallest vortices predicted by theory. For the results from DNS to be meaningful, sufficient spatial resolution must be acquired to ensure that flow phenomena taking place at scales smaller than the grid resolution are negligible. For these reasons, DNS in many cases should be regarded as "highly accurate turbulent calculations without any use of turbulent models."

Because turbulence is essentially chaotic, it is not practical to predict the state of the flow at a particular time and position. This, however, does not defy the importance of DNS in any way. As illustrated in Fig. 6.4, the two solutions are based on the same governing equations and exhibit similar overall behavior over time, although local predictions from turbulent simulation can be influenced greatly even by small perturbations. Hence, it can be expected that DNS is capable of correctly capturing turbulence statistics computed over a long-time interval. Furthermore, for each individual vortex, its motion is expected to be accurately reproduced over its characteristic time scale. This means that correctly performed DNS can reproduce vortex motion and time average flow quantities that characterize the flow field. With appropriate grid resolution and numerical accuracy, DNS can be a trustworthy research tool for obtaining fundamental turbulence data. For DNS, the aforementioned high-order accurate finite-difference methods and spectral methods [4, 13] are often used with appropriate grid resolutions.

6.2.4 Turbulence Simulation with Low Grid Resolution

Let us discuss what the consequence is for performing DNS on a coarse grid. Under this circumstance, the influence from vortices smaller than the mesh width would not be accounted for, although vortices over a range of spatial scales should interact in turbulent flow. The flow simulation in this case would not correctly capture the flow physics at scales larger than grid resolution due to the missing interaction. Without the small-scale vortices responsible for dissipating kinetic energy, the produced turbulent kinetic energy must be dissipated at scales resolvable on the grid. Thus, turbulence represented over the wave numbers resolvable by the simulation would be distorted from reality.[2] Consequently, the Reynolds stress would be affected and in turn may alter the average velocity profile as well. Since larger scale vortices are not as dissipative as smaller scale vortices, the energy over the computed wave numbers would likely be higher in comparison to actual turbulent flows. If grid resolution is grossly insufficient, the energy accumulated over time can crash the simulation.

As an example, let us consider the influence of grid resolution on simulation of turbulent flow between two parallel flat plates (channel flow). Turbulent channel flow is one of the most fundamental flows with wall turbulence and has been studied extensively as a standard problem to validate numerical calculations. LES performed by Deardorff [7], Schumann [42], Moin and Kim [34] as well as DNS by Kim, Moin, and Moser [25] have highlighted the value of numerical simulation and are regarded as milestones in the field of CFD.

Let us consider the fully-developed turbulent flow between two parallel flat plates with constant pressure gradient in the x-direction, as illustrated in Fig. 6.5. If we view the pressure gradient required to maintain the flow as external forcing in the momentum equation, only the pressure fluctuation needs to be solved. Setting the channel width to be 2δ and the *friction velocity* to be $u_\tau = \sqrt{\tau_w/\rho}$, where τ_w is the average shear stress at the wall, the average pressure gradient is $\partial \overline{p}/\partial x = -\tau_w/\delta$. Non-dimensionalizing the governing equations using u_τ, δ, and ν, we obtain

$$\frac{\partial u_i^*}{\partial x_i^*} = 0, \quad \frac{\partial u_i^*}{\partial t} + u_j^* \frac{\partial u_i^*}{\partial x_j^*} = \delta_{i1} - \frac{\partial p'^*}{\partial x_i} + \frac{1}{Re_\tau} \frac{\partial^2 u_i^*}{\partial x_j^* \partial x_j^*}, \qquad (6.9)$$

where the superscript $*$ denotes non-dimensional quantities. For the simulation, the only physical parameter to be specified is the Reynolds number $Re_\tau = u_\tau \delta / \nu$. For each of the velocity components and the pressure fluctuation, we can apply the

---

[2]The lowest wave number is determined by the computational domain size and the highest cutoff wave number is determined by grid resolution.

Fig. 6.5 Turbulent flow
between parallel flat plates
(channel flow)

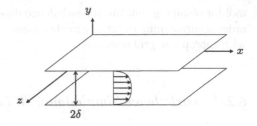

periodic boundary conditions in the streamwise (x) and spanwise (z) directions. At the wall, we prescribe the no-slip boundary condition of $u^* = v^* = w^* = 0$. In what follows, we drop the superscript $*$ to denote non-dimensional variables for simplicity.

Most DNS performed for channel flow in early days of CFD has been for Reynolds number Re_τ in the range of 150–180. Recently, there has been a simulation with $Re_\tau = 1020$ [1], but the Reynolds numbers examined are still low for industrial applications. For this reason, the use of most DNS at this moment is not intended for direct applications in practical engineering problems, but for supplying fundamental data to validate turbulence models. Hence, there are ongoing efforts to obtain turbulence data at higher Reynolds number so that researchers can gain insight into turbulent flows without low Reynolds number effects. A database of DNS results from spectral method calculations is available on the Internet for Reynolds number of $Re_\tau = 180$, 395, and 590 [35].

Next, let us estimate the number of grid points necessary to perform a DNS of turbulent channel flow. For the x and z directions, in which we can apply periodic boundary conditions, the size of the domain must be chosen such that two-point correlations of any fluctuations can be negligible. Furthermore, the grid resolution should be sufficiently finer than the smallest scale of turbulent structures. For example, streak-like structures where regions of fast and low flows appear alternatively in the vicinity of wall have average length scales of $\lambda_x^+ \approx 1000$ and $\lambda_z^+ \approx 100$. Here, the superscript $+$ represents the wall coordinate non-dimensionalized by ν/u_τ, which yields $\delta^+ = Re_\tau$.

The grid resolution in the spanwise direction can influence the solution significantly. If we need to compute up to the second-order moment $\overline{u_i' u_j'}$, it would be sufficient to choose $\Delta z^+ \approx 10$ for the second-order central-difference schemes. If accurate predictions for the third or fourth-order moments (i.e., $\overline{u_i' u_j' u_k' u_l'}$) are necessary, one should set $\Delta z^+ < 5$. These guidelines can be relaxed for higher-order central-difference schemes or spectral methods. For the spanwise extent H_z of the computational domain, at least two to three times the width of the channel (i.e., 4δ to 6δ) should be chosen.

The grid size in the streamwise direction Δx can be larger than Δz, but excessive difference in their sizes can lead to anisotropy in the mesh which should be avoided. Hence, it would be desirable to set Δx to be around $2\Delta z$ to $4\Delta z$. In the streamwise direction, the extent H_x should be about 8δ to 16δ. While this choice in not large compared to λ_x, it does not appear to show much influence on the turbulence

Table 6.1 Grid setup for turbulent channel flow simulation ($Re_\tau = 150$)

N_x	N_y	N_z	H_x/δ	H_y/δ	H_z/δ	Δx^+	Δy^+	Δz^+
4	64	4	7.68	2	3.84	288	0.9 $\sim$ 9	144
8	64	8	7.68	2	3.84	144	0.9 $\sim$ 9	72
16	64	16	7.68	2	3.84	72	0.9 $\sim$ 9	36
32	64	32	7.68	2	3.84	36	0.9 $\sim$ 9	18
64	64	64	7.68	2	3.84	18	0.9 $\sim$ 9	9
128	64	128	7.68	2	3.84	9	0.9 $\sim$ 9	4.5

statistics. For the wall-normal direction, there should be a few points to resolve the viscous sublayer generated by the no-slip boundary condition at the wall. Hence, the minimum grid size there should be $\Delta y^+_{\min} < 1$. At the center of the channel, the wall-normal grid size Δy can be made larger to about Δz to $1.5\Delta z$.

For resolving turbulence with DNS, the required number of grid points are the ratios of the necessary extent of the domains and sizes of the grid. For all directions, the ratios are estimated to be $Re_\tau/3$ to Re_τ, which result in the total number of points be between $(Re_\tau/3)^3$ and Re_τ^3 for three-dimensional simulations. We should note that the discussed grid size is not smaller than the Kolmogorov scale.

To study the influence of grid resolution, we perform DNS of channel flow at Re_τ using second-order central-difference method on a staggered grid. Simulations are performed with the parameters listed in Table 6.1. While fixing the computational domain size, the number of grids points N_x and N_z are varied in the streamwise and spanwise directions, respectively. To prescribe the no-slip boundary condition, the number of points in the wall-normal direction N_y is kept constant.

The first simulation considered is performed for a long period of time with a 64^3 grid to obtain fully-developed turbulence. From a snapshot of the developed turbulent flow field, data are spatially extracted or interpolated for different grids and are in turn used as initial conditions for a set of longtime simulations to find the time-averaged flow quantities.

Let us show the average velocity profile in Fig. 6.6 where $N = N_x = N_z$. In the figure is the result from DNS using spectral method with a highly resolved mesh [24], shown for reference. For the second-order accurate central-difference scheme, we employ a scheme with conservation properties for the kinetic energy so that the simulation does not blow up even with coarse grids. It can be noticed that for the most resolved case with $N = 128$, the result is in good agreement with that from the spectral calculation. However, we should point out that the velocity flux decreases in the order of $N = 8, 4, 16, 128, 64, 32$. Even if we exclude cases of $N_x \leq 16$ that are grossly under-resolved, the convergence of the average velocity profile based solely on refining grid resolution is not monotonic. For DNS, examining grid dependence by simply changing the mesh size is not an easy task.

For this flow field, the turbulence production rate is $-\overline{u'v'}(\partial \overline{u}/\partial y)$. The generated fluctuation is fed into the streamwise velocity component, distributed to the other two velocity components, and is then transferred to small-scale vortices for dissipation.

Fig. 6.6 Effect of grid
resolution on the average
velocity profile of turbulent
channel flow

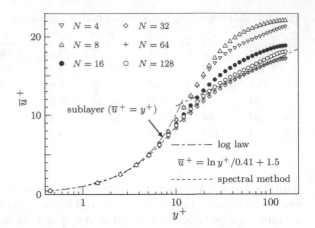

We illustrate in Fig. 6.7 the strength of the velocity fluctuation (turbulence intensity) in the streamwise $u_{rms} = \sqrt{\overline{u'^2}}$ and wall-normal $v_{rms} = \sqrt{\overline{v'^2}}$ components. For these quantities, the convergence is mostly monotonic for decreasing the mesh size. When the grid is coarse, fluctuation in the streamwise direction increases while the fluctuation in the wall-normal direction decreases. The spanwise fluctuation is observed to have the same trend as in the wall-normal fluctuation (not shown in figure).

Under-resolved simulations exhibit higher energy and anisotropy than what are expected. The energy to be dissipated by the small-scale vortices are retained by the larger vortices for the unresolved cases resulting in higher energy in the flow. Furthermore, the energy passed on from the streamwise direction to the other directions is inhibited due to the unaccounted vortex dynamics on the unresolved scales that would have been responsible for the energy transfer.

To determine the average velocity and turbulence intensity for Reynolds number of $Re_\tau = 150$, it appears that a mesh of 64^3 is sufficient based on Fig. 6.7. For correctly capturing higher-order moments, a finer grid would be required. As an example of the third-order moment of velocity fluctuation, the skewness of wall-normal velocity fluctuation $v_{skew} = \overline{v'^3}/\overline{v'^2}^{3/2}$ is shown in Fig. 6.8. Here, we require at least $N = 128$ to match the results from the spectral calculation near the wall.

Next, let us discuss the influence of the Reynolds number. Increasing the Reynolds number without modifying the mesh width results in lowering the grid resolution. We now fix the grid size at 64^3 and vary the Reynolds number as $Re_\tau = 150, 300, 450,$ and 600. The computational domain size and the mesh widths are the same as in Table 6.1 such that $H_x = 7.68\delta$, $H_z = 3.84\delta$, $\Delta x = 0.12\delta$, and $\Delta z = 0.06\delta$. The domain size and grid resolution change when represented in the wall coordinate according to the change in Reynolds number as shown in Table 6.2.

The average velocity profiles are shown for varied Reynolds numbers in Fig. 6.9 along with the results from spectral DNS calculations performed at $Re_\tau = 150$ [24] and 395 [35]. The Reynolds number based on the average centerline velocity u_c is

Fig. 6.7 Effect of grid resolution on the RMS values of u^+ and v^+ velocity profiles of turbulent channel flow. Velocity fluctuations in **a** streamwise and **b** wall-normal directions

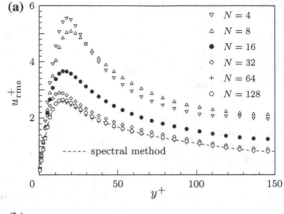

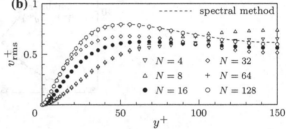

Fig. 6.8 Effect of grid resolution on the skewness of fluctuating wall-normal velocity in turbulent channel flow

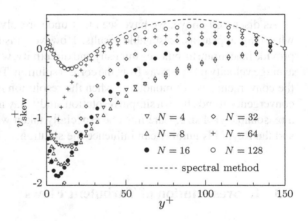

$Re_c = \delta u_c/\nu = 2604, 5266, 8473$, and $12{,}014$ for $Re_\tau = 150, 300, 450$, and 600, respectively. With the second-order central-difference method, the velocity profile has the largest deficit in the log-law region at $Re_\tau = 300$ for the considered values of Re_τ. On the other hand, the velocity in the log law region is high for $Re_\tau = 150$ due to low Reynolds number effect and for $Re_\tau = 450$ and 600 due to the lack of grid resolution. This tells us that the influence of Reynolds number and the effect of low grid resolution appear in a mixed manner. We therefore must choose the grid resolution with care to correctly assess the effect of Reynolds number.

Table 6.2 Changes in grid resolution for turbulent channel flow with varied Reynolds number

Re_τ	H_x^+	H_y^+	H_z^+	Δx^+	Δy^+	Δz^+
150	1152	300	576	18	$0.9 \sim 9$	9
300	2304	600	1152	36	$1.8 \sim 18$	18
450	3456	900	1728	54	$2.8 \sim 27$	27
600	4608	1200	2304	72	$3.7 \sim 36$	36

Fig. 6.9 Effect of Reynolds number on average velocity profile of turbulent channel flow

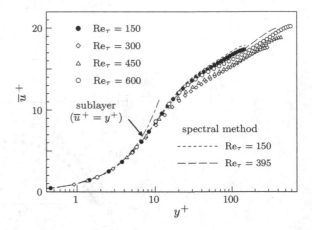

As discussed above, we have seen that under-resolved simulations can produce what may appear as reasonable results. However, insufficient grid resolution can provide inaccurate turbulence intensity and directivity, which may noticeably shift the average velocity profile away from the correct solution. To make matters complicated, the convergence is not monotonic when the resolution is increased. For this reason, convergence trends from a simple resolution study may not be sufficient to determine a reasonable grid size. Care must be especially taken when both the grid resolution and the Reynolds number can influence the solution.

6.3 Representation of Turbulent Flows

6.3.1 Turbulence Models

In most applications, the large-scale behavior of turbulent flow is sought, instead of the full turbulence data in detail. In general, it is known from experimental findings and theory that large-scale vortices in the flow are strongly influenced by the boundary geometry, while the small-scale vortices are more universal, isotropic, and dissipative. For the purpose of engineering calculations, capturing the large-scale vortical fluctuations is important since they affect the average flow quantities. Therefore, it

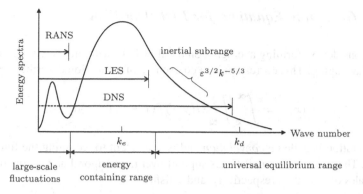

Fig. 6.10 Length-scale coverage by different approaches of turbulence simulation

becomes useful to numerically solve the averaged governing equation that contains the influence of turbulence that is not resolved by the grid. The equations to be solved are obtained by averaging the Navier–Stokes equations in some manner. In the averaged equations, we must relate the averaged flow and the turbulent vortices, which calls for the use of *turbulence models*. Illustrated in Fig. 6.10 is an example energy spectra for a given flow field that contains large-scale (non-turbulent) fluctuations. We describe here the range of wave numbers for which different turbulence simulation methods are employed. Approximate wave number range for each turbulence modeling approach follows Hinze [14].

If we only need to solve for the steady flow (or quasi-steady flow, where the unsteadiness is slow compared to turbulence), the whole turbulence fluctuation needs to be modeled. In such situations, the Navier–Stokes equations are *Reynolds averaged*, which will be discussed in detail in Chap. 7. Solving for the Reynolds-averaged flow field is referred to as the *Reynolds-averaged Navier–Stokes* computation or the *Reynolds-averaged numerical simulation* (RANS). It should be noted that there can be large-scale fluctuations caused by coherent flow structures that RANS needs to capture, as depicted in Fig. 6.10.

Another approach for solving turbulent flow is to resolve the large-scale vortices and model the influence from the small-scale vortices, which is called the *Large-Eddy Simulation* (LES). The governing equations for LES are derived by filtering the equations to separate the large and small-scale structures in the flow. In LES, we can define the larger scale vortices to have sizes larger than the grid resolution since one can capture structures with the period being twice the grid size (as illustrated in Fig. 2.10). We therefore refer to the filtering operation in LES as *grid averaging*, as opposed to the Reynolds averaging.

6.3.2   Governing Equations for Turbulent Flow

Let us consider performing averaging on the flow field without specifying the exact
type of averaging. This corresponds to coarsening of a function described by

$$\bar{f}(x, t) = \int_{-\infty}^{\infty} \int_{-\infty}^{\infty} G(y, s) f(x - y, t - s) \, dy \, ds. \tag{6.10}$$

This operation is called a *convolution* and is equivalent to weighing the function f
with G. The function (*kernel*) G has support over the temporal and spatial domains
with scales of T and L, respectively, and satisfies

$$\int_{-\infty}^{\infty} \int_{-\infty}^{\infty} G(y, s) \, dy \, ds = 1. \tag{6.11}$$

With this convolution kernel, the flow field is coarsened (smoothed out over a coarser
grid) over with scales of T and L. We denote the fluctuation removed from such
operation as $f' = f - \bar{f}$.

Performing a weighted average of the continuity equation, we find

$$\frac{\partial \bar{u}_j}{\partial x_j} = 0, \tag{6.12}$$

which is in the same form as the original continuity equation. The momentum equation, Eq. (6.1), with weighted averaging becomes

$$\frac{\partial \bar{u}_i}{\partial t} + \frac{\partial \overline{u_i u_j}}{\partial x_j} = -\frac{1}{\rho} \frac{\partial \bar{p}}{\partial x_i} + \frac{\partial}{\partial x_j} (2\nu \bar{D}_{ij}), \tag{6.13}$$

where

$$\bar{D}_{ij} = \frac{1}{2} \left(\frac{\partial \bar{u}_i}{\partial x_j} + \frac{\partial \bar{u}_j}{\partial x_i} \right) \tag{6.14}$$

is the velocity gradient tensor based on the averaged velocity field. We have assumed
here that the averaging and derivative operations are commutative

$$\frac{\overline{\partial f}}{\partial t} = \frac{\partial \bar{f}}{\partial t}, \quad \frac{\overline{\partial f}}{\partial x} = \frac{\partial \bar{f}}{\partial x}. \tag{6.15}$$

If the weight function (kernel) G is not uniform in space or time, the above formulation carries commutation errors.

Solving Eqs. (6.12) and (6.13) as the governing equations, the variables to be
solved are the averaged velocity $\bar{u}_i$ and the pressure $\bar{p}$, which are four variables in
three-dimensional calculations. There are four equations to be solved, consisting of

the three momentum equations and the continuity equation. Now, we point out that Eq. (6.13) has a tensor $\overline{u_i u_j}$, which can be expanded as

$$\overline{u_i u_j} = \overline{\bar{u}_i \bar{u}_j} + \overline{\bar{u}_i u'_j} + \overline{u'_i \bar{u}_j} + \overline{u'_i u'_j}. \tag{6.16}$$

We can accordingly rewrite Eq. (6.13) using $\bar{u}_i$ and $\bar{p}$ in a form of the Navier–Stokes equations to arrive at

$$\frac{\partial \bar{u}_i}{\partial t} + \frac{\partial (\bar{u}_i \bar{u}_j)}{\partial x_j} = -\frac{1}{\rho}\frac{\partial \bar{p}}{x_i} + \frac{\partial}{\partial x_j}(-\tau_{ij} + 2\nu \overline{D}_{ij}). \tag{6.17}$$

Utilizing Eq. (6.15), the above equation can alternatively be expressed as

$$\frac{\overline{D}\bar{u}_i}{\overline{D}t} = -\frac{1}{\rho}\frac{\partial \bar{p}}{\partial x_i} + \frac{\partial}{\partial x_j}\left(-\tau_{ij} + 2\nu \overline{D}_{ij}\right), \tag{6.18}$$

where

$$\frac{\overline{D}}{\overline{D}t} = \frac{\partial}{\partial t} + \bar{u}_j \frac{\partial}{\partial x_j} \tag{6.19}$$

is the material derivative based on the averaged velocity. In Eqs. (6.17) and (6.18), the term τ_{ij} is generated due to the averaging operation and appears in a form of a stress tensor

$$\begin{aligned}\tau_{ij} &= \overline{u_i u_j} - \bar{u}_i \bar{u}_j \\ &= \overline{\bar{u}_i \bar{u}_j} - \bar{u}_i \bar{u}_j + \overline{\bar{u}_i u'_j} + \overline{u'_i \bar{u}_j} + \overline{u'_i u'_j}\end{aligned} \tag{6.20}$$

In order to close the system of equations, τ_{ij} needs to be related to variables $\bar{u}_i$ and $\bar{p}$ in some manner, so that the number of equations and the number of unknowns are matched.

6.3.3 Turbulence Modeling Approaches

In the early days of turbulence modeling, most models consisted of relations based on intuition or experience with a set of coefficients that were tuned based on empirical data. Since then, theoretical turbulence research has further supported the development of turbulence models. One such model based on the continuum mechanics approach is discussed by Speziale [44] (instead of the statistical approach). This method derives the model by incorporating the constraint in continuum mechanics into the Taylor series expansion or dimensional analysis of the turbulence fluctuation. Methods that are based on the theory of turbulent statistics are the *two-scale direct interaction approximation* (TS-DIA) [52, 53] and the *renormalization group* (RNG) [51]. With the use of theory, not only the modeling methodology is provided but the constants in the models also can be estimated in some cases.

With the availability of turbulence databases from DNS, development and validation of improved turbulence models have become possible. It is still not easy to measure the high-order turbulent statistics needed for the Reynolds-averaged models or the scale-separated turbulence statistics required for validating LES models. In the model equations, not all terms may necessarily offer clear physical interpretations. In such scenario, DNS can become a powerful tool to gain further insights into the characteristics of the physical quantity being modeled by providing access to the sufficiently resolved flow field data. Based on the knowledge from DNS, we are able to examine whether the components of the turbulence models are physically appropriate, rather than only checking if the computational results match the measurements from experiments.

6.3.4 Visualization of Vortical Structures

To characterize turbulent flow, statistical quantities such as the time-average, fluctuation amplitude, skewness, and flatness are used. For studying the spatial structures in turbulent flow, visualizing isocontours and plotting streamlines are useful. Furthermore, virtual particles can be traced in the numerical flow field to visualize the particle paths and streaklines, which can be validated against experimental visualizations.

Because turbulent flow is characterized by a large number of vortices interacting in a highly complex manner, visualization of the vortical structures can facilitate the understanding of vortex dynamics in turbulent flow and provide insights into the development of turbulence models. In this section, we discuss some of the techniques used to visualize vortices.

The strength of the vortices can be quantified by the vorticity vector

$$\omega_k = \epsilon_{kmn} \frac{\partial u_n}{\partial x_m}, \tag{6.21}$$

where ϵ_{ijk} is the permutation symbol (see Eq. (A.11)). It however does not necessarily represent the strength of rotational motion. As illustrated in Fig. 6.11, vorticity can be present in flows with shear or rotation. Region with shear contains vorticity due to the change in relative angles and stretching between fluid elements as shown in Fig. 6.11a. Fluid moving as rigid-body rotation also possesses vorticity, as illustrated in Fig. 6.11b. In order to capture vortices in the flow, it would hence be sensible to extract flow features that contain higher level of rotation in comparison to shear.

We can decompose the velocity gradient tensor into symmetric and antisymmetric components

$$\frac{\partial u_i}{\partial x_j} = D_{ij} + W_{ij}, \tag{6.22}$$

where the symmetric tensor $D_{ij} = \frac{1}{2}(\partial u_i/\partial x_j + \partial u_j/\partial x_i)$ is the rate-of-strain tensor and the antisymmetric tensor $W_{ij} = \frac{1}{2}(\partial u_i/\partial x_j - \partial u_j/\partial x_i)$ is the rotation (vorticity)

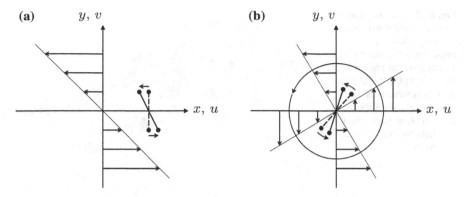

Fig. 6.11 Flows that contain vorticity $\omega_z = \frac{\partial v}{\partial x} - \frac{\partial u}{\partial y} > 0$ and the motion of fluid element. **a** Simple shear flow ($-u \propto y$). **b** rigid-body rotation ($u_\theta \propto r$)

tensor. We define the norms of the rate-of-strain tensor, rate-of-rotation tensor, and vorticity vector in the following manner

$$|\mathbf{D}| = \sqrt{2D_{ij}D_{ij}}, \quad |\mathbf{W}| = \sqrt{2W_{ij}W_{ij}}, \quad |\boldsymbol{\omega}| = \sqrt{\omega_k\omega_k}. \tag{6.23}$$

Note that the rate-of-rotation tensor consists of vorticity elements and are related by

$$W_{ij} = -\epsilon_{ijk}\frac{\omega_k}{2}. \tag{6.24}$$

With the above norm definitions, we have $|\mathbf{W}| = |\boldsymbol{\omega}|$.

Let us express the difference between the rotation and strain present in the flow with

$$|\mathbf{W}|^2 - |\mathbf{D}|^2 = |\boldsymbol{\omega}|^2 - |\mathbf{D}|^2 = -2\frac{\partial u_i}{\partial x_j}\frac{\partial u_j}{\partial x_i}. \tag{6.25}$$

It should be observed that the pressure Poisson equation that is derived from taking the divergence of the momentum equation, Eq. (6.1), is

$$\frac{1}{\rho}\frac{\partial^2 p}{\partial x_i\partial x_i} = -\frac{\partial u_i}{\partial x_j}\frac{\partial u_j}{\partial x_i}, \tag{6.26}$$

which is half of Eq. (6.25), with the assumption of constant density and viscosity and absence of external forcing. We can use the above quantity to characterize vortices to be present in regions where rotational motion is dominant ($|\mathbf{W}|^2 \gg |\mathbf{D}|^2$) with a local pressure minimum. For vortical flows, the low-pressure region usually corresponds to region with large magnitude of rotational velocity. For these reasons, the second

Fig. 6.12 Visualization of laminar vortical structures forming around an impulsively translating low aspect ratio flat plate wing at $Re = 300$ [45]. *Light* and *dark gray* structures correspond to isosurfaces of $|\omega|$ and Q, respectively, to highlight the vortex sheet and cores

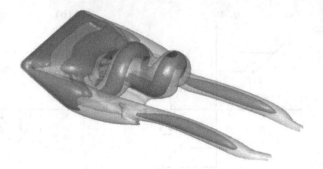

invariant[3] of the velocity gradient tensor, which is also referred to as the Q-value (criterion):

$$Q = \frac{1}{4}(|W|^2 - |D|^2) = -\frac{1}{2}\frac{\partial u_i}{\partial x_j}\frac{\partial u_j}{\partial x_i} \tag{6.27}$$

is often used to capture regions that correspond to the core of vortices [16]. For the simple shear flow example shown in Fig. 6.11a, we have $Q = 0$ with $|W| = |D|$. On the other hand for the rigid-body rotation case shown in Fig. 6.11b, $Q > 0$ with $|W| > |D| = 0$.

By visualizing the vortices using the Q-value isocontour, it is possible to gain the dynamic features of vortical and turbulent flows. One can also consider showing the isocontour of the Q-value colored by other physical quantities such as $|\omega|$, $|u|$, or by their components in particular directions. It should be kept in mind that low-pressure regions in the flow do not necessarily correspond to vortex cores and hence the Q-criterion may misidentify some regions of flow as vortex cores. These issues in capturing vortical structures are considered in Jeong and Hussain [19]. Another commonly employed vortex identification scheme called Λ_2 is also discussed and is compared to the Q-value in their paper. Generally speaking, the use of Q and Λ_2 isocontours leads to similar vortex visualizations.

Visualization Examples

Let us consider two examples for which the Q-value isosurface contours are utilized to visualize the flow field. The first example is the formation of laminar vortices around a low-aspect-ratio rectangular flat plate wing in impulsive translation at $Re = 300$, as shown in Fig. 6.12. The Q-value isosurface is used to capture the vortex cores.[4] The isosurface of $|\omega|$ is also shown with transparency to highlight the behavior of the vortex sheet. The combination of these two isosurfaces shows how the vortex sheets roll up around the wing and form the leading-edge and tip vortices shortly after the wing undergoes the impulsive motion [45].

---

[3] See Problem 6.4.

[4] The 3D printed model of the vortices around a pitching wing shown in Fig. 1.1 is generated with the Q-value isosurface.

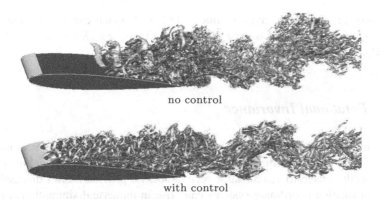

no control

with control

Fig. 6.13 Visualization of vortical structures in turbulent flow over a NACA0012 airfoil at $Re = 23,000$. *Top* Baseline flow and *Bottom* Controlled flow. The Q-value isosurfaces are colored by the streamwise vorticity to highlight the effect of flow control actuators in pulling in high-momentum flow toward the airfoil surface to prevent flow separation (shown by *darker shade of gray*)

The second example uses the Q-value to represent the changes in turbulent flow over a NACA0012 airfoil from the application of flow control. Small rotating jets are introduced near the natural separation point to reattach the flow to achieve lift enhancement and drag reduction. The Q-value isosurface is colored by the streamwise vorticity to show that the large spanwise vortices in the uncontrolled case can be broken up by adding streamwise perturbations and enhancing mixing between the outer high-momentum fluid with near-wall low-momentum fluid, as presented in Fig. 6.13. By pulling the high-momentum fluid closer the wing surface, separation can be mitigated.

6.3.5 Coherent Structure Function

To obtain a sense of magnitude of the Q-value, we can consider the companion quantity

$$E = \frac{1}{4}(|\mathbf{W}|^2 + |\mathbf{D}|^2) = \frac{1}{2}\frac{\partial u_i}{\partial x_j}\frac{\partial u_i}{\partial x_j}. \tag{6.28}$$

This quantity E is related to the dissipation rate of kinetic energy Φ defined in Eq. (3.173) through $\Phi = 2\nu E$. The nondimensional Q-value normalized by E is defined as

$$F = \frac{Q}{E} \tag{6.29}$$

and is called the *coherent structure function*. A vortex with relatively large F corresponds to a columnar vortex that has strong rotation with low kinetic energy dissipation. In general, F takes a value of $|F| < 1$ which can be handled easily without special scaling and is used in turbulence modeling (*Kobayashi model*, see Sect. 8.6.2) [26].

6.3.6 Rotational Invariance

A quantity or function is called rotationally invariant if its value is not influenced by the rotation of the coordinate system. While the concept of rotational invariance is not directly related to visualization techniques, we provide a brief discussion as the use of rotating coordinate systems can arise in numerical simulations of fluid flow (e.g., turbomachinery). For a coordinate system under rotation, the rate-of-strain tensor remains rotationally invariant while the velocity gradient tensor, rate-of-rotation tensor, and vorticity vector do not. If these tensors are used for constructing turbulence models, the influence of rotation should be treated with care.

Let us consider observing the motion of material point X using two different coordinate systems, x and x^*, that are related by $x^* = R(t)x + b(t)$. Here, x and x^* are in inertial and non-inertial reference frames, respectively. The orthogonal tensor representing rigid-body rotation $R(t)$ and an arbitrary vector $b(t)$ are both functions solely of time t.

How physical phenomenon is perceived is dependent on the motion of the observer. However, material properties are *objective* or independent of the motion of the observer. This principle is known as the *principle of material frame indifference* or the *principle of material objectivity*. For a scalar f, vector q, and tensor T to be objective, they must satisfy

$$f^* = f, \quad q^* = Rq, \quad T^* = RTR^T, \tag{6.30}$$

where R^T is the transpose of R with R satisfying $R^T = R^{-1}$ for an orthogonal tensor. Again the superscript of asterisk denotes the non-inertial frame.

Denoting the velocity vectors with $v = \dot{x}$ and $v^* = \dot{x}^*$ for the two frames of reference, we find that $v^* = Rv + \dot{R}x + \dot{B}$. Thus, velocity vectors are clearly not objective.

Next, let us examine the velocity gradient tensors $L = \partial v / \partial x$ and $L^* = \partial v^* / \partial x^*$ for inertial and non-inertial systems, respectively. Noticing that $\partial x / \partial x^* = R^T$, we find

$$L^* = RLR^T + \dot{R}R^T = RLR^T + Q, \tag{6.31}$$

where the tensor Q represents the rotation of the system and is associated with the angular velocity vector $\dot{\theta}$ as an alternating tensor. From the above equation, we observe that the velocity gradient tensor is not objective. Let us also examine the rate-of-strain tensor $D = (L + L^T)/2$ and $D^* = (L^* + L^{*T})/2$ for inertial and non-inertial systems, respectively. Noting that $d(RR^T)/dt = dI/dt = 0$, we have

$$D^* = RDR^T. \tag{6.32}$$

Hence, we see that the rate-of-strain tensor is objective. At last, let us consider the rate-of-rotation tensors $W = (L - L^T)/2$ and $W^* = (L^* - L^{*T})/2$ for the inertial and non-inertial systems, respectively. It can be shown that

$$W^* = RWR^T + Q, \tag{6.33}$$

which tells us that the rate-of-rotation tensor as well as the associated vorticity vector are not objective.

In summary, we find that

$$L^* = L' + Q, \quad D^* = D', \quad W^* = W' + Q, \tag{6.34}$$

where we denote a tensor under rotational transform R as $T' = RTR^T$ for a tensor T.

Using the rotational velocity $\dot{\theta}$ of the coordinate system, we can express in indicial notation

$$W_{ij}^* = W_{ij}' + Q_{ij} = W_{ij}' - \epsilon_{ijk}\dot{\theta}_k^*. \tag{6.35}$$

Therefore, we can note that the following relationship with the rate-of-rotation tensor (angular velocity)

$$W_{ij}' = W_{ij}^* + \epsilon_{ijk}\dot{\theta}_k^* \tag{6.36}$$

should be utilized if non-objective tensors are considered as part of a physical model in a non-inertial reference frame.

For additional details, readers can refer to detailed discussion on transformation properties in Sect. 2.9 of Pope [37].

6.3.7 Modal Decomposition of Turbulent Flows

Post-processing a large volume of data generated from simulation of turbulent fluid flow to gain physical insights can be a challenge. The three-dimensional complex multi-scale dynamics and the spatial resolution requirement for turbulent flows lead to the generation of large data sets. In this section, we discuss data-based modal decomposition techniques that can extract spatial structures in complex flow fields using systematic approaches based on statistics and operator theory. In particular, we can examine spatial and temporal correlations or dynamical properties of the flow field and extract spatial modal structures that highlight regions where important flow physics take place. Modal decompositions [27] can be performed with techniques, such as the *Proper Orthogonal Decomposition*[5] (POD) [3, 5, 15, 20] and the *Dynamic Mode Decomposition* (DMD) [33, 39, 40]. Here, we only provide a brief discussion

---

[5]Also known as the *principal component analysis* (PCA) and the *Karhunen-Loève expansion*.

on POD and DMD but offer Appendix C for additional details. While we discuss modal decomposition methods here in the context of analyzing unsteady fluid flow data from numerical simulation, it should be noted that these techniques can be applied to various types of data, including those from experimental fluid mechanics.

Proper Orthogonal Decomposition (POD)

Lumley first used POD to analyze coherent structures in turbulent flows [29]. Since then this decomposition technique has been used extensively to study a variety of fluid flows [2, 3, 8, 50]. POD finds spatial modes that capture the coherent structures with dominant fluctuations in the flow based on spatiotemporal correlations. The instantaneous flow field can be represented in vector form $x(t)$ and their collection over time can be arranged in a matrix form

$$X = [x(t_1), \cdots, x(t_m)] \in \mathbb{R}^{n \times m} \tag{6.37}$$

(e.g., velocity or vorticity field; see Sect. 1.1 for details). Here, the data matrix can include weights to take temporal weights into consideration [38] but is omitted for simplicity. The POD modes are determined by solving for the eigenvectors of the correlations matrix XX^T

$$XX^T \phi_k = \lambda_k \phi_k, \quad \lambda_1 \geq \ldots \lambda_n \geq 0. \tag{6.38}$$

Here, the eigenvectors ϕ_k have corresponding eigenvalues λ_k which represent the amount of fluctuations the modes capture. If the velocity field is chosen for x, the POD modes capture kinetic energy in an optimal manner.

The original (classical) POD method requires a solution to a large eigenvalue problem of size $n \times n$, where n is the degrees of freedom of the data (the number of data points in the spatial domain times the number of physical variables). There is however an inexpensive alternative to performing the POD analysis called the *method of snapshots* [43], which instead solves an eigenvalue problem using $X^T X \in \mathbb{R}^{m \times m}$, where m is the number of instantaneous flow field data (snapshots). Another approach to determine the POD modes is by way of *singular value decomposition* (SVD). While the use of SVD approach can be computationally expensive compared to the method of snapshots, it is known to be more robust in finding the POD modes. Detailed discussion on these approaches can be found in Appendix C.

The POD reveals the optimal set of spatial modes $\{\phi_j(x)\}_{j=1}^r$ that can reproduce the full flow field using the minimal number of modes r ($\ll n$) with

$$u(x, t) \approx \bar{u}(x) + \sum_{j=1}^{r} a_j(t)\phi_j(x), \tag{6.39}$$

where $\bar{u}$ is the time-average velocity field, $\phi_j(x)$ are the POD modes, and $a_j(t)$ are the time-varying coefficients for the modes. The number of modes r used in the above series are commonly determined by evaluating how close $\sum_{j=1}^r \lambda_j / \sum_{j=1}^n \lambda_j$

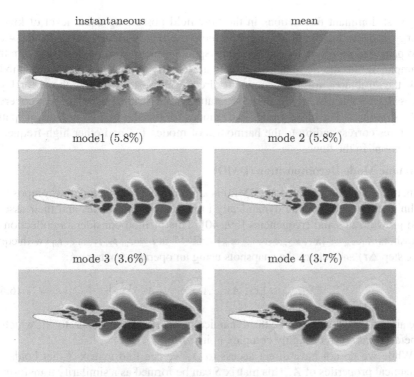

Fig. 6.14 POD analysis of turbulent flow over a NACA0012 airfoil at $Re = 23,000$ (see Fig. 6.13 (*top*)). Shown are the instantaneous and time average streamwise velocity fields and the associated four most dominant POD modes. Data made available by courtesy of A. G. Nair and P. M. Munday

reaches unity. These POD modes are orthogonal and can not only identify regions that exhibit unsteady fluctuations but also can serve as basis vectors (functions) to construct reduced-order models of the flow. A derivation of a reduced-order model using POD modes as a basis is given in Sect. 1.4 and can be used to model the behavior of unsteady fluid flows with significant reduction in computational effort.

POD Analysis Example

Let us revisit the turbulent flow over an airfoil (without control) considered in Sect. 6.3.4 and Fig. 6.13 (top) using the POD analysis. Here, we have visualized the instantaneous and time average streamwise velocity on a spanwise slice in Fig. 6.14. We can observe that there are large-scale vortical structures in the wake from von Kármán shedding yielding spatial and temporal fluctuations about the mean flow. There are also finerscale turbulent flow structures present in the flow.

Performing POD on the flow field data, we can find the dominant modes. Here, the first four dominant POD modes with the percentage of kinetic energy held by the modes are shown in Fig. 6.14. The shown four modes together capture approximately 19 % of the unsteady fluctuations over the examined domain. Modes 1 and 2 represent the most dominant fluctuations in the flow field possessing equal level of kinetic energy, constituting that the pair are periodic oscillatory modes. Modes 3 and 4 are also pairs and represent the subharmonic spatial structures of modes 1 and 2 in this example. While the dominant structures are clearly visible in the wake of modes 1–4, the structures are less prominent directly above the wing as the shear layer necessitates some distance from the separation point (leading edge) to form coherent wake structures. The higher-order POD modes for this flow are comprised of spatial structures corresponding to the harmonics of modes 1–4 and other high-frequency oscillations in the flow.

Dynamic Mode Decomposition (DMD)

Dynamic mode decomposition (DMD) is another data-based modal decomposition technique that captures the dynamically important spatial modes and their associated growth rates and frequencies [39, 40]. This method considers a collection of snapshots $X_{1 \to m} = [x_1, x_2, x_3, \ldots, x_m] \in \mathbb{R}^{n \times m}$ at $t = t_1, t_2, t_3, \ldots, t_m$ (with equal time step Δt) and relates the snapshots using an operator (matrix) A:

$$X_{1 \to m} \approx \left[x_1, A x_1, A^2 x_1 \ldots, A^{m-1} x_1 \right]. \tag{6.40}$$

The aim of DMD is to capture the dynamical properties of this operator A, which in general is unknown or highly complex in its form.

While we do not have direct access to A, we can find matrix $\tilde{S}$ that holds the dynamical properties of A. This matrix $\tilde{S}$ can be formed as a similarity transform of the matrix A based on the POD modes $\Phi = [\phi_1, \ldots, \phi_n]$ and be computed with

$$\tilde{S} \equiv \Phi^T A \Phi = \Phi^T X_{2 \to m+1} X_{1 \to m}^\dagger \Phi = \Phi^T X_{2 \to m+1} \Psi \Sigma^{-1}, \tag{6.41}$$

where $X_{2 \to m+1} = [x_2, x_3, \ldots, x_{m+1}]$ is the shifted data matrix and Φ, Ψ, and Σ are obtained from SVD [12, 48] of the data matrix $X_{1 \to m} = \Phi \Sigma \Psi^T$. In the above equation, the superscript T and dagger $\dagger$ denote the transpose and the Moore–Penrose inverse [48], respectively.

Based on this matrix $\tilde{S}$, we can then find the eigenvalues μ_k and eigenvectors ξ_k by solving an eigenvalue problem of

$$\tilde{S} \xi_k = \mu_k \xi_k. \tag{6.42}$$

The eigenvalues μ_k and the eigenvectors ξ_k are also referred to as the *Ritz values* and *Ritz vectors*, respectively, which are used to approximate the eigenvalues and eigenvectors A. These Ritz vectors ξ_k can be transformed to be the *dynamic modes* (DMD modes)

$$\zeta_k = \Phi \xi_k, \tag{6.43}$$

which highlight spatial structures of dynamical importance. As it can be seen from the transform, Eq. (6.43), DMD modes ζ_k are comprised of the superposition of POD modes Φ.

We can also take the Ritz values μ_k and determine the frequencies and growth rates of these modes $\sigma_k = \text{Re}(\sigma_k) + i\text{Im}(\sigma_k) = \ln(\mu_k)/\Delta t$. Since DMD was introduced by Schmid [40], there have been a number of extensions and modifications to the DMD algorithm [49]. Let us also mention in passing that the finite-dimensional operator A in the DMD analysis can be considered as an approximation to the infinite-dimensional *Koopman operator*, which appears in dynamical systems theory [32, 33, 39, 49].

In comparison to the POD analysis, DMD modes reveal dynamically important structures in the flow field and they can be considered as approximations to the global modes from stability analysis [6, 47]. The DMD modes are not orthonormal and cannot be ranked by the Ritz values in terms of energy content as in the case of POD modes. Nonetheless, DMD can reveal the growth rates and frequencies associated with the modes, which are not directly available in the POD analysis. As such, DMD has been used to study a number of different flows based on data from CFD and experiments to capture dynamical features of the flows [22, 39–41]. Additional details and an example of DMD analysis is offered in Appendix C.

Other Decomposition Methods

While we limit the discussion here to only POD and DMD-based analyses, there are a number of other modal decomposition techniques, including balanced POD [18, 30, 38], global stability analysis [46, 47], resolvent analysis [21, 31], and independent component analysis [17]. We ask readers to refer to the references cited herein for additional details.

Development of modal decomposition techniques is an active area of research and should become ever more important as the data size from CFD-based analysis increases in the future. This trend is not only seen in CFD but also in various areas of computational science. Handling of massive data (*big data*) is a research topic that is tackled in an interdisciplinary manner across different fields. Compressed sampling/sensing [10, 27] should be one of the important fields of research that plays an important role in analyzing massive data, including those from CFD.

6.4 Remarks

Let us summarize our discussion in this chapter as shown in Table 6.3. The simulation methods listed higher in the table have less assumption in the calculation but requires large computational resources. As we go lower on the list, the dependence on turbulence model becomes higher since we must model vortices over a wider range of scales. At this moment, there is not a universal model for turbulent flows.

The utilization of DNS requires care since underresolved simulations can provide results that may appear convincing. We have discussed in Sect. 6.2.4 that "direct numerical simulation" of the Navier–Stokes equations is still possible even with insufficient (under-resolved) grid resolution. As an example, we considered turbulent channel flow which provides guidance in assessing numerical results from other

Table 6.3 Simulation methods for turbulent flow

FTS	Full Turbulence Simulation
DNS	Direct Numerical Simulation
LES	Large-Eddy Simulation
RANS	Reynolds-Averaged Navier–Stokes (numerical simulation)

turbulent flow simulations. For flow problems that have not been studied in the past, demonstrating the validity of the results is not a simple task.

When we are forced to use insufficient grid resolution, it would be necessary to introduce turbulence models to represent the flow physics that occur at scales smaller than the mesh size. Oftentimes, there are uses of artificial viscosity by employing high-order upwinding schemes either as a substitute for the turbulence model or as an addition to the turbulence model. This practice, however, should be avoided as much as possible. Artificial viscosity expressed in terms of higher-order derivatives in particular should not be used since it can yield unphysical results. There should be no need for numerical viscosity, except for flows with discontinuity present, such as shock waves, flames, gas–liquid interface, phase transformation, or ill-formed grid arrangements. One should avoid introducing artificial viscosity to the flow solvers, especially for the case of single phase incompressible flows. If grid resolution is insufficient even with appropriate discretization, we should seek a physically meaningful turbulence model.

While turbulent channel flow has been studied extensively, there still exists many challenges that need to be addressed. In the pursuit of simulating higher Reynolds number flows and handling complex grid structures often used in the industrial applications, it is crucial to develop advanced simulation methods and turbulence models. At the current moment, many industrial problems are tackled with coarse "simulation." Results from such computations should be handled cautiously.

6.5 Exercises

6.1 Consider a homogeneous turbulent flow field in equilibrium where the supply of turbulent kinetic energy is balanced by dissipation.

1. For turbulent motion at the smallest scale, viscous dissipation becomes the dominant physics. This means that turbulent motion can be characterized by the energy dissipation rate ε and kinematic viscosity ν. Using dimensional analysis, estimate the characteristic scales of the smallest turbulent motion (*Kolmogorov scales*) for velocity u_K, length l_K, time t_K, and wave number k_K based on ε and ν.
2. For turbulent flows at sufficiently high Reynolds number, scales at which turbulent kinetic energy is supplied and viscous dissipation takes place are far apart. Between the two scales, there is a range called the *inertial subrange*, in which we can assume that there is only energy transfer ε from large to small scales

without kinetic energy supply or dissipation. Determine the expression for the energy spectra $E(k)$ for such case using dimensional analysis based on ε and wavenumber k.

6.2 Consider performing a DNS of channel flow (between two parallel flat plates)

1. Determine the average pressure gradient required to sustain a channel flow between the parallel plates with spacing of 2δ and an average wall shear stress of τ_w.
2. Non-dimensionalize the governing equations using the density ρ, friction velocity u_τ, and the channel half-width δ for constant viscosity incompressible flow and derive Eqs. (6.9).
3. Given the non-dimensional time average velocity u_m (centerline velocity) from simulation, find the friction coefficient $C_f \equiv \tau_w/(\frac{1}{2}\rho u_m^2)$.

6.3 Numerically solve the differential equations for the Lorenz system, Eq. (6.8), using the fourth-order Runge–Kutta method for the initial conditions of

$$(x_0, y_0, z_0) = (10, 10, 10 + \epsilon).$$

Demonstrate for $\epsilon = 10^{-4}, 10^{-3}, 10^{-2}, 10^{-1}$, and 10^0 the mean and standard deviation of x, y, and z over a longtime are independent of the initial conditions for this chaotic system.

6.4 For a rank-2 (3×3) tensor A, find the three tensor invariants, I_1, I_2, and I_3. One of these invariants I_2 is the Q-criteria from Sect. 6.3.4. Show this by noting that invariants are independent of the choice of orthogonal axes (coordinate transform).

6.5 Given a flow field data X that numerically satisfies the incompressible Navier–Stokes equations, show that each individual POD modes satisfy the incompressibility and no-slip boundary conditions (if there are any no-slip walls in the domain of the original simulation).

References

1. Abe, H., Kawamura, H., Matsuo, Y.: Surface heat-flux fluctuations in a turbulent channel flow up to $Re_\tau = 1020$ with $Pr = 0.025$ and 0.71. J. Heat Fluid Flow **25**, 404–419 (2004)
2. Aubry, N., Holmes, P., Lumley, J.L., Stone, E.: The dynamics of coherent structures in the wall region of a turbulent boundary layer. J. Fluid Mech. **192**, 115–173 (1988)
3. Berkooz, G., Holmes, P., Lumley, J.L.: The proper orthogonal decompsotion in the analysis of turbulent flows. Annu. Rev. Fluid Mech. **25**, 539–575 (1993)
4. Canuto, C., Hussaini, M.Y., Quarteroni, A., Zang, T.A.: Spectral Methods in Fluid Dynamics. Springer, New York (1988)
5. Chatterjee, A.: An introduction to the proper orthogonal decomposition. Curr. Sci. **78**(7), 808–817 (2000)
6. Chomaz, J.M.: Global instabilities in spatially developing flows: non-normality and nonlinearity. Annu. Rev. Fluid Mech. **37**, 357–392 (2005)

7. Deardorff, J.W.: A numerical study of three-dimensional turbulent channel flow at large Reynolds numbers. J. Fluid Mech. **41**(2), 453–480 (1970)
8. Delville, J., Ukeiley, L., Cordier, L., Bonnet, J.P., Glauser, M.: Examination of large-scale structures in a turbulent plane mixing layer. Part 1. Proper orthogonal decomposition. J. Fluid Mech. **391**, 91–122 (1999)
9. Dimotakis, P.E.: The mixing transition in turbulent flows. J. Fluid Mech. **409**, 69–98 (2000)
10. Eldar, Y.C., Kutyniok, G. (eds.): Compressed Senseing: Theory and Applications. Cambridge Univ. Press (2012)
11. Gleick, J.: Chaos-Making a New Science. Penguin Books, New York (2008)
12. Golub, G.H., Loan, C.F.V.: Matrix Computations, 3rd edn. Johns Hopkins Univ. Press (1996)
13. Gottlieb, D., Orszag, S.A.: Numerical Analysis of Spectral Methods: Theory and Applications. SIAM, Philadelphia (1993)
14. Hinze, J.O.: Turbulence. McGraw-Hill, New York (1975)
15. Holmes, P., Lumley, J.L., Berkooz, G., Rowley, C.W.: Turbulence, Coherent Structures, Dynamical Systems and Symmetry, 2nd edn. Cambridge Univ. Press (2012)
16. Hunt, J.C.R., Wray, A.A., Moin, P.: Eddies, streams, and convergence zones in turbulent flows. In: Proceedings of the Summer Program, Center for Turbulence Research, pp. 193–208. Stanford, CA (1988)
17. Hyvärinen, A., Karhunen, J., Oja, E.: Independent Component Analysis. Wiley (2001)
18. Ilak, M., Rowley, C.W.: Modeling of transitional channel flow using balanced propoer orthogonal decomposition. Phys. Fluids **20**, 034,103 (2008)
19. Jeong, J., Hussain, F.: On the identification of a vortex. J. Fluid Mech. **285**, 69–94 (1995)
20. Jolliffe, I.T.: Principal Component Analysis. Springer (2002)
21. Jovanovic, M.R., Bamieh, B.: Componentwise energy amplification in channel flows. J. Fluid Mech. **534**, 145–183 (2005)
22. Jovanović, M.R., Schmid, P.J., Nichols, J.W.: Sparsity-promoting dynamic mode decomposition. Phys. Fluids **26**, 024, 103 (2014)
23. Kaneda, Y., Ishihara, T., Yokokawa, M., Itakura, K., Uno, A.: Energy dissipation rate and energy spectrum in high resolution direct numerical simulations of turbulence in a periodic box. Phys. Fluids **15**(2), L21–24 (2003)
24. Kasagi, N., Tomita, Y., Kuroda, A.: Direct numerical simulation of passive scalar field in a turbulent channel flow. J. Heat Transf. **114**(3), 598–606 (1992)
25. Kim, J., Moin, P., Moser, R.: Turbulence statistics in fully developed channel flow at low Reynolds number. J. Fluid Mech. **177**, 133–166 (1987)
26. Kobayashi, H.: The subgrid-scale models based on coherent structures for rotating homogeneous turbulence and turbulent channel flow. Phys. Fluids **17**, 045, 104 (2005)
27. Kutz, J.N.: Data-Driven Modeling and Scientific Computation. Oxford Univ. Press (2013)
28. Lorenz, E.N.: Deterministic nonperiodic flow. J. Atmos. Sci. **20**, 130–141 (1963)
29. Lumley, J.L.: The structure of inhomogeneous turbulent flows. In: Yaglom, A.M., Tatarsky, V.I. (eds.) Atmosphetic Turbulence and Radio Wave Propagation, pp. 166–178. Nauka, Moscow (1967)
30. Ma, Z., Ahuja, S., Rowley, C.W.: Reduced-order models for control of fluids using the eigensystem realization algorithm. Theo. Comp. Fluid Dyn. **25**, 233–247 (2011)
31. McKeon, B.J., Sharma, A.S.: A critical-layer framework for turbulent pipe flow. J. Fluid Mech. **658**, 336–382 (2010)
32. Mezić, I.: Spectral properties of dynamical systems, model reduction and decompositions. Nonlin. Dyn. **41**, 309–325 (2005)
33. Mezić, I.: Analysis of fluid flows via spectral properties of the Koopman operator. Annu. Rev. Fluid Mech. **45**, 357–378 (2013)
34. Moin, P., Kim, J.: Numerical investigation of turbulent channel flow. J. Fluid Mech. **118**, 341–377 (1982)
35. Moser, R.D., Kim, J., Mansour, N.N.: Direct numerical simulation of turbulent channel flow up to $Re_\tau = 590$. Phys. Fluids **11**(4), 943–945 (1999)

36. Ohta, T., Kajishima, T.: Analysis of non-steady separated turbulent flow in an asymmetric plane diffuser by direct numerial simulations. J. Fluid Sci. Tech. **5**(3), 515–527 (2010)
37. Pope, S.B.: Turbulent Flows. Cambridge Univ. Press (2000)
38. Rowley, C.W.: Model reduction for fluids, using balanced proper orthogonal decomposition. Int. J. Bif. Chaos **15**(3), 997–1013 (2005)
39. Rowley, C.W., Mezić, I., Bagheri, S., Henningson, D.S.: Spectral analysis of nonlinear flows. J. Fluid Mech. **641**, 115–127 (2009)
40. Schmid, P.J.: Dynamic mode decomposition of numerical and experimental data. J. Fluid Mech. **656**, 5–28 (2010)
41. Schmid, P.J., Li, L., Juniper, M.P., Pust, O.: Applications of the dynamic mode decomposition. Theor. Comput. Fluid Dyn. **25**, 249–259 (2011)
42. Schumann, U.: Subgrid scale model for finite difference simulations of turbulent flows in plane channels and annuli. J. Comput. Phys. **18**(4), 376–404 (1975)
43. Sirovich, L.: Turbulence and the dynamics of coherent structures, Parts I-III. Q. Appl. Math. **XLV**, 561–590 (1987)
44. Speziale, C.G.: Analytical methods for the development of Reynolds-stress closures in turbulence. Annu. Rev. Fluid Mech. **23**, 107–157 (1991)
45. Taira, K., Colonius, T.: Three-dimensional flows around low-aspect-ratio flat-plate wings at low Reynolds numbers. J. Fluid Mech. **623**, 187–207 (2009)
46. Theofilis, V.: Advances in global linear instability analysis of nonparallel and three-dimensional flows. Prog. Aero. Sci. **39**, 249–315 (2003)
47. Theofilis, V.: Global linear instability. Annu. Rev. Fluid Mech. **43**, 319–352 (2011)
48. Trefethen, L.N., Bau, D.: Numerical Linear Algebra. SIAM (1997)
49. Tu, J.H., Rowley, C.W., Luchtenburg, D.M., Brunton, S.L., Kutz, J.N.: On dynamic mode decomposition: Theory and applications. J. Comput. Dyn. **1**(2), 391–421 (2014)
50. Ukeiley, L., Cordier, L., Manceau, R., Delville, J., Glauser, M., Bonnet, J.P.: Examinatino of large-scale structures in a turbulent plane mixing layer. Part 2. Dynamical systems model. J. Fluid Mech. **441**, 67–108 (2001)
51. Yakhot, V., Orszag, S.A.: Renormalization group analysis of turbulence. I. Basic Theory. J. Sci. Comput. **1**(1), 3–51 (1986)
52. Yoshizawa, A.: Statistical analysis of the deviation of the Reynolds stress from its eddy-viscosity representation. Phys. Fluids **27**(6), 1377–1387 (1984)
53. Yoshizawa, A.: Derivation of a model Reynolds-stress transport equation using the renormalization of the eddy-viscosity-type representation. Phys. Fluids A **5**(3), 707–715 (1993)

Chapter 7
Reynolds-Averaged Navier–Stokes Equations

7.1 Introduction

The constitutive equations used in the Reynolds-averaged Navier–Stokes (RANS) equations are referred to as turbulence models. Although a large number of studies have been performed on the development of turbulence models, there has not been a universal turbulence model that is applicable to all turbulent flows. However, we in general suggest the use of the k-ε model for "simple" flows and the Large-Eddy Simulation (LES) for more complex flows (Chap. 8) found in many practical engineering applications. In this book, we refer to "simple flows" as stationary flows that have average streamlines that are relatively straight on the absolute coordinate system with low level of acceleration. Flows that impinge on walls, separate from corners, pass through a curved channel, and is in a rotational field would not be considered simple. In this and the next chapters, discussions on the strengths and limitations of these methods are offered. Since our objective is not to introduce all turbulence models available, we ask readers to refer to [3, 6, 28] for comprehensive reviews on RANS.

7.2 Reynolds-Averaged Equations

7.2.1 Reynolds Average

Averaging can be performed to extract the large-scale dynamics of the flow field. The key to simulating such large-scale dynamics is to average over the small-scale fluctuations and model the nonlinear influence from the small-scale fluctuations, in the governing equations, that can alter the large-scale fluid motion. To average a velocity field u_i that has fluctuating components, we can use time averaging for a statistically steady flow or spatial averaging for a statistically uniform flow. More generally, we can consider the use of ensemble averaging [8]. In any of these cases,

© Springer International Publishing AG 2017
T. Kajishima and K. Taira, *Computational Fluid Dynamics*,
DOI 10.1007/978-3-319-45304-0_7

the average quantity varies slowly and there is some level of correlation between the average $\bar{u}_i$ and the fluctuation u_i' $(=u_i - \bar{u}_i)$. Thus, strictly speaking we have

$$\overline{u_i' \bar{u}_j} \neq 0, \quad \bar{\bar{u}}_i \neq \bar{u}_i. \tag{7.1}$$

For the discussion in this chapter, let us redefine the averaging operation such that it satisfies

$$\overline{u_i'} = 0, \quad \overline{u_i' \bar{u}_j} = 0, \quad \bar{\bar{u}}_i = \bar{u}_i. \tag{7.2}$$

These relations in Eq. (7.2) are referred to as the Reynolds-averaging laws. The ensemble average that satisfies these laws is called the *Reynolds average*. This conceptual averaging operation conveniently removes fluctuating components from the flow field variables without explicitly defining the spatial length scale used in the averaging operation. In what follows, we assume that the averaging and derivative operations are commutative for Reynolds averaging (see Eq. (6.15)).

The continuity equation for incompressible flow

$$\frac{\partial u_k}{\partial x_k} = 0 \tag{7.3}$$

with Reynolds averaging being performed becomes

$$\frac{\partial \bar{u}_k}{\partial x_k} = 0. \tag{7.4}$$

Taking the Reynolds average of the momentum equation

$$\frac{\partial u_i}{\partial t} + \frac{\partial (u_i u_j)}{\partial x_j} = -\frac{1}{\rho}\frac{\partial p}{\partial x_i} + \frac{\partial}{\partial x_j}(2\nu D_{ij}), \tag{7.5}$$

we obtain

$$\frac{\partial \bar{u}_i}{\partial t} + \frac{\partial (\bar{u}_i \bar{u}_j)}{\partial x_j} = -\frac{1}{\rho}\frac{\partial \bar{p}}{\partial x_i} + \frac{\partial}{\partial x_j}(-\tau_{ij} + 2\nu \bar{D}_{ij}), \tag{7.6}$$

which can be rewritten as

$$\frac{\overline{D}\bar{u}_i}{\overline{D}t} = -\frac{1}{\rho}\frac{\partial \bar{p}}{\partial x_i} + \frac{\partial}{\partial x_j}\left(-\tau_{ij} + 2\nu \bar{D}_{ij}\right) \tag{7.7}$$

using Eq. (7.4). Here we can observe that Eqs. (7.4), (7.6), and (7.7) are in the same forms as in Eqs. (6.12), (6.17), and (6.18).

Because the nonlinear term under Reynolds averaging satisfies

$$\overline{u_i u_j} = \bar{u}_i \bar{u}_j + \overline{u_i' u_j'}, \tag{7.8}$$

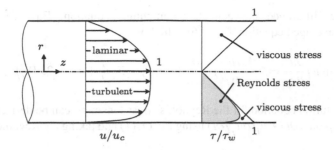

Fig. 7.1 Velocity and stress profiles in fully-developed pipe flow. The centerline velocity and the wall shear stress are denoted by u_c and τ_w, respectively

we find that the tensor τ_{ij} simplifies to

$$\tau_{ij} = \overline{u_i' u_j'} \tag{7.9}$$

instead of the full expression in Eq. (6.20). The symmetric tensor $\rho\tau_{ij}$ represents the correlation of velocity fluctuation and has the same units as the molecular viscous stress tensor $-2\mu\overline{D}_{ij}$. We therefore call $\rho\tau_{ij}$ the *turbulent stress* or the *Reynolds stress*. When the density is uniform, τ_{ij} may be simply referred to as the Reynolds stress also.

The objective here is to solve the Reynolds-averaged continuity and momentum equations, Eqs. (7.4) and (7.6), to find the averaged flow field ($\overline{u}_i$ and $\overline{p}$). What must be provided to these equations is a turbulence model for the Reynolds stress τ_{ij}.

Now, let us consider the fully developed velocity and stress profiles in a circular pipe, as illustrated in Fig. 7.1. For laminar flow, the velocity profile is parabolic and the molecular viscous stress balances the pressure gradient. For turbulent flow, the viscous stress becomes large only in the vicinity of the wall and the Reynolds stress becomes dominant in the entire flow field. It is therefore critical to provide an accurate model for the Reynolds stress.

7.2.2 Reynolds Stress Equation

The governing equation for the Reynolds stress is derived from the equations for the velocity fluctuations. Taking the difference between Eq. (7.3) and its Reynolds-averaged equation, Eq. (7.4), the continuity equation for the fluctuating components can be derived as

$$\frac{\partial u_k'}{\partial x_k} = 0. \tag{7.10}$$

The momentum equation for the fluctuating component of velocity is determined by taking the difference between the momentum equation, Eq. (7.5) and its Reynolds-averaged equation, Eq. (7.6), which is

$$\frac{\partial u'_i}{\partial t} + \bar{u}_j \frac{\partial u'_i}{\partial x_j} = -\frac{1}{\rho} \frac{\partial p'}{\partial x_i} + \frac{\partial}{\partial x_j} \left(-u'_i u'_j - \bar{u}_i u'_j + \tau_{ij} + 2\nu D'_{ij} \right). \tag{7.11}$$

The evolution equation for the Reynolds stress tensor τ_{ij} can be derived by reexpressing $u'_j (\partial u'_i / \partial t) + u'_i (\partial u'_j / \partial t)$ using Eq. (7.11) and taking the Reynolds average to yield

$$\frac{\overline{D}\tau_{ij}}{\overline{D}t} = P_{ij} + \Pi_{ij} - \varepsilon_{ij} + \frac{\partial J_{ijk}}{\partial x_k} \tag{7.12}$$

which is called the *Reynolds stress equation*. Here, the first term on the right-hand side represents the production of Reynolds stress

$$P_{ij} = -\tau_{ik} \frac{\partial \bar{u}_j}{\partial x_k} - \tau_{jk} \frac{\partial \bar{u}_i}{\partial x_k}, \tag{7.13}$$

the second term denotes the pressure-strain correlation tensor

$$\Pi_{ij} = \frac{\overline{p'}}{\rho} \left(\frac{\partial u'_i}{\partial x_j} + \frac{\partial u'_j}{\partial x_i} \right), \tag{7.14}$$

the third term is the dissipation

$$\varepsilon_{ij} = 2\nu \overline{\frac{\partial u'_i}{\partial x_k} \frac{\partial u'_j}{\partial x_k}}, \tag{7.15}$$

and in the fourth term, $J_{ijk} = J_{(T)ijk} + J_{(P)ijk} + J_{(V)ijk}$ represents the diffusive flux with

$$J_{(T)ijk} = -\overline{u'_i u'_j u'_k}, \tag{7.16}$$

$$J_{(P)ijk} = -\frac{1}{\rho} (\overline{p' u'_i} \delta_{jk} + \overline{p' u'_j} \delta_{ik}), \tag{7.17}$$

$$J_{(V)ijk} = \nu \frac{\partial \tau_{ij}}{\partial x_k}. \tag{7.18}$$

Each of these flux terms is related to velocity fluctuation, pressure fluctuation, and viscosity, respectively. The diffusive flux in Eq. (7.12) is expressed in the divergence form, which tells us that the net increase or decrease can only be achieved through the computational boundary. This can be shown by integrating $\partial J_{ijk} / \partial x_k$ over the whole computational domain and invoking Gauss' theorem.

In the governing equation for the second-order correlation $\tau_{ij} = \overline{u_i' u_j'}$, higher order correlations $\overline{u_i' u_j' u_k'}$ and $\overline{u_j' p'}$ appear. These higher order terms appear due to the nonlinearity in the Navier–Stokes equations. We can derive the governing equations for such high-order terms, but they would always result in containing even higher order correlations *ad infinitum*. It thus becomes necessary to truncate the high-order correlations and represent their effects with a model using low-order correlations. This procedure is called *closure* and the model for higher-order correlation is referred to as *turbulence model*.

To solve the stress equation Eq. (7.12) along with the continuity and momentum equations, the dissipation term ε_{ij}, the third-order correlations of pressure-strain correlation term, Π_{ij}, and the turbulent diffusion flux, $J_{(T)ijk} + J_{(P)ijk}$, would require some form of modeling because they are not expressed in terms of $\overline{u}_i$, $\overline{p}$, and $\overline{u_i' u_j'}$. Such model is referred to as the *Reynolds stress model* (RSM), the *stress equation model*, or the *second moment closure* and is distinguished from low-order models that find the Reynolds stress without the use of the stress equation.

7.3 Modeling of Eddy Viscosity

One can consider the stress created by turbulent motion of vortices to be analogous to the viscous stress caused by the random motion of molecules. The viscous stress is the product of the viscosity and the velocity gradient. If we are to model the turbulent stress as the product of a coefficient representing diffusion of vortices and the average velocity gradient, we can write

$$\rho \tau_{ij} = \frac{2}{3} \delta_{ij} \rho k - 2 \mu_T \overline{D}_{ij}. \tag{7.19}$$

We require the first term on the right-hand side of Eq. (7.19), because the velocity gradient tensor is traceless ($\overline{D}_{ii} = 0$). The term $k = \overline{u_i' u_i'}/2 = \tau_{ii}/2$ is the kinetic energy of the velocity fluctuation.

The model represented by Eq. (7.19) is known as the *eddy-viscosity model*. This approach was proposed by Boussinesq in 1877 and is called the *Boussinesq approximation*.[1] In analogy to the dynamic viscosity coefficient μ, the coefficient μ_T is referred to as the *eddy-viscosity coefficient*. The first term on the right-hand side of Eq. (7.19) represents the isotropic component of the Reynolds stress τ_{ij} with $2k/3$ ($=\tau_{mm}/3$). Moving this term to the left-hand side and expressing $\tau_{ij}^a = \tau_{ij} - \delta_{ij} \tau_{mm}/3$, we have

$$\rho \tau_{ij}^a = -2 \mu_T \overline{D}_{ij}. \tag{7.20}$$

---

[1] There are also different Boussinesq approximations that appear in the analysis of flows with buoyancy and water waves.

We can thus observe that the eddy-viscosity model relates the anisotropic component $\rho\tau_{ij}^a$ to the mean strain rate $\overline{D}_{ij}$ in a proportional manner.

With the use of the eddy-viscosity model, we can write the Reynolds averaged momentum equation as

$$\frac{\overline{D\overline{u}_i}}{\overline{D}t} = -\frac{1}{\rho}\frac{\partial \overline{P}}{\partial x_j} + \frac{\partial}{\partial x_j}\left[2(\nu + \nu_T)\overline{D}_{ij}\right], \tag{7.21}$$

where $\nu_T = \mu_T/\rho$ is the kinematic eddy viscosity. In comparison to the original momentum equation, the Reynolds-averaged momentum equation has the sum of the kinematic viscosity ν and the kinematic eddy viscosity ν_T appearing in the viscous term. Another point to note is that the pressure variable $\overline{P}$ is comprised of the dynamic pressure $\overline{p}$ and the isotropic component of the Reynolds stress:

$$\overline{P} = \overline{p} + \frac{2}{3}\rho k. \tag{7.22}$$

Through Taylor series expansion of the fluctuating components and dimensional analysis, Eq. (7.19) can be understood as the first-order Taylor series approximation of the Reynolds stress in t/T, where t and l denote the temporal and spatial scales of the fluctuating components and T and L represent the temporal and spatial scales of the average flow field [26] (also recall our discussion on temporal and spatial length scales from Sect. 6.2.1). The corresponding eddy viscosity is $\nu_T = l^2/t$. The kinematic viscosity can be derived dimensionally to be $\nu = l^2/t$ in a similar manner if we instead set the molecular scales to be (t, l) and the continuum flow scales to be (T, L). It should however be noted that the assumptions made for the (molecular) viscosity and the eddy viscosity are completely different. The ratio of the molecular motion scales (t, l) to the continuum flow scales (T, L) is extremely small $t/T \leq \mathcal{O}(10^{-6})$, which results in the kinematic viscosity being characterized only by (t, l) [26]. This implies that the viscous term in the Navier–Stokes equations very accurately represents the viscous physics. On the other hand, the fluctuating velocity scales (t, l) and the mean flow scales (T, L) are not as clearly distinguished in the eddy-viscosity model as the scales of the molecular and macroscopic physics. There is not a single characteristic scale that captures the eddy dynamics in such case. Since the largest vortex in the flow represents the characteristic length scale of the flow, it can be possible to observe $t/T = \mathcal{O}(1)$.

In any case, the unknown stress tensor with six independent components is now condensed into a single scalar ν_T by relating the Reynolds stress and the average velocity gradient as stated in Eq. (7.19). Once the eddy viscosity ν_T is provided in some manner, the Reynolds-averaged equations are closed (note that we also need k to determine the average pressure $\overline{p}$ from Eq. (7.22)).

The eddy viscosity ν_T is determined by simultaneously solving the transport equations for the relevant turbulent quantities along with the continuity and momentum equations. Based on the number of transport equations added to the governing equations, models are referred to as the n-equation model. Some of the well-known models in use are the zero-, one-, and two-equation models that we discuss below.

Zero-Equation Model

A model that computes the eddy viscosity solely from the Reynolds-averaged velocity field is called the zero-equation model. In the *mixing length model* proposed by Prandtl in 1925, the eddy viscosity is modeled as

$$\nu_T = l_m^2 \left| \frac{\partial \overline{u}}{\partial y} \right|, \qquad (7.23)$$

where l_m represents the mixing length over which a fluid element may travel while maintaining its momentum. Within the turbulent boundary layer, we recover the logarithmic velocity profile by setting $l_m = \kappa y$ (see Sect. 7.4.1).

In order to generalize the mixing length model, Eq. (7.23), the model needs to be extended to turbulent flows that have velocity gradients other than $\partial \overline{u}/\partial y$. We can choose the norm of the rate of strain tensor $\overline{D}_{ij}$ or the magnitude of averaged vorticity $\overline{\omega}_i$. Using the rate-of-strain tensor, we have

$$\nu_T = l_m^2 |\overline{D}|, \quad |\overline{D}| = \sqrt{2 \overline{D}_{ij} \overline{D}_{ij}}, \qquad (7.24)$$

which is often employed in LES discussed later. For the Reynolds-averaged formulation, the Baldwin–Lomax model [2] uses the norm of the vorticity vector

$$\nu_T = l_m^2 |\overline{\omega}|, \quad |\overline{\omega}| = \sqrt{\overline{\omega}_i \overline{\omega}_i} \qquad (7.25)$$

to model the eddy viscosity in the inner layer. The expressions in Eqs. (7.24) and (7.25) are general descriptions of the zero-equation models. Note however that vorticity is not an objective quantity as discussed in Sect. 6.3.6.

In any case, one can determine ν_T by providing l_m and consequently compute the average flow field without the burden of solving an extra equation. For some flows, the zero equation models show good agreement with experiments. Nonetheless, these models are not likely to be a universal turbulence model, because the choice of l_m usually requires some prior knowledge of the flow.

One-Equation Model

The one-equation model refers to a family of turbulence models that utilize a transport equation that is solved simultaneously with the Reynolds-averaged Navier–Stokes equations to determine the eddy viscosity. Earlier versions of one-equation models chose the transport equation for the kinetic energy k, but required the turbulence length scale to be specified based on experience to determine the eddy viscosity and to provide closure for the kinetic energy equation. For such reason, significant improvements over the zero-equation model were generally not achieved.

Presently, it is mainstream to use the transport equation for eddy viscosity or a closely related variable for the one-equation model. Utilizing the eddy viscosity ν_T as

the transport variable may not be a physically intuitive approach, but we can express
the transport of $\nu_T = k/\omega$ as

$$
\begin{aligned}
\frac{\overline{D}\nu_T}{\overline{D}t} &= \frac{\partial}{\partial t}\frac{k}{\omega} + \bar{u}_j\frac{\partial}{\partial x_j}\frac{k}{\omega} \\
&= \frac{1}{\omega}\left\{(1-\alpha)P_k - (\beta^* - \beta)k\omega + \frac{\partial}{\partial x_j}\left[\left(\frac{\nu_T}{\sigma_\omega} + \nu\right)\frac{\partial\nu_T}{\partial x_j}\right]\right. \\
&\quad\left. + 2\left(\frac{\nu_T}{\sigma_\omega} + \nu\right)\left(\frac{1}{\omega}\frac{\partial k}{\partial x_j}\frac{\partial\omega}{\partial x_j} - \frac{1}{\omega^2}\frac{\partial\omega}{\partial x_j}\frac{\partial\omega}{\partial x_j}\right)\right\}
\end{aligned}
\tag{7.26}
$$

based on the k-ω model discussed later in Sect. 7.5. This form of the equation for
ν_T has resemblance to the form of two-equation models. Needless to say, the above
eddy-viscosity equation must be closed without introducing additional variables (k
and ω or k and ϵ for the two-equation models).

One of the most well-known one-equation model is the *Spalart–Allmaras model*
(SA) model [25], which computes the eddy viscosity with

$$
\nu_T = \tilde{\nu}\frac{(\tilde{\nu}/\nu)^3}{(\tilde{\nu}/\nu)^3 + 7.1^3}
\tag{7.27}
$$

and the transport equation

$$
\frac{\overline{D}\tilde{\nu}}{\overline{D}t} = \tilde{S}\tilde{\nu} - c_{w1}f_w\left(\frac{\tilde{\nu}}{d}\right)^2 + \frac{1}{\sigma_\nu}\left[\frac{\partial}{\partial x_j}(\tilde{\nu} + \nu)\frac{\partial\tilde{\nu}}{\partial x_j} + c_{b2}\frac{\partial\tilde{\nu}}{\partial x_j}\frac{\partial\tilde{\nu}}{\partial x_j}\right].
\tag{7.28}
$$

Similar to the classical one-equation models, the one-equation model based on the
eddy-viscosity transport equation contains a number of constants and functions,
which requires some prior knowledge of the flow. It should however be noted that
the constants in the SA model and its newer variants have been well tuned. With
appropriate choices of model parameters, the solutions from the one-equation model
can be expected to provide reliable results. The SA model and its variants have been
widely used in the aerodynamics community.

Two-Equation Model

The eddy viscosity ν_T can be modeled by using two out of three parameters of length
(l), time (t), and velocity (q) to yield

$$
\nu_T = \frac{l^2}{t} = lq = q^2t.
\tag{7.29}
$$

To evaluate the above relation, we would not need to directly derive the equations
for l, t, or q but choose two parameters to dimensionally match ν_T. While there is
some arbitrariness to this approach, one can choose the two parameters such that the
terms in their transport equations have physical significance.

For one of the parameters, we can select the turbulent kinetic energy k and use its square root $\sqrt{k}$ as the characteristic velocity q. The governing equation for energy $k = \tau_{ii}/2$ can be derived by taking a tensor contraction of the stress equation, Eq. (7.12),

$$\frac{\overline{D}k}{\overline{D}t} = P_k - \varepsilon + \frac{\partial J_{(k)j}}{\partial x_j}, \tag{7.30}$$

where the first term P_k on the right-hand side represents the production of turbulent energy based on the average velocity gradient

$$P_k = \frac{1}{2}P_{ii} = -\tau_{ij}\frac{\partial \overline{u}_i}{\partial x_j}. \tag{7.31}$$

The second term ε on the right hand side

$$\varepsilon = \frac{1}{2}\varepsilon_{ii} = \nu\overline{\frac{\partial u'_i}{\partial x_j}\frac{\partial u'_i}{\partial x_j}} \tag{7.32}$$

denotes the rate of kinetic energy dissipation per unit mass. The third term on the right-hand side $J_{(k)j} = J_{(Tk)j} + J_{(Pk)j} + J_{(Vk)j}$ is the diffusive flux term, where

$$J_{(Tk)j} = \frac{1}{2}J_{(T)iij} = -\overline{u'_j u'_i u'_i} \tag{7.33}$$

$$J_{(Pk)j} = \frac{1}{2}J_{(P)iij} = -\frac{1}{\rho}\overline{p'u'_j} \tag{7.34}$$

$$J_{(Vk)j} = \frac{1}{2}J_{(V)iij} = \nu\frac{\partial k}{\partial x_j} \tag{7.35}$$

correspond to the fluxes due to velocity fluctuation, pressure fluctuation, and viscous effect, respectively.

Comparing the transport equations for the Reynolds stress, Eq. (7.12), and the turbulent energy, Eq. (7.30), we can notice that the pressure-strain correlation term is eliminated in the later equation due to the use of the continuity equation. While this term accounts for the interactions amongst velocity fluctuation components, it does not appear in the energy equation since it is derived by taking the contraction.

The production of turbulent energy can be evaluated with

$$P_k = 2\nu_T\overline{D}_{ij}\overline{D}_{ij} \tag{7.36}$$

for a given ν_T, since $P_k = -\tau_{ij}\overline{D}_{ij} = \tau_{ij}^a\overline{D}_{ij}$. Thus, we need models for ε, $J_{(Tk)j}$, and $J_{(Pk)j}$ to close the k-equation. The diffusive fluxes from turbulent and pressure fluctuations are often combined into a single diffusive gradient term

$$J_{(Tk)j} + J_{(Pk)j} = \frac{\nu_T}{\sigma_k}\frac{\partial k}{\partial x_j}, \tag{7.37}$$

where σ_k is a non-dimensional constant that provides the ratio between the eddy-viscosity coefficient ν_T and the turbulent diffusion coefficient for k. The dissipation rate ε will be discussed further in the next section.

There are numerous two-equation models proposed depending on how the parameters are chosen. While the k-ε model is the most widely used model, there is no concrete reason why this model is superior to other models that combines k with another variable. The k-ε model appears to have become widespread with its implementation being available in commercial CFD solvers. One of the strengths of the k-ε model is that the solution of the ε-equation directly gives the sink term in the k-equation.

7.4 k-ε Model

The eddy viscosity ν_T can be dimensionally related to the turbulent kinetic energy k and the kinetic energy dissipation rate ε through

$$\nu_T = C_\mu \frac{k^2}{\varepsilon}, \tag{7.38}$$

where C_μ is a non-dimensional constant. Now, let us consider the physical significance of k/ε. In the context of uniform and isotropic turbulent flow, there should be no production or diffusion of turbulent kinetic energy, which leads Eq. (7.30) to be $\partial k/\partial t = -\varepsilon$. Assuming that ε is constant $\varepsilon = \varepsilon_0$, the solution to this differential equation is $k = k_0 - \varepsilon_0 t$, where $k = 0$ for $t = k_0/\varepsilon_0$. Through this simple analysis, we can interpret k/ε as the characteristic time scale for turbulence to be sustained. The k-ε model is based on these scaling arguments and chooses $q = \sqrt{k}$, $t = k/\varepsilon$, and $l = k^{3/2}/\varepsilon$ in Eq. (7.29).

We can derive the exact governing equation for the dissipation rate ε. By taking the partial derivative of Eq. (7.11) with respect to x_m, evaluating $2\nu(\partial u_i'/\partial x_m) \cdot \partial(\partial u_i'/\partial x_m)/\partial t$, and performing Reynolds averaging, we find

$$\frac{\overline{D}\varepsilon}{\overline{D}t} = P_\varepsilon - \Phi_\varepsilon + \frac{\partial J_{(\varepsilon)j}}{\partial x_j}. \tag{7.39}$$

The right-hand side terms may be difficult to directly relate to physics, but their behaviors have become understood based on DNS studies [15]. The term P_ε is

$$P_\varepsilon = -2\nu \left(\overline{\frac{\partial u_i'}{\partial x_k} \frac{\partial u_j'}{\partial x_k}} + \overline{\frac{\partial u_k'}{\partial x_i} \frac{\partial u_k'}{\partial x_j}} \right) \frac{\partial \overline{u}_j}{\partial x_i}$$

$$- 2\nu \overline{u_i' \frac{\partial u_j'}{\partial x_k} \frac{\partial^2 \overline{u}_j}{\partial x_i \partial x_k}} - 2\nu \overline{\frac{\partial u_j'}{\partial x_i} \frac{\partial u_j'}{\partial x_k} \frac{\partial u_i'}{\partial x_k}}, \tag{7.40}$$

which can be viewed as the production term of ε. The first and third terms are the production terms due to the average velocity gradient and vortex stretching, respectively. However, the second term has no clear physical interpretation. In P_ε, note that there are terms that can decrease ε. Next, the term Φ_ε is expressed as

$$\Phi_\varepsilon = 2\nu^2 \overline{\frac{\partial^2 u_i'}{\partial x_j \partial x_k} \frac{\partial^2 u_i'}{\partial x_j \partial x_k}} \tag{7.41}$$

and, by analogy to the k-equation, Φ_ε represents dissipation of ε. The last term in Eq. (7.39) describes diffusion with $J_{(\varepsilon)j} = J_{(T\varepsilon)j} + J_{(P\varepsilon)j} + J_{(V\varepsilon)j}$, where

$$J_{(T\varepsilon)j} = -\nu \overline{\frac{\partial u_i'}{\partial x_k} \frac{\partial u_i'}{\partial x_k} u_j'} \tag{7.42}$$

$$J_{(P\varepsilon)j} = -\frac{\nu}{\rho} \overline{\frac{\partial u_j'}{\partial x_i} \frac{\partial p'}{\partial x_i}} \tag{7.43}$$

$$J_{(V\varepsilon)j} = \nu \frac{\partial \varepsilon}{\partial x_j} \tag{7.44}$$

correspond to the diffusive fluxes associated with velocity fluctuation, pressure fluctuation, and viscosity, respectively.

For modeling the ε-equation, the production and destruction terms are combined and approximated as

$$P_\varepsilon - \Phi_\varepsilon = (C_{\varepsilon 1} P_k - C_{\varepsilon 2} \varepsilon) \frac{\varepsilon}{k}, \tag{7.45}$$

where $C_{\varepsilon 1}$ and $C_{\varepsilon 2}$ are non-dimensional constants. Note that the production and destruction terms are not separated as $P_\varepsilon = C_{\varepsilon 1} P_k \varepsilon / k$ and $\Phi_\varepsilon = C_{\varepsilon 2} \varepsilon^2 / k$, because for flows with no average velocity gradient, $P_k = 0$ but $P_\varepsilon \neq 0$ due to the last term in Eq. (7.40).

The diffusive flux attributed to turbulence is approximated in an analogous manner to k

$$J_{(T\varepsilon)j} + J_{(V\varepsilon)j} = \frac{\nu_T}{\sigma_\varepsilon} \frac{\partial \varepsilon}{\partial x_j}, \tag{7.46}$$

where σ_ε is a non-dimensional constant.

Standard k-ε Model

We summarize the standard k-ε model [12] below. The governing equations consist of the continuity equation

$$\frac{\partial \overline{u}_i}{\partial x_i} = 0 \qquad (7.47)$$

and the momentum equation

$$\frac{\overline{D}\overline{u}_i}{\overline{D}t} = -\frac{1}{\rho}\frac{\partial \overline{P}}{\partial x_i} + \frac{\partial}{\partial x_j}\left[2(\nu + \nu_T)\overline{D}_{ij}\right], \qquad (7.48)$$

where the eddy viscosity is evaluated as

$$\nu_T = C_\mu \frac{k^2}{\varepsilon}. \qquad (7.49)$$

The turbulent kinetic energy and the energy dissipation are determined by solving the following equations

$$\frac{\overline{D}k}{\overline{D}t} = P_k - \varepsilon + \frac{\partial}{\partial x_j}\left[\left(\frac{\nu_T}{\sigma_k} + \nu\right)\frac{\partial k}{\partial x_j}\right] \qquad (7.50)$$

$$\frac{\overline{D}\varepsilon}{\overline{D}t} = (C_{\varepsilon 1} P_k - C_{\varepsilon 2}\varepsilon)\frac{\varepsilon}{k} + \frac{\partial}{\partial x_j}\left[\left(\frac{\nu_T}{\sigma_\varepsilon} + \nu\right)\frac{\partial \varepsilon}{\partial x_j}\right], \qquad (7.51)$$

where the constants take the following values

$$C_\mu = 0.09, \quad \sigma_k = 1.0, \quad \sigma_\varepsilon = 1.3, \quad C_{\varepsilon 1} = 1.44, \quad C_{\varepsilon 2} = 1.92. \qquad (7.52)$$

We briefly describe how the above constants are derived following Chap. 2 of Ref. [17] (see also [28]). First, we estimate the constant C_μ. Let us assume a two-dimensional turbulent shear flow in which the production and dissipation of energy are in equilibrium, $-\tau_{12}(\partial \overline{u}_1/\partial x_2) - \varepsilon = 0$. By expressing the Reynolds stress with the eddy-viscosity model $\tau_{12} = -\nu_T(\partial \overline{u}_1/\partial x_2)$ and combining these two equations to eliminate $\partial \overline{u}_1/\partial x_2$, we obtain $\nu_T = \tau_{12}^2/\varepsilon$. Since the value for $|\tau_{12}|/k \approx 0.3$ is known from experiments, we obtain $C_\mu = 0.09$.

Next, we estimate the values for $C_{\varepsilon 1}$ and $C_{\varepsilon 2}$. For isotropic turbulence, turbulence kinetic energy decays as $\partial k/\partial t = -\varepsilon$ and $\partial \varepsilon/\partial t = -C_{\varepsilon 2}(\varepsilon^2/k)$. By combining these two equations, we find $\partial^2 k/\partial t^2 = C_{\varepsilon 2}(1/k)(\partial k/\partial t)^2$. For the initial state of turbulence decay, experiments have revealed that $k \propto t^{-1}$, which tells us $\partial k/\partial t \propto -t^{-2}$ and $\partial^2 k/\partial t^2 \propto 2t^{-3}$, leading us to obtain $C_{\varepsilon 2} = 2$. For flat-plate boundary layers with no average pressure gradients, we can assume $D\varepsilon/Dt = 0$ as well as local equilibrium ($P_k = \varepsilon$). Neglecting viscous diffusion, we have $(C_{\varepsilon 1} - C_{\varepsilon 2})\varepsilon^2/k + \partial\{(\nu_T/\sigma_\varepsilon)(\partial \varepsilon/\partial y)\}/\partial y = 0$, where y is the wall-normal direction. Utilizing a relation corresponding to the log law $\varepsilon = C_\mu^{3/4}k^{3/2}/\kappa y$ and letting k be uniform in the direction of y, we obtain $C_{\varepsilon 2} - C_{\varepsilon 1} = \kappa^2/(\sigma_\varepsilon\sqrt{C_\mu})$. For $C_{\varepsilon 2} = 2$, we have $C_{\varepsilon 1} = 1.5$. Past applications of this model to various types of flows have shown that values of $C_{\varepsilon 2} = 1.92$ and $C_{\varepsilon 1} = 1.44$ provide reasonable predictions of turbulent flows. Comparison with experimental results have yielded $\sigma_k = 1.0$ and $\sigma_\varepsilon = 1.3$.

7.4.1 Treatment of Near-Wall Region

When there is separation between the fluctuation scale of turbulent kinetic energy k and that of the dissipation rate ε, the k- and ε-equations are independent. This is an assumption for the k-ε model to be meaningful as a two-equation model. Such condition is met for high-Reynolds number flows as discussed in Sect. 6.2.1. For this reason, we can say that the standard k-ε model is suitable for high-Reynolds number flows. For near-wall turbulent flows or stably stratified turbulent flows, the aforementioned turbulent scales are not clearly separable. Hence, the use of $k^{3/2}/\varepsilon$ and k/ε for the length and time scales, respectively, is no longer appropriate in these regions.

In the near-wall regions, there are two effects that confront the assumptions made for the k-ε model. First is the low-Reynolds number viscous effect. Second is the anisotropic effect that results from the suppression of fluctuations in the wall-normal direction compared to the wall-tangential directions. In order to address these issues, there are two approaches that can be taken. We can either eliminate the application of the k-ε model in the wall region, or incorporate a correction to the k-ε model that can be applied to the wall region.

Application of the Wall Law

By utilizing the law of the wall, we can perform turbulent flow simulation without resolving the flow near the wall. Now, let us denote the wall-normal direction by y and the velocity tangential to the wall by u. We take the first grid point from the wall to be y_p (approximately $y_p^+ = 30$ to 200). For the region between the wall and y_p, we do not solve the governing equations but instead apply the logarithmic law at y_p

$$\overline{u}_p^+ = \frac{1}{\kappa} \ln y_p^+ + B, \tag{7.53}$$

where $y^+ \equiv y u_\tau / \nu$ and $u^+ \equiv u/u_\tau$ are the wall coordinate and the velocity non-dimensionalized by the friction velocity u_τ. For a flat plate, values around $\kappa = 0.4$ and $B = 5.5$ are selected.

Solving for the flow is effortless when the friction velocity u_τ is known. For example, the friction velocity $u_\tau = \sqrt{\tau_w/\rho}$ can be determined by equating the pressure gradient and the wall shear stress for fully developed turbulence under constant pressure gradient in pipe flow or channel flow. Once $y_p^+ = y_p u_\tau / \nu$ is verified to lie within the log-law region, the velocity $\overline{u}_p$ can be specified using Eq. (7.53) at the first grid point.

For a more general case, in which the wall shear stress is determined from the flow, an iterative scheme is necessitated as presented below. We can define a function F for an instance in time with a given u_p at y_p:

$$F(u_\tau) = \frac{\overline{u}_p}{u_\tau} - \frac{1}{\kappa}\ln\frac{y_p u_\tau}{\nu} - B = 0, \tag{7.54}$$

where the objective is to determine the root u_τ. The Newton–Raphson method

$$u_\tau^{m+1} = u_\tau^m - \frac{F(u_\tau^m)}{F'(u_\tau^m)}, \quad m = 0, 1, 2, \cdots, \tag{7.55}$$

can be utilized to solve the Eq. (7.54). Here, $F'(u_\tau) = -(\overline{u}_p/u_\tau + 1/\kappa)/u_\tau$. When the iterative process has reached convergence, we have the frictional velocity u_τ. With this found solution, it should be checked whether y_p^+ for the found u_τ is within the log-law region. In the wall region, k and ε at y_p are provided by

$$k_p = \frac{u_\tau^2}{\sqrt{C_\mu}}, \quad \varepsilon_p = \frac{u_\tau^3}{\kappa y_p} \tag{7.56}$$

by assuming the local equilibrium between the production and dissipation of turbulent energy ($P_k \approx \varepsilon$) near the wall and the Reynolds stress ($-\overline{u'v'} \approx u_\tau^2$) to be constant.

We can further generalize the application of the law of the wall by replacing Eq. (7.53) with Spalding's law of the wall:

$$y_p^+ = \overline{u}_p^+ + \exp(-\kappa B)\left[\exp(\kappa\overline{u}_p^+) - 1 - \kappa\overline{u}_p^+ - \frac{(\kappa\overline{u}_p^+)^2}{2} - \frac{(\kappa\overline{u}_p^+)^3}{6}\right] \tag{7.57}$$

so that there would not be an issue with y_p^+ being in the buffer or viscous sub-layer. However, it would not be advisable to have the first grid point lie in a region where it would be appropriate to specify a boundary condition for the higher-Reynolds-number k-ε model.

For flows for which the logarithmic law does not hold, we can use a similar approach to solve for the flow if there is a relation that replaces the logarithmic law. However, we must be able to provide the boundary conditions for the k and ε equations, in addition to u_p at the first grid point y_p.

Low-Reynolds Number k-ε Model

The wall law is not universal for all turbulent flows. For flows in which the log-wall does not hold, the iterative scheme mentioned above would diverge or provide unphysical results. For problems where the various turbulent quantities are especially important near the near-wall regions, such as flows with heat transfer physics, it is necessary to prescribe the no-slip condition at the wall (for a stationary wall, $k = 0$ and $\overline{u}_i = 0$) and compute all the way up to the wall boundary. A model that takes the low-Reynolds number effect near the wall into account and allow for computations of the overall flow field is referred to as the low-Reynolds number *k-ε* model.

The common formulation of the low-Reynolds number *k-ε* model involves the use of the damping function f_μ for ν_T and the correction functions f_1 and f_2 for the production and dissipation terms in the *ε*-equation, and the wall correction terms D and E in the k and *ε*-equations, respectively [21]. We in general can write the low-Reynolds number *k-ε* model as

$$\nu_T = C_\mu f_\mu \frac{k^2}{\varepsilon}, \tag{7.58}$$

$$\frac{\overline{D}k}{\overline{D}t} = P_k - \varepsilon + D + \frac{\partial}{\partial x_j}\left[\left(\frac{\nu_T}{\sigma_k} + \nu\right)\frac{\partial k}{\partial x_j}\right], \tag{7.59}$$

$$\frac{\overline{D}\varepsilon}{\overline{D}t} = (C_{\varepsilon 1} f_1 P_k - C_{\varepsilon 2} f_2 \varepsilon)\frac{\varepsilon}{k} + E + \frac{\partial}{\partial x_j}\left[\left(\frac{\nu_T}{\sigma_k} + \nu\right)\frac{\partial \varepsilon}{\partial x_j}\right]. \tag{7.60}$$

Since the Jones–Launder model [9] was introduced, there have been many low-Reynolds number models that have been proposed.

Because the low-Reynolds number *k-ε* model has been introduced to represent the behavior of turbulence near the wall, it is necessary to accurately model the asymptotic behavior of the flow, as described below. Consider a flow over a flat plate with no-slip applied at the boundary. By expanding the velocity fluctuations with Taylor series in the wall-normal direction (y) and utilizing the no-slip boundary condition and the continuity equation, the velocity fluctuation in each direction can be expanded in the following manner

$$\left.\begin{array}{ll} u'(y) = a_1 y + a_2 y^2 + a_3 y^3 + \cdots & \text{(streamwise)} \\ v'(y) = b_2 y^2 + b_3 y^3 + \cdots & \text{(vertical)} \\ w'(y) = c_1 y + c_2 y^2 + c_3 y^3 + \cdots & \text{(spanwise)} \end{array}\right\} \tag{7.61}$$

The turbulence quantities near the wall ($y \to 0$) become

$$k = (a_1^2 + c_1^2)y^2/2, \quad \varepsilon = \nu(a_1^2 + c_1^2), \quad \overline{u'v'} = a_1 b_2 y^3, \tag{7.62}$$

which consequently yields $\nu_T \propto y^3$. For a proper model, all of the above conditions need to be satisfied.

There are models that are known to reproduce the asymptotic turbulent flow behavior near the wall, such as the model by Myong and Kasagi [18] and the one by Nagano et al. [19]. The difference between these two models lies in how the characteristic length scales are provided in the wall region. For regions away from the wall, the representative length scale that determines the eddy viscosity is derived from size of the relatively large vortices that possess the majority of the turbulent kinetic energy. In the vicinity of the wall these models switch to different characteristic length scales. The Myong–Kasagi model uses the Taylor micro scale $\lambda = (\nu k/\varepsilon)^{1/2}$ while the Nagano–Tagawa–Niimi model utilizes the Kolmogorov scale $\eta = (\nu^3/\varepsilon)^{1/4}$. Switching the length scales from the near to far-wall regions is closely related to how dissipation and diffusion are balanced above the wall.

Only flows in close neighborhood of the wall show behavior described by Eqs. (7.61) and (7.62). Therefore, sufficient grid resolution is required near the wall to accurately reproduce turbulent flows. For high-Reynolds-number flows, it may be a challenge to produce (and perform computation on) such a large grid. Even if the grid resolution may not be as fine as one would like, results can be obtained with the no-slip boundary condition and a low-Reynolds number k-ε model (the grid size still needs to be reasonable). For models that satisfy the asymptotic behavior near the wall, the results do not appear to be significantly affected by the use of coarse grids, as long as $y^+ \leq 2$ to 4 for the first grid point [27].

It has been pointed out that the use of the wall coordinate $y^+ = yu_\tau/\nu$ often leads to problems in the conventional low-Reynolds number k-ε models. It becomes inappropriate to use y^+ for cases where u_τ becomes very small near the separation point or for flows around a complex-shaped body in which case, it may be difficult to evaluate the wall-normal distance y. Abe et al. [1] proposed a model that used the Kolmogorov scale velocity $u_\varepsilon \equiv (\nu\varepsilon)^{1/4}$ and the turbulent Reynolds number $Re_t \equiv k^2/\nu\varepsilon$ instead of u_τ. This model uses

$$
\left.
\begin{aligned}
&C_\mu = 0.09, \quad C_{\varepsilon 1} = 1.5, \quad C_{\varepsilon 2} = 1.9, \quad \sigma_k = 1.4, \quad \sigma_\varepsilon = 1.4 \\[4pt]
&f_\mu = \left[1 - \exp\left(-\frac{y^*}{14}\right)\right]^2 \left\{1 + \frac{5}{Re_t^{3/4}} \exp\left[-\left(\frac{Re_t}{200}\right)^2\right]\right\} \\[4pt]
&f_1 = 1 \\[4pt]
&f_2 = \left[1 - \exp\left(-\frac{y^*}{3.1}\right)\right]^2 \left\{1 - 0.3\exp\left[-\left(\frac{Re_t}{6.5}\right)^2\right]\right\} \\[4pt]
&D = E = 0,
\end{aligned}
\right\}
\tag{7.63}
$$

where $y^* \equiv y/\eta$. In this formulation, the need for u_τ has been eliminated. This model, Eq. (7.63), also satisfies the asymptotic behavior near the wall. There has been efforts to improve the models based on examining DNS results.

7.4.2 Computational Details of the k-ε Model

Comparing the Reynolds-averaged equations using the eddy-viscosity model, Eq. (7.21), with the Navier–Stokes equations, the only difference is the viscosity ν being changed to $\nu + \nu_T$. Thus, we essentially need to solve for the flow field as if it is laminar flow with non-constant viscosity. Equations (7.21) and (7.4) must be solved in a coupled manner to determine $\overline{u}_i$ and $\overline{P}$. Because it would be a challenge to implicitly treat the non-constant eddy viscosity ν_T term, we can practically handle only the constant kinematic viscosity term implicitly and treat the eddy-viscosity term explicitly. For the k-ε model, we can find $\overline{p}$ from $\overline{P}$ through Eq. (7.22) since we have access to k.

Time Stepping of the k-ε Equation

For discretizing the k- and ε-equations, there is not much difference from how the other equations are discretized. Because the variables k and ε are scalar quantities, it would be sensible to place them at the cell centers when using staggered or collocated grids. The production term can be evaluated from the available velocity and Reynolds stress. The advective and diffusive terms can be treated in a similar way to how the analogous terms in the momentum equation are discretized, as discussed in Chap. 3.

It has been reported that implicitly handling the sink term increases numerical stability [20]. For example, we can time step the k- and ε-equations using the second-order Adams–Bashforth method for the production, advection, and diffusive terms and the Crank–Nicolson method for the sink terms

$$\frac{k^{n+1} - k^n}{\Delta t} = \frac{3H_k^n - H_k^{n-1}}{2} - \left(\frac{\varepsilon}{k}\right)^n \frac{k^{n+1} + k^n}{2}, \tag{7.64}$$

$$\frac{\varepsilon^{n+1} - \varepsilon^n}{\Delta t} = \frac{3H_\varepsilon^n - H_\varepsilon^{n-1}}{2} - C_{\varepsilon 2} \left(\frac{\varepsilon}{k}\right)^n \frac{\varepsilon^{n+1} + \varepsilon^n}{2}, \tag{7.65}$$

where H_k and H_ε represent the collection of the production, advection, and diffusive terms in the respective equations.

Boundary Condition for Dissipation Rate

It is necessary to prescribe the wall boundary condition for the dissipation rate ε when we use the low-Reynolds number k-ε model. Denoting the wall-normal direction by y, we can specify

$$\varepsilon_w = \nu \left[\frac{\partial^2 k}{\partial y^2}\right]_w \tag{7.66}$$

or

$$\varepsilon_w = 2\nu \left[\frac{\partial \sqrt{k}}{\partial y}\right]_w^2. \tag{7.67}$$

Because one-sided finite-difference scheme often has reduced accuracy (see Sect. 2.3.1), the expression

$$\varepsilon_p = 2\nu \frac{k_p}{y_p^2} \tag{7.68}$$

is often used instead to avoid differentiation (where y_p, k_p, and ε_p are values at the point next to the wall). For this boundary condition, k and ε would be placed at the center of the grid cells. Since ε would not be defined on the wall in such case, the variable placements would be consistent with how we evaluate the variables in computations.

7.4.3 Features and Applications of the k-ε Model

The k-ε model is widely used in engineering and industrial applications that are often a part of larger engineering systems. Therefore, it is quite important to understand the properties of the model and its limitations. The strengths of the k-ε model are

- The model parameters have been well tuned and widely accepted for many applications.
- There are techniques proposed to use reduced number of grid points near the wall by employing the wall law. Hence the calculations can be performed with reasonable computational cost.
- The eddy viscosity is always positive which provides stability to numerical calculations.

There are cases when the k-ε model cannot correctly predict the behavior of turbulent flow. The k-ε model has difficulty predicting the following types of flows:

- Flows for which the eddy-viscosity approximation does not hold.
- Flows with its average profile being greatly influenced by the anisotropy in Reynolds stress.
- Flows that are not in local equilibrium ($P_k \not\approx \varepsilon$).
- Flows for which the approximation k/ε and $k^{3/2}/\varepsilon$ for the time and length scales are not appropriate.

Let us consider specific examples of flows for which the k-ε model would not predict the flow behavior correctly, as depicted in Figs. 7.2 and 7.3. The discussion below does not only apply to the k-ε model but also to the eddy-viscosity models in general. First, we consider a flow that hits a wall as illustrated in Fig. 7.2a, which models wind hitting a building. By combining Eq. (7.31) and the continuity equation ($\frac{\partial \overline{u}}{\partial x} + \frac{\partial \overline{v}}{\partial y} = 0$) the production of energy can be written as

$$P_k = -(\overline{u'u'} - \overline{v'v'})\frac{\partial \overline{u}}{\partial x}. \tag{7.69}$$

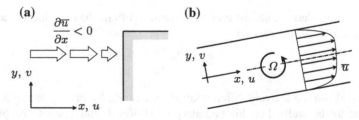

Fig. 7.2 Examples of turbulent flows for which k-ε model does not perform well. **a** Flow hitting a wall. **b** Flow within a rotating channel

Expressing the above using the eddy viscosity, we have

$$P_k = 4\nu_T \left(\frac{\partial \overline{u}}{\partial x}\right)^2 . \tag{7.70}$$

In actual turbulent flows, the difference between $\overline{u'u'}$ and $\overline{v'v'}$ should contribute to the production of energy. However the use of the eddy-viscosity model makes the sum of those terms artificially contribute to the production of turbulence [17]. For this reason, the turbulent kinetic energy k becomes unrealistically large where the flow hits the wall and accordingly makes the prediction of the overall flow field difficult.

Next, let us take an example of a flow in a two-dimensional channel that rotates at an angular velocity of Ω. The high-pressure side (bottom wall in the figure) shows increases in wall stress and turbulent fluctuations. The low-pressure side (top wall in the figure) exhibits the opposite trend. Such increase or decrease can be explained by looking at the production term P_{ij} and the rotational term R_{ij}

$$P_{11} + R_{11} = -\overline{u'v'}\frac{\partial \overline{u}}{\partial y} + 4\Omega \overline{u'v'} \tag{7.71}$$

$$P_{22} + R_{22} = -4\Omega \overline{u'v'} \tag{7.72}$$

$$P_{33} + R_{33} = -\overline{v'v'}\frac{\partial \overline{u}}{\partial y} - 2\Omega \left(\overline{u'u'} - \overline{v'v'}\right) \tag{7.73}$$

in the equation for each component of the Reynolds stress. For a rotating turbulent channel flow, there is redistribution of turbulent kinetic energy between the velocity fluctuation components $\overline{u'u'}$ and $\overline{v'v'}$ due to the Coriolis force, which consequently alters the Reynolds shear stress. The eddy viscosity model only takes k into account and does not include the effects from the aforementioned redistribution. For that reason, the model is not able to capture the features in rotational turbulent flow (the high-pressure side destabilizing and the low-pressure side laminarizing) [13].

Let us consider another example of fully developed turbulent flow inside a square duct. The primary velocity profile in the duct is shown in Fig. 7.3, which is determined from LES (see next chapter) [10]. For turbulent flow in noncircular pipes, it is known that streamwise velocity contours protrude into convex regions (corners) of the cross section as seen in Fig. 7.3a. There is also *secondary flow* comprised of the velocity components orthogonal to the primary streamwise velocity. The corresponding cross-sectional view of the secondary flow for a square duct is illustrated in Fig. 7.3b. In general, there are two types of cross-stream secondary flow

- *Prandtl's secondary flow of the first kind*—secondary flow that appears in curved pipes or pipes under rotation
- *Prandtl's secondary flow of the second kind*—secondary flow that appears in stationary non-circular pipes

What we have in Fig. 7.3 is Prandtl's secondary flow of the second kind. Although the magnitude of the secondary flow is of order 1/100 of the streamwise velocity and 1/10

Fig. 7.3 Depiction of the velocity profile inside a square duct [10].
a Streamwise flow profile ($\overline{u}_1$). **b** Cross-sectional secondary flow ($\overline{u}_2$, $\overline{u}_3$)

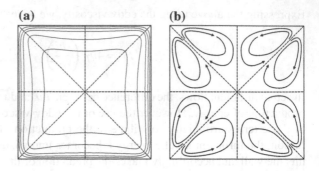

(a)   **(b)**

of the turbulent velocity fluctuation, secondary flow modifies the primary velocity profile and greatly influences frictional drag. Thus, it is important to accurately predict the secondary flow for analyzing flows in ducts and designing engineering systems with ducts. To capture the turbulent flow profile, as shown in Fig. 7.3, it is necessary to use a turbulence model that can correctly evaluate the near-wall behavior including the difference between the normal components of the Reynolds stress $\overline{u'_2 u'_2} - \overline{u'_3 u'_3}$ as well as the shear component $\overline{u'_2 u'_3}$. The eddy-viscosity model such as the k-ε model can usually capture the primary shear components $\overline{u'_1 u'_2}$ and $\overline{u'_1 u'_3}$ but has difficulty predicting the aforementioned secondary components.

As we have discussed above, the eddy-viscosity models including the k-ε model, is not appropriate for flows characterized by the redistribution of velocity fluctuations amongst different directional components. It is often the case that the normal components of the Reynolds stress ($\overline{u'u'}$, $\overline{v'v'}$, and $\overline{w'w'}$) determined from the eddy viscosity approximation, Eq. (7.19), is quite different from actual physics. Average flow fields that can be represented accurately with the k-ε model are flows where one component of the shear stress (e.g., $\overline{u'v'} = -\nu_T \partial \overline{u}/\partial y$) is dominant and the other normal components do not contribute much to the average flow.

7.5 Other Eddy-Viscosity Models

k-ω Model

Another model that is used as much as the aforementioned k-ϵ model is the k-ω model. While there are several variations of the k-ω model, let us present the standard form of the model following Wilcox [28] for constant density ρ. We present here the standard k-ω model following the standard k-ε model form of Eqs. (7.49), (7.50), and (7.51) from the k-ε model

$$\nu_T = \frac{k}{\omega}, \tag{7.74}$$

$$\frac{\overline{D}k}{\overline{D}t} = P_k - \beta^* k\omega + \frac{\partial}{\partial x_j}\left[\left(\frac{\nu_T}{\sigma_k^*} + \nu\right)\frac{\partial k}{\partial x_j}\right], \tag{7.75}$$

$$\frac{\overline{D}\omega}{\overline{D}t} = \alpha\frac{\omega}{k}P_k - \beta^*\omega^2 + \frac{\partial}{\partial x_j}\left[\left(\frac{\nu_T}{\sigma_\omega} + \nu\right)\frac{\partial \omega}{\partial x_j}\right], \tag{7.76}$$

$$\alpha = \frac{5}{9}, \quad \beta = \frac{3}{40}, \quad \beta^* = 0.09, \quad \sigma_k^* = 2, \quad \sigma_\omega = 2, \tag{7.77}$$

where we note that ω has the dimension of inverse time as in the case of ε/k. By comparing Eqs. (7.49), (7.50) and Eqs. (7.74), (7.75), the parameters $\beta^* = C_\mu = 0.09$ match for the relationship of $\varepsilon = \beta^* k\omega = C_\mu k\omega$ to hold. However, there is difference in the turbulent eddy-viscosity coefficients for k in Eqs. (7.50) between (7.75) with $\sigma_k^* \neq \sigma_k$.

Now let us consider flow over a flat plate and examine Eqs. (7.75) and (7.76) near the wall. In the limit of $y \to 0$, we find these equations become

$$\nu\frac{\partial^2 k}{\partial y^2} = \beta^* k\omega, \quad \nu\frac{\partial^2 \omega}{\partial y^2} = \beta\omega^2. \tag{7.78}$$

Solving these equations, we find that $\omega \propto y^{-2}$ and $k \propto y^{3.23}$. On the other hand, the near-wall behavior of the k-ϵ model from Eq. (7.62) shows that $k \propto y^2$ and $\omega \propto \varepsilon/k \propto y^{-2}$. While the behavior of ω agrees with that from Eq. (7.76), the value of ω becomes infinite which needs to be controlled, for example by providing bounds on ω near the wall. For k, Eq. (7.75) itself does not behave in an manner consistent with the desired asymptotic behavior and hence necessitates the use of wall corrections.

We can also examine the relationship between the k-ε and k-ω models. Substituting $\beta^* k\omega$ into ε in Eq. (7.51) and splitting the right-hand side, we find

$$\frac{\overline{D}\varepsilon}{\overline{D}t} = \beta^*\omega\left(\frac{\partial k}{\partial t} + \overline{u}_j\frac{\partial k}{\partial x_j}\right) + \beta^* k\left(\frac{\partial \omega}{\partial t} + \overline{u}_j\frac{\partial \omega}{\partial x_j}\right). \tag{7.79}$$

Now, using Eqs. (7.75) and (7.76) from the standard k-ω model with the above equation, we arrive at

$$\frac{\overline{D}\varepsilon}{\overline{D}t} = \frac{\varepsilon}{k}(C_{\varepsilon1}^* P_k - C_{\varepsilon2}^* \varepsilon) + \frac{\partial}{\partial x_j}\left[\left(\frac{\nu_T}{\sigma^*} + \nu\right)\frac{\partial \varepsilon}{\partial x_j}\right] + S_\omega, \tag{7.80}$$

where $C_{\varepsilon1}^* = 1 + \alpha$, $C_{\varepsilon2}^* = 1 + \beta/\beta^*$, and both σ_k^* and σ_ω are set to σ^*. The difference between the k-ε and k-ω models, except for the parameters, can be consolidated into a single term S_ω

$$S_\omega = -2\left(\frac{\nu_T}{\sigma^*} + \nu\right)\frac{\partial k}{\partial x_j}\frac{\partial \omega}{\partial x_j}$$

$$= 2\omega\left(\frac{\nu_T}{\sigma^*} + \nu\right)\left(\frac{1}{k}\frac{\partial k}{\partial x_j}\frac{\partial k}{\partial x_j} - \frac{1}{\varepsilon}\frac{\partial k}{\partial x_j}\frac{\partial \varepsilon}{\partial x_j}\right). \tag{7.81}$$

To evaluate the eddy viscosity, there are choices other than ε and ω to be combined with k. In those cases, we should be cautious with the possibility of the viscous coefficient becoming negative, which is not desirable from the view point of numerical stability. If a different variable is to be used in a two-equation model, it is beneficial to choose a variable that is physically and numerically appropriate for finding the eddy viscosity.

Shear Stress Transport (SST) Model

In general, the k-ω model performs well in predicting boundary layer flows with adverse pressure gradient and separation. However, it is reported that the model has difficulty predicting spatially developing free shear flows. To address this shortcoming, Menter [16] proposed the *shear stress transport (SST) model* that combines the strengths of the k-ϵ and k-ω models.

The SST model is expressed in terms of the k-ω model and consists of Eq. (7.75) and a ω-equation

$$\frac{\overline{D}\omega}{\overline{D}t} = \gamma\frac{\omega}{k}P_k - \beta^*\omega^2 + \frac{\partial}{\partial x_j}\left[\left(\frac{\nu_T}{\sigma_\omega} + \nu\right)\frac{\partial \omega}{\partial x_j}\right]$$
$$+ 2(1 - F_1)\sigma_{\omega 2}\frac{1}{\omega}\frac{\partial k}{\partial x_j}\frac{\partial \omega}{\partial x_j} \tag{7.82}$$

that has a source term similar to Eq. (7.81). In this model, the k-ϵ and k-ω models are blended through function F_1 (where $0 \le F_1 \le 1$). When $F_1 = 0$ we recover the k-ϵ model and when $F_1 = 1$ we retrieve the k-ω model. Two sets of parameters are provided for each k-ϵ and k-ω model. Denoting the model parameters from the k-ω and k-ϵ models with ϕ_ω and ϕ_ϵ, respectively, the SST model uses the weighted average $\phi = F_1\phi_\omega + (1 - F_1)\phi_\epsilon$ to determine its parameters. For full details on the model, readers should consult with Menter [16]. Similar to the SA model, the SST model has also been used widely in the field of aerodynamics.

Nonlinear Eddy-Viscosity Model

The ratio of the Reynolds stress to the strain rate tensor of the average velocity field should actually be described by a tensor instead of a single scalar ν_T. The eddy-viscosity model can be modified to include such variation. It is also possible to generalize the model by not only using the first-order strain rate tensor but also using higher-order terms in the expansion. There are nonlinear eddy-viscosity models that are proposed based on such ideas (high-order eddy-viscosity model and anisotropic eddy-viscosity model) [26].

In the nonlinear eddy-viscosity model, the Reynolds stress is expressed with the first-order term in $\overline{D}_{ij}$, Eq. (7.20), as well as the second and third-order terms in

$\overline{D}_{ij}$ and $\overline{W}_{ij} = (\partial \overline{u}_i/\partial x_j - \partial \overline{u}_j/\partial x_i)/2$. As the expansion may not be convergent, there is no guarantee that the computational results are to improve. Furthermore, one should be cautious in utilizing the model for non-inertial systems because $\overline{W}_{ij}$ is not an objective quantity (see Sect. 6.3.6).

There are theoretical studies that relate the nonlinear eddy-viscosity models and the stress equations models to explain the physical contributions from the higher-order expansion terms. Furthermore, there are a number of studies reporting that the nonlinear eddy viscosity model requires less computation resource than the stress equation model for predicting complex turbulent flows which would not have been accurately captured with the k-ε model.

7.6 Reynolds Stress Equation Model

The stress equation model closes the system of equations comprised of the governing equations for the average flow and a model constructed at the level of the Reynolds stress equation, Eq. (7.12). As mentioned in Sect. 7.1, for turbulent flows that cannot be predicted satisfactorily with the standard k-ε model, we would in general recommend the use of LES models instead of the stress equation model. We therefore will only discuss the Reynolds stress equation model briefly for the completeness of the discussion of turbulence modeling.

Because the Reynolds stress is a second-order velocity correlation tensor, we cannot determine the length or time scale by itself. We need at least one additional parameter related to turbulence. Usually, the turbulent kinetic energy dissipation rate ε is used to supplement the system of equations. Thus, in a similar manner to the k-ε model, the time and length scales are k/ε and $k^{3/2}/\varepsilon$, respectively. The parameters employed for the Reynolds stress equation model are the Reynolds stress, ε, and the average velocity field.

7.6.1 Basic Form of the Stress Equation

There is no need to model the production and the viscous diffusion terms (Eqs. (7.13) and (7.18), respectively), in the stress equation, Eq. (7.12), because they are expressed in terms of the Reynolds stress and the average velocity gradient. Below we describe how the rest of the terms can be modeled. One major advantage to use the stress equation model is that the production term requires no eddy viscosity.

Pressure-Strain Correlation Model

The pressure-strain term Π_{ij} satisfies $\Pi_{ii} = 0$ due to the continuity equation. For this reason, we can interpret that the pressure-strain correlation transfers the Reynolds stress among its components. Hence this term is also referred to as the *redistribution term*. Commonly, the model for Π_{ij} consists of three parts

$$\Pi_{ij} = \Pi_{(1)ij} + \Pi_{(2)ij} + \Pi_{(w)ij}, \tag{7.83}$$

which we describe below.

Let us examine the pressure-strain correlation terms analytically. The pressure fluctuation p' can be determined from the Poisson equation that is derived by taking the divergence of Eq. (7.11) and invoking incompressibility

$$\frac{1}{\rho}\nabla^2 p' = -2\frac{\partial u'_i}{\partial x_j}\frac{\partial \overline{u}_j}{\partial x_i} - \frac{\partial^2 (u'_i u'_j - \tau_{ij})}{\partial x_i \partial x_j}. \tag{7.84}$$

The above equation can be solved with Green's theorem to yield

$$\frac{p'}{\rho} = \frac{1}{2\pi}\int_V \frac{\partial u'_i}{\partial x_j}\frac{\partial \overline{u}_j}{\partial x_i}\frac{dV_y}{|\boldsymbol{x} - \boldsymbol{y}|} + \frac{1}{4\pi}\int_V \frac{\partial^2 (u'_i u'_j - \tau_{ij})}{\partial x_i \partial x_j}\frac{dV_y}{|\boldsymbol{x} - \boldsymbol{y}|}$$
$$+ \frac{1}{4\pi}\int_S \left(\frac{1}{|\boldsymbol{x} - \boldsymbol{y}|}\frac{\partial p'}{\partial n} - p'\frac{\partial}{\partial n}\frac{1}{|\boldsymbol{x} - \boldsymbol{y}|}\right)\frac{dS_y}{\rho}, \tag{7.85}$$

where V is the volume of interest, S is its surface, and n is the unit normal vector at the surface.

We can derive the three-part representation of the pressure-strain (redistribution) term, Eq. (7.83), by multiplying $(\partial u'_i/\partial x_j + \partial u'_j/\partial x_i)$ to the above equation and taking the Reynolds average. The first term in Eq. (7.83) is

$$\Pi_{(1)ij} = \frac{1}{4\pi}\int_V \overline{\left(\frac{\partial u'_i}{\partial x_j} + \frac{\partial u'_j}{\partial x_i}\right)\frac{\partial^2 (u'_k u'_l)}{\partial y_k \partial y_l}}\frac{dV_y}{|\boldsymbol{x} - \boldsymbol{y}|} \tag{7.86}$$

and is called the *slow term*, which corresponds to the component that changes through the velocity fluctuation. The second term is

$$\Pi_{(2)ij} = \frac{1}{2\pi}\int_V \frac{\partial \overline{u}_k}{\partial x_l}\overline{\left(\frac{\partial u'_i}{\partial x_j} + \frac{\partial u'_j}{\partial x_i}\right)\frac{\partial u'_l}{\partial y_k}}\frac{dV_y}{|\boldsymbol{x} - \boldsymbol{y}|} \tag{7.87}$$

and is referred to as the *rapid term*, which represents the component that is immediately affected by the change in the average flow field. The third term is known as the *wall-reflection term*. This term is written in a form proportional to $k^{3/2}/(\varepsilon y_d)$, where $k^{3/2}/\varepsilon$ and y_d are the characteristic length scale and the distance from the wall, respectively [4]. For locations far enough from the wall, this term can be neglected.

Terms that arise from pressure are dependent on the whole flow field for incompressible flow (due to the elliptic nature). Thus, Eqs. (7.86) and (7.87) are in the integral forms. Reproducing these integrals in a model makes its use unattractive, and has often led to approximating these expressions locally at a single point.

For the slow term, the model proposed by Rotta [22] incorporates isotropy in the simplest fashion as follows

$$\Pi_{(1)ij} = -C_1 \frac{\varepsilon}{k} \left(\tau_{ij} - \frac{2}{3}\delta_{ij}k \right), \tag{7.88}$$

where C_1 is a positive modeling constant. The term $\Pi_{(1)ij}$ acts to relax the anisotropic fluctuation $\tau_{ij}^a = \tau_{ij} - \frac{2}{3}\delta_{ij}k$, which is based on the following idea. If we consider a problem of anisotropic turbulence decay in uniform flow without mean velocity gradient, what are left in the stress equation are the time derivative term, dissipation term and the slow term (in the pressure-strain correlation). If we perform such experiment, the fluctuation should decay and reach isotropy. At least for such example, the slow term should be able to steer turbulence to become isotropic. One can also model the difference between $\Pi_{(1)ij}$ and the anisotropic component of ε_{ij} as the *return (return-to-isotropy) term* $\Pi_{(1)ij} - \varepsilon_{ij}^a$.

Because the rapid term contains the average velocity gradient, its change quickly acts to perform redistribution in the turbulence. In general the rapid term is written as

$$\Pi_{(2)ij} = -\frac{C_2 + 8}{11} \left(P_{ij} - \frac{2}{3}\delta_{ij}P \right) - \frac{30C_2 - 2}{55}k \left(\frac{\partial \overline{u}_i}{\partial x_j} + \frac{\partial \overline{u}_i}{\partial x_i} \right)$$
$$- \frac{8C_2 - 2}{11} \left(Q_{ij} - \frac{2}{3}\delta_{ij}P \right), \tag{7.89}$$

where

$$P_{ij} = -\tau_{ki}\frac{\partial \overline{u}_j}{\partial x_k} - \tau_{kj}\frac{\partial \overline{u}_i}{\partial x_k}, \quad Q_{ij} = -\tau_{ki}\frac{\partial \overline{u}_k}{\partial x_j} - \tau_{kj}\frac{\partial \overline{u}_k}{\partial x_i}, \tag{7.90}$$

$P = \frac{1}{2}P_{kk} = \frac{1}{2}Q_{kk}$, and C_2 is a model parameter. The above model is known as the *Launder–Reece–Rodi* (LRR) *model* [11]. In most cases, the first term in Eq. (7.89) is dominant. Based on this observation, the rapid term is sometimes simplified as

$$\Pi_{(2)ij} = -C_2 \left(P_{ij} - \frac{2}{3}\delta_{ij}P_k \right) \tag{7.91}$$

with the anisotropic component of the production term having a negative sign. While in reality the Reynolds stress is determined through the process of production, redistribution, and dissipation, it can be seen that this model quickly responds to the anisotropy of production and leads the flow to become isotropic. This is the reason for which this term is called the rapid term.

The slow term responds to the anisotropy of the Reynolds stress components and the rapid term is proportional to the anisotropy in the production term. These terms guide the flow towards isotropy. When the flow field has a boundary surface, the terms with area integral play an important role. In fact, with only $\Pi_{(1)ij}$ and $\Pi_{(2)ij}$, we observe isotropization effects regardless of the constraints imposed by the boundary surface. However, the existence of the solid wall selectively reduces the wall-normal fluctuation, effectively acting as a guide towards anisotropy. This is the wall reflective effect in the pressure fluctuation field which corresponds to the area integral in Eq. (7.85). This term also consists of $\Pi_{(w1)ij}$ and $\Pi_{(w2)ij}$ that are due to velocity fluctuation and the influence from the average flow field. We note in passing that numerous models exist for each of these terms.

Dissipation Term Model

When the turbulence Reynolds number $k^2/\nu\varepsilon$ is high enough, it can be thought that the dissipation of the Reynolds stress is carried out solely by the small eddies, similar to the dissipation of the turbulent kinetic energy. Turbulence at such scale can be viewed as isotropic so that

$$\varepsilon_{ij} = \frac{2}{3}\delta_{ij}\varepsilon \tag{7.92}$$

is a reasonable approximation. For the return term $\Pi_{(1)ij} - \varepsilon_{ij}^a$, the isotropic component $\frac{2}{3}\delta_{ij}\varepsilon$ is left as the dissipation term in the stress equation. If we have access to the value of ε, there is no need for a model.

The ε-equation coupled with the stress equation is

$$\frac{\overline{D}\varepsilon}{\overline{D}t} = -C_{\varepsilon 1}\frac{\varepsilon}{k}\tau_{ij}\frac{\partial \bar{u}_i}{\partial x_j} - C_{\varepsilon 2}\frac{\varepsilon^2}{k} + \frac{\partial}{\partial x_i}\left[\left(C_\varepsilon\frac{k}{\varepsilon}\tau_{ij} + \nu\right)\frac{\partial\varepsilon}{\partial x_j}\right]. \tag{7.93}$$

Except for the diffusion term, this equation shares the same basic formulation of the k-ε model. Note however that there is a difference between this model and the ε-equation in the k-ε model. The Reynolds stress is utilized for τ_{ij}, instead of utilizing the eddy-viscosity model in the above expression.

Diffusion Term Model

For the diffusive flux due to turbulence $J_{(T)ijk}$, there is the *Daly–Harlow model* [5] that takes different effects into account through a coefficient in the gradient diffusivity model

$$J_{(T)ijk} = -\overline{u_i'u_j'u_k'} = C_s\frac{k}{\varepsilon}\tau_{kl}\frac{\partial\tau_{ij}}{\partial x_l}, \tag{7.94}$$

where C_s is a constant. However, Eq. (7.94) does not satisfy the symmetry property of $\overline{u_i'u_j'u_k'} = \overline{u_i'u_k'u_j'}$. The *Hanjalić–Launder model* [7] resolves this issue by setting

$$J_{(T)ijk} = -\overline{u_i'u_j'u_k'} = C_s'\frac{k}{\varepsilon}\left(\tau_{kl}\frac{\partial\tau_{ij}}{\partial x_l} + \tau_{il}\frac{\partial\tau_{jk}}{\partial x_l} + \tau_{jl}\frac{\partial\tau_{ki}}{\partial x_l}\right), \tag{7.95}$$

where C_s' is a constant. The diffusive flux related to pressure fluctuation $J_{(P)ijk}$ is often neglected or lumped together with $J_{(T)ijk}$.

We leave the above formulation to be the basic form of the stress equation model since there is not a well-established standard model as in the k-ε models. The near-wall region still requires correction for anisotropic turbulence or low-Reynolds number effects, similar to the k-ε models. The stress equation models with such corrections are proposed by Hanjalić and Launder [7], Shima [23], and So [24].

7.6.2 Features of the Stress Equation Model

The stress equation model has been shown to perform better than the k-ε model for rotating flows, flows with curved streamlines, slowly varying unsteady flows, secondary flows of the second kind, impinging flows, rough-wall channel flow, wall jets, and other flows. Although it is possible to implement modifications in the k-ε by improving upon the complexity of the model for each type of flow, it is not clear whether the modifications can make any improvements for flows with combined effects. The stress equation model has provided good results for the aforementioned cases of flows without requiring users to choose a model or tune parameters. The reason for its ability to predict turbulent flows well is attributed to modeling each component of the Reynolds stress production term and considering the stress redistribution through the pressure-strain correlation term.

Nonetheless, the stress equation models have not been widely used. Some of the reasons for not being utilized can be listed below:

- Due to the added complexity, the number of equations in the model is increased which requires added computation time compared to the k-ε model.
- Each part of the model has not been fully validated.
- Providing the initial and boundary conditions for all terms becomes cumbersome.
- The Reynolds equation is not diffusive and can be susceptible to numerical instability.

The issues related to computational expense have been somewhat resolved with the advancement of computer hardware technology. The rest of the problems are related to the numerical simulation techniques for the stress equation model, especially the stability issue. For actual implementations, often times experience-based upper and lower bounds on the magnitude of each term are specified such that negative diffusion coefficients or extreme nonequilibrium conditions are not encountered during the computations [29]. At the moment, most practical calculations require numerical alternations (e.g., artificial viscosity, specifications of bounds, etc.), which make it difficult to separate the influence from the use of models and the numerical techniques.

7.7 Remarks

For "simple flows," the k-ε eddy-viscosity models can predict the Reynolds-averaged flow field quite well. What we mean by "simple flows" are stationary flows that have average streamlines that are relatively straight on the absolute coordinate system with low level of acceleration. Flows that impinge on walls, separate from corners, pass through a curved channel, and are in a rotational field would not be considered simple flows. The rate-of-strain tensor and the Reynolds stress tensor are symmetric tensors with six independent elements. For simple flows, a pair of elements representing shear becomes dominant (e.g., $\partial \overline{u}/\partial y$ and $\overline{u'v'}$). In such cases, the k-ε model is well tuned to predict the average flow field. One such representative case is the developed turbulent flow in circular pipe as illustrated in Fig. 7.1.

The question is how one should solve for the turbulent flow field when the eddy-viscosity model is not applicable. For certain types of problems, we can make small changes to isotropic eddy-viscosity models (k-ε, k-ω, SST models) by incorporating an anisotropic eddy-viscosity contribution to better predict the flow. However, there are no set guidelines for choosing a model for flows for which we cannot predict what kind of vortical structures appear. Letting commercial software select appropriate models for complex flows is not advisable.

For flows that are not simple, one can give up on the use of the eddy-viscosity model and instead choose the stress equation model or LES, which is discussed in the next chapter. Rather than using a sophisticated stress equation mode, we recommend the use of LES, even if it is a low-order LES model. At the moment, the stress equation models are not as well tuned as the k-ε and k-ω models. Moreover, the implementation of the stress equation model is not straightforward. As the stress equation model becomes further developed, we may start to see the limitation on the use of the one-point closure model (both the k-ε and stress equation models only handle the velocity fluctuation moment at a single point). If we insist on the use of Reynolds averaging, derived models may become further complex. Such a trend would not be welcomed by the users.

Let us also comment on the ambiguity of Reynolds averaging. As we have discussed in Sect. 7.2.1, the Reynolds average is defined as an averaging operation that satisfies the Reynolds averaging laws and is not defined as a particular average over a spatial or temporal scale. For example, there is no clear distinction between the vortical structures and the turbulent fluctuations that appear in jets, wakes, and mixing layers. Thus, for unsteady flow calculations, it becomes unclear whether the vortical structures are regarded as unsteadiness in the Reynolds-averaged flow or as fluctuating component that is averaged out to result in a steady Reynolds-averaged flow. There is also influence from the spatial and temporal resolution in computing the average flow. If we further consider examples of two-phase flows with particles, bubbles, or droplets, there is an open question on how the scale from Reynolds averaging and the scale used to compute phase-average (e.g., number density, volume fraction, etc.) would interact. For turbulent flows comprised of many physical scales, it is easier to incorporate various physical components into LES, in which scales that are computed and modeled are defined in a relatively clear manner, as opposed to the Reynolds-averaged equations in which scale separation is not so obvious.

In this chapter, we have kept the discussions on turbulence models brief with the exception of the k-ε model. The details of turbulence models, including the model parameters, empirical functions, and switching criteria for hybrid models, can vary from one turbulence solver program to the other. Furthermore, such details are constantly being adjusted. Hence, we have not attempted to list the vast set of turbulence models in this book. For additional details on the models or their comparisons, we recommend readers to refer to Wilcox [28] and Durbin and Pettersson Reif [6] or resources available on the internet:

Fig. 7.4 Invariant map of the second and third invariants of the anisotropic component of the Reynolds stress tensor

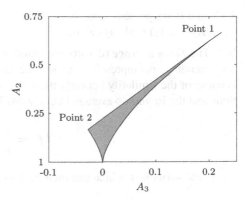

- Turbulence Modeling Resource website[2] hosted by NASA Langley Research Center (C. Rumsey)
- CFD Online website[3]

7.8 Exercises

7.1 Given the Reynolds stress tensor, let us define its anisotropic components by

$$a_{ij} = \frac{1}{2k}\overline{u_i' u_j'} - \frac{1}{3}\delta_{ij},$$

where $k = \frac{1}{2}\overline{u_k' u_k'}$ is the turbulent kinetic energy. The second and third invariants of this tensor are provided by

$$A_2 = a_{ij}a_{ji} \quad \text{and} \quad A_3 = a_{ij}a_{jk}a_{ki},$$

respectively. The state of the Reynolds stress can be represented by the *invariant map* [14], which plots A_3 on the horizontal axis and A_2 on the vertical axis, as shown in Fig. 7.4. The state of the flow can be found to reside inside the gray region defined by Points 1, 2, and the origin. (Also see Problem 6.4)

1. Show that isotropic turbulence is represented on the plot at the origin.
2. Show that Point 1 represents flows having velocity fluctuations in only one direction. Find the coordinates of Point 1.
3. Show that Point 2 corresponds to flows that have isotropic two-component fluctuations. Determine the coordinates of Point 2.

---

[2]http://turbmodels.larc.nasa.gov.
[3]http://www.cfd-online.com.

4. When flows possess velocity fluctuations in two directions, show that $A = 1 - \frac{9}{2}(A_2 - 2A_3)$ is always zero.

7.2 The *Favre average* (density weighted average) is considered as a valuable tool for extending incompressible turbulence models to compressible turbulent flows, because of the similarity between the forms of the Favre-averaged governing equations and the Reynolds-averaged equations. The Favre average is defined as

$$\langle f \rangle = \frac{1}{\overline{\rho}}\,\overline{\rho f}$$

for a field variable f which can then be decomposed as

$$f = \langle f \rangle + f'',$$

where f'' denotes the fluctuating component.

1. Similar to the Reynolds average, prove that the Favre average satisfies

$$\langle fg \rangle = \langle f \rangle \langle g \rangle + \langle f'' g'' \rangle.$$

2. Show that the difference between the Favre and Reynolds averages is

$$\langle f \rangle - \overline{f} = \frac{1}{\overline{\rho}}\,\overline{\rho' f'}.$$

3. Replace the velocity variable in the Reynolds-averaged continuity equation

$$\frac{\partial \overline{\rho}}{\partial t} + \frac{\partial}{\partial x_k}\overline{\rho u_k} = 0$$

by its Favre average and state the resulting equation.

4. The Reynolds-averaged momentum equation is given by

$$\frac{\partial}{\partial t}\overline{\rho u_i} + \frac{\partial}{\partial x_k}\overline{\rho u_i u_k} = \frac{\partial}{\partial x_k}(-\delta_{ik}\overline{p} + \overline{\tau}_{ik})$$

where $\tau_{ik} = \mu\left(\dfrac{\partial u_i}{\partial x_k} + \dfrac{\partial u_k}{\partial x_i}\right) - \dfrac{2}{3}\delta_{ik}\mu\dfrac{\partial u_m}{\partial x_m}$ is the viscous stress. Show by replacing the velocity variable with its Favre average that the momentum equation can be expressed as

$$\frac{\partial}{\partial t}\left(\overline{\rho}\,\langle u \rangle_i\right) + \frac{\partial}{\partial x_k}\left(\overline{\rho}\,\langle u \rangle_i \langle u \rangle_k\right) = \frac{\partial}{\partial x_k}(-\delta_{ik}\overline{p} + \overline{\tau}_{ik} - \overline{\rho}\,\langle u_i'' u_k'' \rangle).$$

5. Derive the transport equation for the second-order tensor of the Favre fluctuation velocity

$$\langle R \rangle_{ij} = \langle u_i'' u_j'' \rangle \quad \left(= \langle u_i u_j \rangle - \langle u \rangle_i \langle u \rangle_j \right).$$

7.3 For a generalized coordinate system, derive the transport equations for the turbulent kinetic energy k

$$\frac{\partial k}{\partial t} + \overline{U}^k \frac{\partial k}{\partial \xi^k} = P_K - \varepsilon + \frac{1}{\sqrt{\hat{g}}} \frac{\partial}{\partial \xi^k} \left\{ \left(\nu + \frac{\nu_T}{\sigma_k} \right) \sqrt{\hat{g}} \hat{g}^{kl} \frac{\partial k}{\partial \xi^l} \right\}$$

and the energy dissipation ε

$$\frac{\partial \varepsilon}{\partial t} + \overline{U}^k \frac{\partial \varepsilon}{\partial \xi^k} = \left(C_{\varepsilon 1} P_K - C_{\varepsilon 2} \right) \frac{\varepsilon}{k} + \frac{1}{\sqrt{\hat{g}}} \frac{\partial}{\partial \xi^k} \left\{ \left(\nu + \frac{\nu_T}{\sigma_\varepsilon} \right) \sqrt{\hat{g}} \hat{g}^{kl} \frac{\partial \varepsilon}{\partial \xi^l} \right\},$$

where $P_K = -\hat{g}_{ij} \hat{g}_{kl} \left(\frac{2}{3} \hat{g}^{ik} k - 2\nu_T D^{ik} \right) D^{jl} = 2\nu_T \hat{g}_{ij} \hat{g}_{kl} D^{ik} D^{jl}$ represents the production term for the turbulent kinetic energy. Refer to Appendix A for details on deriving the governing equations on a generalized coordinate system.

References

1. Abe, K., Nagano, Y., Kondoh, T.: An improved k-ϵ model for preduction of turbulent flows with separation and reattachment. Trans. Jpan. Soc. Mech. Eng. B **58**(554), 3003–3010 (1992)
2. Baldwin, B.S., Lomax, H.: Thin layer approximation and algebraic model for separated turbulent flows. AIAA Paper 78–257 (1978)
3. Cebeci, T.: Analysis of Turbulent Flows with Computer Programs, 3rd edn. Butterworth-Heinemann (2013)
4. Daiguji, H., Miyake, Y., Yoshizawa, A. (eds.): Computational Fluid Dynamics of Turbulent Flow: Models and Numerical Methods. Univ. Tokyo Press (1998)
5. Daly, B.J., Harlow, F.H.: Transport equations in turbulence. Phys. Fluids **13**(11), 2634–2649 (1970)
6. Durbin, P.A., Reif, B.A.P.: Statistical Theory and Modeling for Turbulent Flows, 2nd edn. Wiley (2011)
7. Hanjalić, K., Launder, B.E.: Contribution towards a Reynolds-stress closure for low-Reynolds-number turbulence. J. Fluid Mech. **74**(4), 593–610 (1976)
8. Hinze, J.O.: Turbulence. McGraw-Hill, New York (1975)
9. Jones, W.P., Launder, B.E.: The prediction of laminarization with a two-equation model of turbulence. Int. J. Heat Mass Trans. **15**(2), 301–314 (1972)
10. Kajishima, T., Miyake, Y.: A discussion on eddy viscosity models on the basis of the large eddy simulation of turbulent flow in a square duct. Comput. Fluids **21**(2), 151–161 (1992)
11. Launder, B.E., Reece, G.J., Rodi, W.: Progress in the development of a Reynolds-stress turbulence closure. J. Fluid Mech. **68**(3), 537–566 (1975)
12. Launder, B.E., Spalding, D.B.: The numerical computation of turbulent flows. Comput. Methods Appl. Mech. Engrg. **3**(2), 269–289 (1974)
13. Launder, B.E., Tselepidakis, D.P., Younis, B.A.: A second-moment closure study of rotating channel flow. J. Fluid Mech. **183**, 63–75 (1987)
14. Lumley, J.L., Newman, G.R.: The return to isotropy of homogeneous turbulence. J. Fluid Mech. **82**(1), 161–178 (1977)

15. Mansour, N.N., Kim, J., Moin, P.: Reynolds-stress and dissipation-rate budgets in a turbulent channel flow. J. Fluid Mech. **194**, 15–44 (1988)
16. Menter, F.R.: Two-equation eddy-viscosity turbulence models for engineering applications. AIAA J. **32**(8), 1598–1605 (1994)
17. Miyata, H. (ed.): Analysis of turbulent flows. In: Computational Fluid Dynamics Series, vol. 3. Univ. Tokyo Press (1995)
18. Myong, H.K., Kasagi, N.: A new approach to the improvement of k-ϵ turbulence model for wall-bounded shear flows. JSME Int. J. **33**(1), 63–72 (1990)
19. Nagano, Y., Tagawa, M., Niimi, M.: An improvement of the k-ϵ turbulence model (the limiting behavior of wall and free turbulence, and the effect of adverse pressure gradient). Trans. Jpn. Soc. Mech. Eng. B **55**(512), 1008–1015 (1989)
20. Patankar, S.: Numerical Heat Transfer and Fluid Flow. Hemisphere, Washington (1980)
21. Patel, V.C., Rodi, W., Scheuerer, G.: Turbulence models for near-wall and low Reynolds number flows - A review. AIAA J. **23**(9), 1308–1319 (1985)
22. Rotta, J.C.: Statistische theorie nichthomogener turbulenz. Zeitschrift für physik **129**(6), 547–572 (1951)
23. Shima, N.: Prediction of turbulent boundary layers with a second-moment closure. Parts I and II. J. Fluids Eng. **115**(1), 56–69 (1993)
24. So, R.M.C., Lai, Y.G., Hwang, B.C., Yoo, G.J.: Low-Reynolds-number modelling of flows over a backward-facing step. ZAMP **39**(1), 13–27 (1988)
25. Spalart, P.R., Allmaras, S.R.: A one-equation turbulence model for aerodynamic forces. AIAA Paper 92–439 (1992)
26. Speziale, C.G.: Analytical methods for the development of Reynolds-stress closures in turbulence. Annu. Rev. Fluid Mech. **23**, 107–157 (1991)
27. Suga, K.: Near-wall grid dependency of low-Reynolds-number eddy viscosity turbulence models. Trans. Jpn. Soc. Mech. Eng. B **64**(626), 3315–3322 (1998)
28. Wilcox, D.C.: Turbulence Modeling for CFD, 3rd edn. DCW Industries (2006)
29. Yamamoto, M.: Time average turbulent flow model. Stress equation model (RSM). Turbomachinery **24**(4), 240–247 (1996)

Chapter 8
Large-Eddy Simulation

8.1 Introduction

Let us briefly reexamine the spatial scales present in turbulent flows before discussing how large-eddy simulation can be formulated. Consider the turbulent energy spectra for various turbulent flows across a wide range of Reynolds numbers. Shown in Fig. 8.1 are the energy spectra for various three-dimensional turbulent flows non-dimensionalized by the Kolmogorov scale [4]. Note that the vortices represented with low wave numbers are dependent on the problem. On the other hand, the small vortices represented by the high wave numbers exhibit a universal behavior (independent of the flow field). This is due to the isotropic nature of turbulence near the Kolmogorov scale. Based upon this observation, we can consider modeling the small-scale vortices that possess universality in their behavior and directly resolving the large scale vortices that are influenced by the setup of the flow field. Note however that it would not be appropriate to simply simulate only the large-scale vortices on a coarse grid with the Navier–Stokes equations, because there are interactions amongst vortices over wide range of scale due to the nonlinearity in turbulent flows. Large-eddy simulation (LES) resolves the large-scale vortices in the turbulent flow field and incorporates the influence of small-scale vortices that are not resolved by the grid through a model. This approach taken by LES has been shown to be successful in simulating a variety of complex turbulent flows.

8.2 Governing Equations for LES

The filtered Navier–Stokes equations are commonly used as the governing equations for LES with its filter width set to be close to the size of the mesh. Thus, the scales that are directly solved for on the grid are referred to as the *grid scales* (GS) or the *resolvable scales*. On the other hand, the small scales not captured by the grid are called the *subgrid scales* (SGS) or the *residual scales*.

© Springer International Publishing AG 2017
T. Kajishima and K. Taira, *Computational Fluid Dynamics*,
DOI 10.1007/978-3-319-45304-0_8

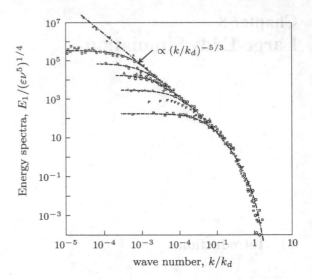

Fig. 8.1 Universality of
turbulent energy spectra
(based on Ref. [4])

8.2.1   Filtering

The governing equations for LES are derived by filtering the conservation equations
for fluid flow [15]. The filtering operation should not be noticeable in the simulated
results for a chosen LES model. The most widely used model, the Smagorinsky
model, is not sensitive to the type of filter function but can be influenced by the
width of the filter. What we describe below on filtering is not relevant for some of
the LES models, but it is important to identify which scales are modeled and which
are directly computed in LES.

Filters
Filtering can be performed through an application of convolution

$$\overline{f}(x) = \int_{-\infty}^{\infty} G(y) f(x - y) \mathrm{d}y, \tag{8.1}$$

where the function $G(y)$ should be positive around $y = 0$ and satisfy $\lim_{y \to \pm\infty}$
$G(y) = 0$. We construct this function to have the property of

$$\int_{-\infty}^{\infty} G(y) \mathrm{d}y = 1 \tag{8.2}$$

so that $\overline{f}(x)$ is the weighted average of $f(x)$ near x. The function G is referred to
as the *filter function*.

Let us take a few examples of the filter functions. First, we can consider the
simplest function shown in Fig. 8.2a known as the *box filter*

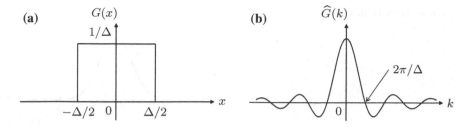

Fig. 8.2 a The box filter and its b Fourier transform

$$G(x) = \begin{cases} 1/\Delta & (|x| < \Delta/2) \\ 0 & (|x| > \Delta/2) \end{cases} \tag{8.3}$$

which has a filter width of Δ and satisfies Eq. (8.2). This filter is also referred to as the *top hat filter*. While it may appear that the fine-scale fluctuations are truncated, the Fourier transform of the filter

$$\widehat{G}(k) = \frac{\sin(\Delta k/2)}{\Delta k/2} \tag{8.4}$$

reveals that the filter function damps high wave number components but in an oscillatory manner. It should be noted with caution that this filter provides negative values for certain wave numbers, as shown in Fig. 8.2b.

Next, consider applying a box filter in the wave space with

$$\widehat{G}(k) = \begin{cases} 1 & (|k| < \pi/\Delta) \\ 0 & (|k| > \pi/\Delta) \end{cases} \tag{8.5}$$

to remove fine-scale (high-frequency) components. This filter is known as the *spectral cutoff filter* and is analogous to the box filter in physical space (see Fig. 8.2a). However, the inverse Fourier transform of this filter reveals that the weighted average in physical space is dependent on an oscillatory filter function of

$$G(x) = 2\frac{\sin(\pi x/\Delta)}{\pi x} \tag{8.6}$$

analogous to what is illustrated in Fig. 8.2b. The box filter does not have a sharp cutoff in the wave space and the spectral cutoff filter does not have a sharp cutoff in the physical space.

We can consider an alternative form of filter based on a Gaussian distribution. While the support of the Gaussian profile is not compact,[1] it retains its Gaussian profile in both the wave and physical spaces. The characteristic length can be set to

---

[1] A Gaussian profile decays fast but is nonzero over the entire space.

Fig. 8.3 Gaussian filter

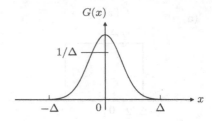

Δ to arrive at a filter function of

$$G(x) = \sqrt{\frac{6}{\pi \Delta^2}} \exp\left(-\frac{6x^2}{\Delta^2}\right), \qquad (8.7)$$

which satisfies Eq. (8.2). This is called the *Gaussian filter* as shown in Fig. 8.3. The Fourier transformed filter in wave space is

$$\widehat{G}(k) = \exp\left(-\frac{\Delta^2 k^2}{24}\right). \qquad (8.8)$$

Regardless of the choice of the filter function, we must define the filter size, Δ, that separates the small and large-scale structures. If we let all fluctuating components that are computed on the grid scale to be the "large-scale" components, we are able to take advantage of the grid resolution most effectively in the numerical simulation. That is, we do not waste any directly computed results. For this reason, the filtering width is usually set to the grid size Δ. We hence refer to the scale extracted from the filtering operation as the grid scale. On the other hand, the fluctuation

$$f' = f - \overline{f}, \qquad (8.9)$$

is called the subgrid scale (SGS) component.

Because LES are performed for three-dimensional turbulent flows, we must represent the filtering operation in a three-dimensional manner. Using a filter function G with its characteristic length being the mesh size, we can define the grid-scale component, $\overline{f}$,

$$\overline{f}(x) = \int_{-\infty}^{\infty} G(y) f(x - y) dy. \qquad (8.10)$$

The convolution represented by Eq. (8.10) can be written as

$$\widehat{\overline{f}}(k) = \widehat{G}(k) \cdot \widehat{f}(k) \qquad (8.11)$$

in Fourier space as a simple product operation. Equation (8.11) tells us that it is equivalent to perform the convolution in real space or to inverse transform the product of the Fourier transformed filter and function. For a Cartesian coordinate system, we can apply the filtering operation in each of the three directions as

$$G(x - y) = \prod_{i=1}^{3} G_i (x_i - y_i) \tag{8.12}$$

using the aforementioned filter function. Note that we can select different filter functions or filter widths for each spatial direction.

Issues Associated with Filtering

Filtering essentially separates flow structures into large and small-scale components based on the grid size. However, this does not mean that all grid-scale components can be captured by numerical computation. Similarly, there can be subgrid-scale components existing over the computational scales. This is due to the fact that filter function cannot separate scales sharply in the physical or wave space (or both). For example, let us depict in Fig. 8.4 the separation of scales using a Gaussian filter. Although the grid-scale component $\bar{u}$ exists over the wave numbers higher than the cutoff wave number ($k > \pi/\Delta$), the grid cannot support such components. On the other hand for wave numbers lower than the cutoff wave number ($k < \pi/\Delta$), it may appear that the difference between the spectra of u and $\bar{u}$ is small. However, such difference may not be negligible when the spectra are shown over the logarithmic scale. In some cases, such difference may not be negligible when considering level of the spectra for u'. When using this type of filter, the "subgrid-scale" components that are larger than mesh width should also be taken into consideration.

The filter function need not be uniform over space and time, in which case the derivative and filtering operations are no longer commutative. That is Eq. (6.15) does not hold. This point is often overlooked when LES models are developed. However, care should be taken when models are developed, especially when filtering can influence the simulated flow in a significant fashion.

Fig. 8.4 Illustration of scale separation using a Gaussian filter (log–log plot)

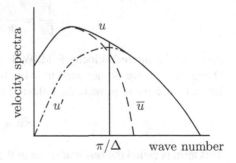

8.2.2 Governing Equations for Large-Eddy Simulation

Under the assumption that the derivative and filtering operations are commutative (i.e., Eq. (6.15)), the filtered continuity and momentum equations for incompressible flow become

$$\frac{\partial \overline{u}_i}{\partial x_i} = 0 \tag{8.13}$$

$$\frac{\partial \overline{u}_i}{\partial t} + \frac{\partial \overline{u_i u_j}}{\partial x_j} = -\frac{1}{\rho}\frac{\partial \overline{p}}{\partial x_i} + \frac{\partial}{\partial x_j}(2\nu \overline{D}_{ij}), \tag{8.14}$$

where $\overline{D}_{ij}$ is the grid-scale rate-of-strain tensor

$$\overline{D}_{ij} = \frac{1}{2}\left(\frac{\partial \overline{u}_i}{\partial x_j} + \frac{\partial \overline{u}_j}{\partial x_i}\right). \tag{8.15}$$

Here, the grid-scale flow field $(\overline{u}, \overline{p})$ is to be determined but Eq. (8.14) contains the term $\overline{u_i u_j}$, which cannot be represented by the grid-scale variables. Rewriting the above momentum equation using the grid-scale variables, we have

$$\frac{\partial \overline{u}_i}{\partial t} + \frac{\partial (\overline{u}_i \overline{u}_j)}{\partial x_j} = -\frac{1}{\rho}\frac{\partial \overline{p}}{\partial x_i} + \frac{\partial}{\partial x_j}(-\tau_{ij} + 2\nu \overline{D}_{ij}) \tag{8.16}$$

or alternatively, through the use of Eq. (8.14),

$$\frac{\overline{D}\overline{u}_i}{\overline{D}t} = -\frac{1}{\rho}\frac{\partial \overline{p}}{\partial x_i} + \frac{\partial}{\partial x_j}(-\tau_{ij} + 2\nu \overline{D}_{ij}), \tag{8.17}$$

where $\overline{D}/\overline{D}t$ denotes the material derivative

$$\frac{\overline{D}}{\overline{D}t} = \frac{\partial}{\partial t} + \overline{u}_j \frac{\partial}{\partial x_j} \tag{8.18}$$

based on the filtered velocity field. With the momentum equation expressed in the above manner, we do not need to solve for the eddies smaller than the grid scale. Instead, we have to provide the influence of the subgrid-scale eddies through

$$\tau_{ij} = \overline{u_i u_j} - \overline{u}_i \overline{u}_j. \tag{8.19}$$

This term is called the *residual stress* or the *subgrid-scale stress* (SGS stress), which is created from filtering the Navier–Stokes equations. As illustrated in Fig. 8.4, this stress term resides over the subgrid-scale wave numbers but may have some minor contributions from the grid-scale wave numbers. Strictly speaking $\rho \tau_{ij}$ has the dimension of stress. However, when ρ is constant, τ_{ij} is often referred to as stress.

The subgrid-scale stress τ_{ij} has traditionally been decomposed in the following manner

$$\tau_{ij} = L_{ij} + C_{ij} + R_{ij}, \tag{8.20}$$

where

• *Leonard term*

$$L_{ij} = \overline{\bar{u}_i \bar{u}_j} - \bar{u}_i \bar{u}_j \tag{8.21}$$

• *Cross term*

$$C_{ij} = \overline{\bar{u}_i u'_j} + \overline{u'_i \bar{u}_j} \tag{8.22}$$

• *SGS Reynolds stress*

$$R_{ij} = \overline{u'_i u'_j} \tag{8.23}$$

Notice that L_{ij} and C_{ij} are not referred to as stresses but terms. This is because these terms do not satisfy invariance under Galilean transformation independently. This can be resolved with the reformulation proposed by Garmano [8][2]:

$$\tau_{ij} = L_{ij}^m + C_{ij}^m + R_{ij}^m, \tag{8.24}$$

where

• *Modified Leonard stress*

$$L_{ij}^m = \overline{\bar{u}_i \bar{u}_j} - \bar{\bar{u}}_i \bar{\bar{u}}_j \tag{8.25}$$

• *Modified cross stress*

$$C_{ij}^m = \overline{\bar{u}_i u'_j} + \overline{u'_i \bar{u}_j} - (\bar{\bar{u}}_i \overline{u'_j} + \overline{u'_i} \bar{\bar{u}}_j) \tag{8.26}$$

• *Modified SGS Reynolds stress*

$$R_{ij}^m = \overline{u'_i u'_j} - \overline{u'_i}\, \overline{u'_j} \tag{8.27}$$

The modified Leonard stress L_{ij}^m is determined with the filtered grid-scale velocity but the evaluation of C_{ij}^m and R_{ij}^m requires the use of models.

---

[2]For the turbulent stress $\tau(u, v) = \overline{uv} - \bar{u}\,\bar{v}$ to be Galilean invariant, $\tau(u+\alpha, v+\beta) = \tau(u, v)$ must hold for a system moving at a constant velocity of (α, β). If we let velocity U to be the constant velocity $\overline{U}$ for $u^* = u + U$, then we have $\bar{u}^* = \bar{u} + U$ and $u^{*\prime} = u'$. Thus the Leonard stress L_{ij} becomes

$$L_{ij} = \overline{\bar{u}_i^* \bar{u}_j^*} - \bar{u}_i^* \bar{u}_j^* - (\overline{\bar{u}_i^*} - \bar{u}_i^*)U_j - (\overline{\bar{u}_j^*} - \bar{u}_j^*)U_i,$$

which is not Galilean invariant since U remains in the expression. The modified L_{ij}^m on the other hand becomes

$$L_{ij}^m = \overline{\bar{u}_i^* \bar{u}_j^*} - \overline{\bar{u}_i^*}\, \overline{\bar{u}_j^*},$$

which is Galilean invariant.

The difference between the modified Leonard stress L_{ij}^m and the Leonard stress L_{ij} is

$$B_{ij} = L_{ij}^m - L_{ij} = \overline{\bar{u}_i \bar{u}_j} - \overline{\bar{u}}_i \overline{\bar{u}}_j. \tag{8.28}$$

This term is known as the *scale-similarity term* [1] and is used in a model discussed later in Sect. 8.4.

8.3   Smagorinsky Model

The Smagorinsky model is one of the most widely used LES models that utilizes the filter width Δ as the characteristic length. This model was originally proposed for its use in high-Reynolds number atmospheric fluid flow simulations for which the employed grid was very coarse [29]. Presently, the Smagorinsky model is applied to analyze a wide variety of academic and industrial turbulent flow problems.

8.3.1   Local Equilibrium and Eddy-Viscosity Assumptions

By filtering the flow field, the kinetic energy $\bar{k} = \overline{u_k u_k}/2$ is decomposed[3] into the grid-scale energy $k_{GS} = \bar{u}_k \bar{u}_k/2$ and the subgrid-scale energy $k_{SGS} = (\overline{u_k u_k} - \bar{u}_k \bar{u}_k)/2$.

The conservation relation for the grid-scale energy k_{GS} becomes

$$\frac{\overline{D}k_{GS}}{\overline{D}t} = \tau_{ij}\overline{D}_{ij} - \bar{\varepsilon}_{GS} + \frac{\partial}{\partial x_j}\left(-\bar{u}_i \tau_{ij} - \frac{\overline{p}\bar{u}_j}{\rho} + \nu \frac{\partial k_{GS}}{\partial x_j} \right), \tag{8.29}$$

where $\tau_{ij}\overline{D}_{ij}$ represents the rate of energy transport to the subgrid-scale energy k_{SGS}. Accordingly, the conservation relation for the subgrid-scale energy k_{SGS} is

$$\begin{aligned}
\frac{\overline{D}k_{SGS}}{\overline{D}t} = &-\tau_{ij}\overline{D}_{ij} - \varepsilon_{SGS} \\
&+ \frac{\partial}{\partial x_j}\left[\bar{u}_i \tau_{ij} - \frac{1}{2}\left(\overline{u_i u_i u_j} + \overline{u_j u_i u_i} \right) - \frac{\overline{p u_j} - \overline{p}\,\bar{u}_j}{\rho} + \nu \frac{\partial k_{SGS}}{\partial x_j} \right],
\end{aligned} \tag{8.30}$$

where $-\tau_{ij}\overline{D}_{ij}$ appears as the rate of energy production. We can assume local equilibrium such that the dissipation rate of subgrid-scale energy ε_{SGS}

---

[3]It is often misinterpreted that the sum of the grid-scale and subgrid-scale energy $k_{GS} + k_{SGS}$ is k. However it should be noted that this sum should be $\bar{k}$. The kinetic energy distribution from experiments or DNS should be filtered when LES results are compared with such results.

$$\varepsilon_{SGS} = \overline{\varepsilon} - \overline{\varepsilon}_{GS} = \nu \overline{\frac{\partial u_i}{\partial x_j} \frac{\partial u_i}{\partial x_j}} - \nu \frac{\partial \overline{u}_i}{\partial x_j} \frac{\partial \overline{u}_i}{\partial x_j} \tag{8.31}$$

is in balance with the production rate to yield

$$\varepsilon_{SGS} = -\tau_{ij} \overline{D}_{ij}. \tag{8.32}$$

For the subgrid-scale stress τ_{ij}, let us employ the eddy-viscosity approximation in analogy to the physical viscosity or the eddy viscosity for Reynolds stress discuss in the previous chapter. Then, we can write

$$\tau_{ij}^a = -2\nu_e \overline{D}_{ij}, \tag{8.33}$$

where ν_e is called the *subgrid-scale (SGS) eddy-viscosity coefficient*. In what follows, we denote the anisotropic component of the stress with the superscript of a for convenience, as in $\tau_{ij}^a = \tau_{ij} - \delta_{ij}\tau_{kk}/3$. Note that from the traceless condition ($\overline{D}_{ii} = 0$), we have $\tau_{ij}^a \overline{D}_{ij} = \tau_{ij} \overline{D}_{ij}$.

Substituting Eq. (8.33) into the momentum equation, Eq. (8.16), we have

$$\frac{\partial \overline{u}_i}{\partial t} + \frac{\partial (\overline{u}_i \overline{u}_j)}{\partial x_j} = -\frac{1}{\rho} \frac{\partial \overline{P}}{\partial x_i} + \frac{\partial}{\partial x_j} \left[2(\nu + \nu_e) \overline{D}_{ij} \right]. \tag{8.34}$$

By providing the SGS eddy-viscosity coefficient ν_e and simultaneously solving Eqs. (8.13) and (8.34), we can determine the filtered velocity field $\overline{u}_i$ and the *modified pressure*[4] $\overline{P} = \overline{p} + \frac{1}{3}\rho\tau_{kk}$. At this point, what remains to be provided is a model for estimating ν_e. Assuming local equilibrium and the eddy-viscosity approximation forms the basis to derive the Smagorinsky model for evaluating this eddy-viscosity coefficient, as described below.

8.3.2 Derivation of the Smagorinsky Model

The eddy viscosity has a dimensional of the product of velocity (q) and length (l), which can be expressed as

$$\nu_e = C_\nu q l. \tag{8.35}$$

With the local equilibrium assumption described by Eq. (8.32), the eddy-viscosity approximation, Eq. (8.33), yields $\varepsilon_{SGS} = 2\nu_e \overline{D}_{ij} \overline{D}_{ij}$. Taking the dimension of the dissipation rate into consideration, we can set

---

[4]The modified pressure here is different from the pressure term $\overline{P} = \overline{p} + \frac{2}{3}\rho k$ that appears in eddy-viscosity models for RANS but shares similarity in how it combines the isotropic stress components (see Sect. 7.3).

$$\nu_e \overline{D}_{ij} \overline{D}_{ij} = C_\varepsilon q^3 / l \tag{8.36}$$

scaled by q and l. By eliminating q using Eqs. (8.35) and (8.36) and setting l to be the filter width Δ, the subgrid-scale eddy-viscosity coefficient becomes

$$\nu_e = (C_s \Delta)^2 |\overline{D}|. \tag{8.37}$$

This model is called the *Smagorinsky model* [29]. Here, $|\overline{D}|$ represents the norm of the rate-of-strain tensor

$$|\overline{D}| = \sqrt{2\overline{D}_{ij}\overline{D}_{ij}}. \tag{8.38}$$

The constant C_s is referred to as the *Smagorinsky constant* and is the only non-dimensional constant that needs to be provided.

From the aforementioned local equilibrium assumption,

$$\varepsilon_{SGS} = (C_s \Delta)^2 |\overline{D}|^3. \tag{8.39}$$

Next by considering turbulent statistics theory, Lilly [18] estimated

$$\frac{1}{2}|\overline{D}|^2 = \int_0^{\pi/\Delta} k^2 E(k) \mathrm{d}k = \frac{3}{4}\alpha\varepsilon^{2/3}\left(\frac{\pi}{\Delta}\right)^{4/3}, \tag{8.40}$$

where the Kolmogorov spectra $E(k) = \alpha\varepsilon^{2/3}k^{-5/3}$ is used. We can take viscous dissipation to be approximately equal to the subgrid-scale dissipation ($\varepsilon = \varepsilon_{SGS}$) when Δ is in the inertial subrange. Accordingly, we find that [18]

$$C_s = \frac{1}{\pi}\left(\frac{3\alpha}{2}\right)^{-3/4} = 0.235\alpha^{-3/4}. \tag{8.41}$$

Substituting the Kolmogorov constant of $\alpha = 1.5$, we obtain $C_s = 0.173$, which is taken to be the theoretical value for the Smagorinsky constant.

In summary, we have assumed local equilibrium and utilized eddy viscosity with a characteristic length scale of Δ to derive the Smagorinsky model. The model can further incorporate the Kolmogorov spectra to estimate the sole non-dimensional parameter C_s.

8.3.3   Properties of the Smagorinsky Model

The subgrid-scale model provided by Eqs. (8.33) and (8.37) can predict the overall energy dissipation well given an appropriate choice of the Smagorinsky constant C_s. However it has been observed that the local subgrid-turbulent behavior is not well reproduced with this model. Comparison of the actual SGS component obtained from decomposing DNS results by filtering and that from the SGS model based on the grid-scale computation can have low correlation.

For isotropic turbulence, it is reported that the use of the theoretical C_s value allows for the LES with the Smagorinsky model to match the experimental measurements quite well. However, the value for C_s needs to be modified to a lower value between 0.10 and 0.15 for shear flows. Thus, we can clearly notice that the Smagorinsky constant is not universal.

The Smagorinsky model transfers energy from k_{GS} to k_{SGS} since the eddy viscosity ν_e is always positive and cannot represent the inverse cascade (backward scatter). In reality, there can exist regions in the flow where the energy in the small scales is transferred to the large scales. The energy transport toward smaller scales and then to dissipative scales occur in an average sense, but without averaging, energy can locally be transferred in the opposite direction. It is however somewhat convenient that the Smagorinsky model has this shortcoming since the representation of the inverse cascade can lead to numerical instability, as we will describe later.

When velocity gradient is nonzero for the grid-scale component, we have $|\overline{D}| > 0$ and correspondingly a positive subgrid-scale eddy viscosity in Eq. (8.37). Therefore, correction needs to be implemented to ensure that τ_{ij} is zero for laminar flow. This leads to the use of *switching* (on and off) the Smagorinsky model for flows with transition or re-laminarization, based on whether the local flow is laminar or turbulent.

For flows in the vicinity of the wall, the energy balance does not satisfy the local equilibrium between production and dissipation. If we are to analyze the flow in such area with the enforcement of the no-slip boundary condition, it becomes necessary to introduce damping to yield $\tau_{ij} = 0$ at the wall.

We have listed the shortcomings of the Smagorinsky model in order to understand its limitations. However, this model works well for many practical simulations. In particular, the *robustness* of the Smagorinsky model should be a welcoming property to provide numerical stability for practical applications.

8.3.4 Modification in the Near-Wall Region

Similar to the k-ε model, LES can treat flows near the wall by prescribing (1) the wall law with coarse grids or (2) the no-slip boundary condition with sufficient grid resolution.

(1) Application of the Wall Law
The implementation of the wall law is performed in a similar manner to how it is incorporated in the k-ε model. The difference is that the LES utilizes the velocity field including the turbulent fluctuation, instead of the averaged velocity field used by the k-ε model. Note that the use of the wall law is founded on an averaged concept. Thus, the validity of applying the wall law for a local instantaneous flow in LES is, strictly speaking, somewhat inappropriate.

Fig. 8.5 Wall-tangential
velocity at a grid point near
the wall

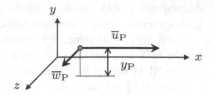

Let us consider a flow along the wall with the streamwise, wall-normal, and spanwise directions denoted by x, y, and z, as depicted in Fig. 8.5. At the center of the cell neighboring the wall, we assume that the grid-scale velocity is available. For a staggered grid, this velocity can be the average value of the adjacent velocity values. For a collocated grid, the velocity value at the cell center can be used. At this moment, the velocity components $\overline{u}_p$ and $\overline{w}_p$ are provided at a point that is y_p away from the wall. This leads to a problem of solving for the shear velocity near the wall using the wall law $F(u_\tau) = 0$.

For the wall law, the use of the log law

$$F(u_\tau) = \frac{\overline{u}_p}{u_\tau} - \frac{1}{\kappa} \ln \frac{y_p u_\tau}{\nu} - B = 0 \qquad (8.42)$$

is common. The shear velocity can be solved by the Newton–Raphson method

$$u_\tau^{m+1} = u_\tau^m - \frac{F(u_\tau^m)}{F'(u_\tau^m)}, \quad m = 0, 1, 2, \ldots \qquad (8.43)$$

in an iterative manner. For the log law, $F'(u_\tau) = -(\overline{u}_p/u_\tau + 1/\kappa)/u_\tau$.

There is some room for choosing the value of $\overline{u}_p$. We can select the instantaneous local mainstream velocity $\overline{u}_p$, the velocity value including the spanwise velocity $(\overline{u}_p^2 + \overline{w}_p^2)^{1/2}$, or the average velocity value $\langle \overline{u}_p \rangle$ taken over an appropriate region. If we focus on capturing the fluctuation in the wall shear stress $\tau_w = \rho u_\tau^2$, the use of the instantaneous local velocity allows for a solution with higher resolution. From the point of view of the wall law holding only in the averaged sense, we should probably use $\langle \overline{u}_p \rangle$ if the statistically uniform directions can be identified (e.g., wall-parallel directions for channel flow). It is difficult to state which choice of representative velocity value is optimal.

As opposed to the Reynolds-averaged Navier–Stokes equations, LES is also concerned with velocity fluctuations. We therefore must prescribe the boundary conditions for these fluctuations. If the first velocity value away from the wall is in the logarithmic region as depicted in Fig. 8.6, the specification of zero velocity at the wall creates difficulty in capturing the correct velocity gradient near the wall. When using a coarse grid, the no-slip boundary condition needs to be relaxed to yield the correct velocity gradient profile. Assuming that the velocity including the fluctuation follows the log law, we can specify the derivatives

Fig. 8.6 Velocity gradient at
a cell adjacent to the wall

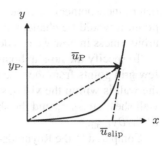

$$\frac{\partial \overline{u}}{\partial y} = \frac{u_\tau}{\kappa y}, \quad \frac{\partial^2 \overline{u}}{\partial y^2} = -\frac{u_\tau}{\kappa y^2} \tag{8.44}$$

along the wall.

In addition to the above approach, if an appropriate wall law $F(u_\tau) = 0$ is known for the flow of interest, that relation can be used to cut down on the number of grid points required from what would be needed if we are to specify the no-slip condition. It should be pointed out that the wall shear stress τ_w can be provided by the model but the utilization of the wall law leads to the problem of not being able to retrieve the wall pressure (later, Eq. (8.108) shows that $\tau_{ii} \neq 0$ for a position away from the wall).

It is possible that the first grid point lies closer to the wall than the log-law region. Even if we can implement a corrected wall model that is valid in that region, the Smagorinsky model would not be valid for its use there. If the first grid point from the wall resides within the sublayer, the approach described below can be taken.

(2) Damping Function Method

Prescribing the no-slip boundary condition to the grid-scale velocity, the fluctuation in the subgrid-scale component should also become zero at the wall also. Nonetheless, as long as there is nonzero gradient in the grid-scale velocity profile, Eq. (8.37) introduces turbulent fluctuation in the subgrid-scale, which needs to be canceled to enforce the no-slip boundary condition. In the Smagorinsky model, a *damping function* f_s is incorporated. For the eddy viscosity

$$\nu_e = (C_s f_s \Delta)^2 |\overline{D}|, \tag{8.45}$$

the *van Driest function*

$$f_s = 1 - \exp\left(-\frac{y^+}{A^+}\right) \tag{8.46}$$

is often employed, where the non-dimensional constant $A^+ \approx 25$.

It may not be possible to uniquely determine $y^+ = y u_\tau / \nu$ in Eq. (8.46). For u_τ, there is room for selection, as to using the local value in the direct neighborhood of the wall or to use some averaged value. Based on the selection for the value of u_τ, there may be some variability in the profile of the turbulent vortices but it is hard to determine whether one choice is superior than another. In the vicinity of a separation point, it would be challenging to determine u_τ. For flow around a corner, there is arbitrariness in how we can define y.

To specify the no-slip boundary condition, it is necessary to smoothly position a few grid points from the viscous sublayer to the buffer zone. If the first point from the wall is within the viscous sublayer, the turbulent stress can be ignored and the wall shear stress τ_w and the shear velocity u_τ can be determined from $\tau_w = \rho u_\tau^2 = \mu |\partial \overline{u}/\partial y|_{\text{wall}}$.

Compared to the Reynolds-averaged models, the subgrid-scale models are not as critically dependent on the definition of the distance from the wall or the near-wall asymptotic velocity behavior. It is because LES calculates the important part of the turbulent fluctuation as the grid-scale component; the subgrid-scale components are residual components of the turbulent fluctuations. Furthermore, as we have discussed earlier, the Smagorinsky model is aimed to express the average energy dissipation by the subgrid-scale fluctuation and is not meant to accurately predict the local subgrid-scale fluctuation. It is inadvisable to attempt local accuracy improvement by solely modifying the damping function.

For flows that are not aligned with the wall, there are dynamic models that apply the no-slip boundary condition without the use of correction functions. It appears that such approach is more appropriate in generalizing the method, instead of adopting the use of damping functions.

8.4   Scale-Similarity Model

8.4.1   Bardina Model

Filtering both sides of the definition of the subgrid-scale component $u_i' = u_i - \overline{u}_i$, we obtain $\overline{u_i'} = \overline{u}_i - \overline{\overline{u}}_i$. Note that both sides of this expression are not zero. The velocity $\overline{u_i'}$ represents the relatively large-scale components extracted by filtering the subgrid-scale velocity. The difference between the grid-scale component and its doubly filtered value (relatively large fluctuation at the grid scale), $\overline{u}_i - \overline{\overline{u}}_i$, denotes the relatively small-scale fluctuation at the grid scale. Thus, the relationship $\overline{u_i'} = \overline{u}_i - \overline{\overline{u}}_i$ implies that there is similarity amongst vortices possessing neighboring scales. Based upon this idea, the cross terms and the subgrid-scale Reynolds stress can be modeled as

$$\overline{u_i' \overline{u}_j} = \left(\overline{u}_i - \overline{\overline{u}}_i \right) \overline{\overline{u}}_j, \quad \overline{\overline{u}_i u_j'} = \overline{\overline{u}}_i \left(\overline{u}_j - \overline{\overline{u}}_j \right) \tag{8.47}$$

$$\overline{u_i' u_j'} = \left(\overline{u}_i - \overline{\overline{u}}_i \right) \left(\overline{u}_j - \overline{\overline{u}}_j \right). \tag{8.48}$$

The sum of the cross term and the subgrid-scale Reynolds stress accordingly becomes

$$C_{ij} + R_{ij} = \overline{u_i'\overline{u}_j + \overline{u}_i u_j'} + \overline{u_i' u_j'} = \overline{\overline{u}_i \overline{u}_j} - \overline{\overline{u}_i}\overline{\overline{u}}_j, \qquad (8.49)$$

which matches with B_{ij} in Eq. (8.28). This is referred to as the *scale-similarity model* or the *Bardina model* [1]. Further addition of L_{ij} to the above equation yields

$$\tau_{ij} = \overline{\overline{u}_i \overline{u}_j} - \overline{\overline{u}_i}\overline{\overline{u}}_j. \qquad (8.50)$$

That is, τ_{ij} provided by the Bardina model matches with the modified Leonard stress L_{ij}^m with the sum $C_{ij}^m + R_{ij}^m$ being set to zero.

The eddy-viscosity model makes τ_{ij}^a to be a scalar multiple of $\overline{D}_{ij}$, which assumes that their principal axes are aligned. The scale-similarity model is not restricted by such assumption. This is considered to be one of the reasons why the Bardina model are observed to have higher correlation with results from the DNS database. This means that the τ_{ij} calculated by the Bardina model using the $\overline{u}$ value obtained by filtering the results in DNS database and the τ_{ij} computed from the u' value from the subfilter scale match locally very well. However, the scale-similarity model is not commonly used by itself, since it does not introduce any dissipation and can destabilize the numerical calculation.

8.4.2 Mixed Model

The subgrid-scale components include vortices (or vortical structures) of all scales that were removed by the filtering operation. The scale-similarity model is suitable for relating the grid-scale and the relatively large-scale structures in the subgrid-scale components. The Smagorinsky model based on the use of eddy viscosity represents the unidirectional effect of dissipation and captures its average property at the smallest scale. In other words, the scale-similarity and eddy-viscosity models have different mechanisms as depicted in Fig. 8.7. Combining the two approaches becomes sensible if we consider the properties of the subgrid-scale components.

Fig. 8.7 Concept of the mixed model

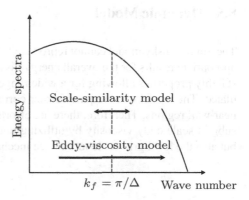

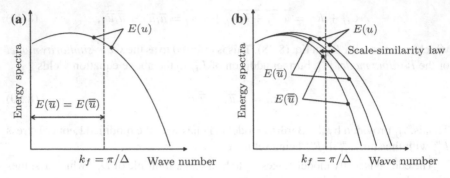

Fig. 8.8 Relationship between the scale-similarity model and filtering. **a** Spectral cutoff filter. **b** Gaussian filter

Based on the above discussion, a *mixed model* [2] can be proposed as

$$\tau_{ij}^a = L_{ij}^{ma} - 2\,(C_s\Delta)^2\,|\overline{\boldsymbol{D}}|\overline{\boldsymbol{D}}_{ij}. \tag{8.51}$$

In this model, L_{ij}^m is directly computed and $C_{ij}^m + R_{ij}^m$ is provided by the Smagorinsky model.

The Bardina model, Eq. (8.50), and the Bardina-based mixed model, Eq. (8.51), are dependent on the choice of filter function. For the spectral cutoff filter, the filtered results are unchanged with multiple filtering operations, as shown in Fig. 8.8a. This implies that $L_{ij}^m = 0$ with the application of double filtering, making the use of scale similarity not being able to rely on the application of the same filter twice. For the Gaussian filter shown in Fig. 8.8b, $\overline{u}$ and u' are distributed over the entire wave space, which leaves some ambiguity in scale separation. It becomes difficult to model actual turbulent flows based on the scale-similarity law as we have initially discussed. The concept of scale similarity is better utilized in dynamic models described below in Sect. 8.5.

8.5   Dynamic Model

The Smagorinsky model cannot reproduce the turbulent flow locally but can provide appropriate results for the overall energy dissipation. It also has attractive numerical stability properties allowing for a wide range of turbulent flow problems to be simulated. The model does require further corrections for anisotropic flows or for the near-wall regions. Therefore, there are efforts to develop models that determine the subgrid-scale eddy viscosity by utilizing not only the grid-scale velocity gradients but also the local grid-scale turbulence mechanism.

8.5.1 Dynamic Eddy-Viscosity Model

Instead of choosing a value for the Smagorinsky constant C_s based on experience, we can consider C_s as a variable to be dynamically determined from the grid-scale velocity field $\overline{u}$ in hope of overcoming some of the shortcomings in the Smagorinsky model. The model that determines C_s dynamically is called the *dynamic Smagorinsky model* (DMS).

(1) Germano Model
Germano et al. [9] introduced the use of a *test filter* $\widetilde{G}$ to extract relatively small scale structures from the grid-scale components. Here, let us define the following test filter operation

$$\widetilde{f}(x) = \int_{-\infty}^{\infty} \widetilde{G}(y) f(x - y) \, dy. \tag{8.52}$$

To distinguish between the test filter and the original filter used in the derivation of the governing equations for LES, we refer to the original filter as the *grid filter*. The test filter width is taken to be larger than that of the grid filter. Applying the test filter on the grid-filtered Navier–Stokes equations, Eq. (8.14), gives

$$\frac{\partial \widetilde{\overline{u}}_i}{\partial t} + \frac{\partial \widetilde{\overline{u}}_i \widetilde{\overline{u}}_j}{\partial x_j} = -\frac{1}{\rho} \frac{\partial \widetilde{\overline{p}}}{\partial x_i} + \frac{\partial}{\partial x_j} \left(-T_{ij} + 2\nu \widetilde{\overline{D}}_{ij} \right), \tag{8.53}$$

where

$$\widetilde{\overline{D}}_{ij} = \frac{1}{2} \left(\frac{\partial \widetilde{\overline{u}}_i}{\partial x_j} + \frac{\partial \widetilde{\overline{u}}_j}{\partial x_i} \right). \tag{8.54}$$

Here the term

$$T_{ij} = \widetilde{\overline{u_i u_j}} - \widetilde{\overline{u}}_i \widetilde{\overline{u}}_j \tag{8.55}$$

represents the residual turbulent stress left after test filtering. This term T_{ij} cannot be calculated but we can evaluate

$$\mathcal{L}_{ij} = \widetilde{\overline{u}_i \overline{u}_j} - \widetilde{\overline{u}}_i \widetilde{\overline{u}}_j \tag{8.56}$$

by applying the test filter on the grid-scale components. The above equation is known as the *Germano identity*. The relationship between $\mathcal{L}_{ij}$ and T_{ij} is provided by

$$\mathcal{L}_{ij} = T_{ij} - \widetilde{\tau}_{ij}. \tag{8.57}$$

With the Smagorinsky model, we can approximate τ_{ij} and T_{ij}

$$\tau_{ij}^a = -2C\overline{\Delta}^2 |\overline{D}| \overline{D}_{ij}, \tag{8.58}$$

$$T_{ij}^a = -2C\widetilde{\Delta}^2 |\widetilde{\overline{D}}| \widetilde{\overline{D}}_{ij} \tag{8.59}$$

using a common constant C for τ_{ij} and T_{ij}. Here, we have replaced C_s^2 with C so that C can also assume a negative value. In the above equations, $\overline{\Delta}$ and $\widetilde{\Delta}$ are the characteristic lengths of the grid filter $\overline{G}$ and the test filter $\widetilde{G}$, respectively, with the ratio $\gamma \equiv \widetilde{\Delta}/\overline{\Delta} > 1$.

By substituting Eqs. (8.58) and (8.59) into Eq. (8.57) and assuming that $\overline{\Delta}$ and C are constant inside the test filter, the anisotropic component of $\mathcal{L}_{ij}$ can be modeled by

$$\mathcal{L}_{ij}^a = -2C\overline{\Delta}^2 \mathcal{M}_{ij}, \tag{8.60}$$

where

$$\mathcal{M}_{ij} = \gamma^2 |\widetilde{\boldsymbol{D}}|\widetilde{\overline{D}}_{ij} - \widetilde{|\boldsymbol{D}|\overline{D}_{ij}}. \tag{8.61}$$

By multiplying $\overline{D}_{ij}$ to both sides of the above equation

$$\mathcal{L}_{mn}\overline{D}_{mn} = -2C\overline{\Delta}^2 \mathcal{M}_{ij}\overline{D}_{ij}, \tag{8.62}$$

where we have $\mathcal{L}_{ij}^a \overline{D}_{ij} = \mathcal{L}_{ij}\overline{D}_{ij}$ because $\overline{D}_{ij}$ is traceless. Therefore, we find

$$C = -\frac{1}{2\overline{\Delta}^2} \frac{\mathcal{L}_{mn}\overline{D}_{mn}}{\mathcal{M}_{kl}\overline{D}_{kl}} \tag{8.63}$$

and, accordingly, we have the SGS eddy-viscosity coefficient to be

$$\nu_e = -\frac{1}{2} \frac{\mathcal{L}_{mn}\overline{D}_{mn}}{\mathcal{M}_{kl}\overline{D}_{kl}} |\overline{\boldsymbol{D}}|. \tag{8.64}$$

Above is the basic form of the *dynamic Smagorinsky model*. It should be kept in mind that $\mathcal{L}_{mn}$ in Eq. (8.63) should be evaluated with Eq. (8.56) and not with Eq. (8.60). The only required input parameter is the ratio of the filter width γ in the dynamic Smagorinsky model.

Determining the subgrid-scale eddy viscosity coefficient dynamically has advantage over the original Smagorinsky model in the following manner. Once the ratio of the filter widths γ is chosen, the eddy viscosity is specified without the use of any empirical parameters. If there are no turbulent scales between the grid and test filters, $\mathcal{L}_{ij}$ vanishes and the eddy viscosity becomes zero. Thus, the dynamic model does not require any special switching for laminar to turbulent transition or damping function near the wall. Furthermore, the subgrid-scale energy dissipation $\varepsilon_{SGS} = \tau_{ij}\overline{D}_{ij}$ can be negative to model the inverse cascade behavior of energy.

On the other hand, numerical instability can occur from the fluctuation of C due to the state of turbulence. The value of C can fluctuate greatly over space and the denominator in Eq. (8.63) can become extremely small. Both of these phenomena can lead to numerical instability. As a remedy, some form of averaging $\langle \rangle$ is performed to relax the local fluctuation of C in actual computations. For example, we can use volume average for isotropic turbulence, wall-normal-plane average for channel flow, or streamwise-plane average for duct flow. Hence in reality Eq. (8.63) is commonly replaced by

$$C = -\frac{1}{2\overline{\Delta}^2} \frac{\langle \mathcal{L}_{mn} \overline{D}_{mn} \rangle}{\langle \mathcal{M}_{kl} \overline{D}_{kl} \rangle}. \tag{8.65}$$

The incorporation of averaging makes the properties of the dynamic Smagorinsky model somewhat different from what was envisioned initially.

(2) Lilly's Least Squares Method

Lilly introduced the use of the least squares method to determine the constant in the dynamic Smagorinsky model [19]. The tensor $\mathcal{L}_{ij}^a$ in Eq. (8.60) has only five independent elements with the enforcement of symmetry and traceless conditions. However, there is not a single scalar C that can satisfy Eq. (8.60) for all five of the components. The approach taken by Lilly is to minimize the error

$$e_{ij} = \mathcal{L}_{ij}^a + 2C\overline{\Delta}^2 \mathcal{M}_{ij} \tag{8.66}$$

in a least squares fashion. The minimization is performed on the squared error of $E = e_{ij}e_{ij}$ by solving

$$\frac{\partial E}{\partial C} = 4\overline{\Delta}^2 (2C\mathcal{M}_{ij}\mathcal{M}_{ij} + \mathcal{L}_{ij}\mathcal{M}_{ij}) = 0 \tag{8.67}$$

for the value C. Because $\frac{\partial^2 E}{\partial C^2} = 8\overline{\Delta}^2 \mathcal{M}_{ij}\mathcal{M}_{ij} \geq 0$, the found C indeed provides the minimum E. From the traceless condition ($\mathcal{M}_{ii} = 0$), we have $\mathcal{L}_{ij}\mathcal{M}_{ij}^a = \mathcal{L}_{ij}\mathcal{M}_{ij}$. The optimal C accordingly becomes

$$C = -\frac{1}{2\overline{\Delta}^2} \frac{\mathcal{L}_{mn}\mathcal{M}_{mn}}{\mathcal{M}_{kl}\mathcal{M}_{kl}}. \tag{8.68}$$

With Lilly's least squares method, the above denominator is always positive unless $|\mathcal{M}| = 0$. This formulation can usually avoid the division by zero which occurs often with Garmano's approach (Eq. (8.63)). Nonetheless, the fluctuation in the value of C is still large and can cause numerical instability. In practice, Lilly's method often necessitates averaging similar to Eq. (8.65) like

$$C = -\frac{1}{2\overline{\Delta}^2} \frac{\langle \mathcal{L}_{mn}\mathcal{M}_{mn} \rangle}{\langle \mathcal{M}_{kl}\mathcal{M}_{kl} \rangle} \tag{8.69}$$

or establishing upper and lower bounds on C.

It is common to use Lilly's least squares method instead of Germano's original formulation not only in the dynamic Smagorinsky model but also in the approaches discussed below.

8.5.2 Extensions of the Dynamic Model

The dynamic Smagorinsky model permits $C(x, t)$ be to negative in the subgrid-scale eddy viscosity model. However, this leads to difficulty in achieving stable computation. The issues with the dynamic Smagorinsky model arise from

1. The use of negative eddy viscosity to represent the inverse cascade; and
2. The mathematical inconsistency in pulling C outside of the test filter during the derivation of Eq. (8.60).

Note that the inverse cascade is not caused by a diffusive process but is due to the nonlinear vortex merging. In order to remedy these problems, we can consider the dynamic mixed model for the first issue and the dynamic localization model for the second one.

The mixed model can turn unstable and the localization model requires increased computational effort. While these models have not yet been widely used, we present the basic concepts behind these models as they provide insights into how corrections can be made to the fundamental construct of the dynamic Smagorinsky model.

(1) Dynamic Mixed Model
By providing a different model that captures the effect of inverse cascade, the eddy-viscosity term can act as the main mechanism for energy dissipation. By using such mixed model as the foundation of the dynamic model, we can expect the value of C to lie within an appropriate range. Moreover, the use of such dynamic model removes the constraint of having the principal axes τ_{ij} and $\overline{D}_{ij}$ aligned. The model based on this approach is called the *dynamic mixed model*.

For τ_{ij} and T_{ij}, we supply different models (h_{ij} for the subgrid-scale and H_{ij} for the subtest-scale) in addition to the subgrid-scale eddy-viscosity model given by Eqs. (8.58) and (8.59). Thus, we have

$$\tau_{ij}^a = -2C\overline{\Delta}^2 |\overline{D}|\overline{D}_{ij} + h_{ij}^a \tag{8.70}$$

$$T_{ij}^a = -2C\widetilde{\Delta}^2 |\widetilde{\overline{D}}|\widetilde{\overline{D}}_{ij} + H_{ij}^a \tag{8.71}$$

for the anisotropic components. The model for the anisotropic part of the Germano identity becomes

$$\mathcal{L}_{ij}^a = -2C\overline{\Delta}^2 \mathcal{M}_{ij} + \mathcal{N}_{ij}^a, \tag{8.72}$$

where $\mathcal{N}_{ij} = H_{ij} - \widetilde{h}_{ij}$. Utilizing Lilly's least squares method to solve for C that minimizes $E = (\mathcal{L}_{ij}^a - \mathcal{N}_{ij}^a + 2C\overline{\Delta}^2 \mathcal{M}_{ij})^2$, we find

$$C = -\frac{1}{2\overline{\Delta}^2} \frac{(\mathcal{L}_{mn} - \mathcal{N}_{mn})\,\mathcal{M}_{mn}}{\mathcal{M}_{kl}\mathcal{M}_{kl}}. \tag{8.73}$$

This is the general representation of the dynamic mixed model. We note that there have been a few extensions made to this model [28, 31, 33].

(2) Dynamic Localization Model

Ghosal et al. [10] have attempted to remedy the mathematical inconsistency of assuming C as an invariant inside the test filter, which is one of the issues associated with the dynamic Smagorinsky and the dynamic mixed models. Here, let us consider the error

$$e_{ij} = \mathcal{L}_{ij}^a + 2\overline{\Delta}^2 \left(\gamma^2 C | \widetilde{\overline{D}} | \widetilde{\overline{D}}_{ij} - \widetilde{C | \overline{D} | \overline{D}_{ij}} \right) \tag{8.74}$$

and the spatial integral of its squared magnitude $E = e_{ij} e_{ij}$

$$F[C] = \int e_{ij}(x) e_{ij}(x) \, dx. \tag{8.75}$$

The proposed approach by Ghosal et al. is to determine C for which $\delta F = 0$. This results in having to solve the Fredholm integral equation of the second kind for $C(x)$

$$C(x) - \int \kappa(x, y) C(y) \, dy = f(x), \tag{8.76}$$

which can be solved iteratively. There is also a solution technique that utilizes localization in an approximate manner to reduce the computation time [24]. The above model that resolves the inconsistency of assuming constant C inside the test filter is called the *dynamic localization model*.

8.6 Other SGS Eddy-Viscosity Models

8.6.1 Structure Function Model

When the subgrid-scale fluctuations can be approximated as isotropic and follow the Kolmogorov spectra, Métais and Lesieur [16, 21] showed that the SGS eddy-viscosity coefficient becomes

$$\nu_e = \frac{2}{3} \alpha^{-\frac{3}{2}} \left[\frac{E_x(k_f)}{k_f} \right]^{\frac{1}{2}}, \tag{8.77}$$

where α is the Kolmogorov constant, $k_f = \pi/\Delta$ is the wave number corresponding to the filter width, and E_x is the energy spectra. This equation however cannot be directly used for LES in physical space.

Now, let us consider the use of the *structure function*

$$F_2(x, r, t) = \langle \|u(x + r, t) - u(x, t)\|^2 \rangle, \tag{8.78}$$

where $r = |r|$ and $\langle \rangle$ denotes the ensemble average. Using the filter width Δ as the representative length scale, F_2 and E_x can be related through

$$F_2(x, \Delta, t) = 4 \int_0^{k_f} E_x(k, t) \left[1 - \frac{\sin(k\Delta)}{k\Delta} \right] dk. \tag{8.79}$$

For a Cartesian grid with uniform spacing, the structure function can be approximated using filtered velocity values at the neighboring six points

$$\begin{aligned}
F_2(x_{i,j,k}, \Delta, t) = \frac{1}{6} \Big(&\|\overline{u}_{i+1,j,k} - \overline{u}_{i,j,k}\|^2 + \|\overline{u}_{i-1,j,k} - \overline{u}_{i,j,k}\|^2 \\
&+ \|\overline{u}_{i,j+1,k} - \overline{u}_{i,j,k}\|^2 + \|\overline{u}_{i,j-1,k} - \overline{u}_{i,j,k}\|^2 \\
&+ \|\overline{u}_{i,j,k+1} - \overline{u}_{i,j,k}\|^2 + \|\overline{u}_{i,j,k-1} - \overline{u}_{i,j,k}\|^2 \Big).
\end{aligned} \tag{8.80}$$

When there is a direction in which the grid is comparably fine (e.g., turbulent shear flow), variable values in that direction can be removed and instead a four-point evaluation can be used.

With the value of the structure function F_2 being available and the assumption of the Kolmogorov spectra $E_x(k, t) = \alpha \varepsilon(t)^{2/3} k^{-5/3}$ holding over the low wave number ($k \le k_f$), we can derive from Eq. (8.79) the SGS eddy viscosity

$$\nu_e(x, t) = 0.105 \alpha^{-\frac{3}{2}} \Delta \sqrt{F_2(x, \Delta, t)}, \tag{8.81}$$

which is known as the *structure function model* [16, 21]. If F_2 is evaluated correctly, this model requires no wall correction because F_2 decays near the wall, unlike the Smagorinsky model. For the background, development, and applications of the structure function model, we refer readers to reference [17].

8.6.2 Coherent Structure Model

Kobayashi proposed the *coherent structure model (Kobayashi model)* [14] to determine the SGS eddy viscosity coefficient based on the turbulent structures in the flow. Let us take the grid-scale velocity field and decompose the velocity gradient tensor $\partial \overline{u}_i / \partial x_j$ into the symmetric and asymmetric components, in a manner similar to Eq. (6.22),

$$\frac{\partial \overline{u}_i}{\partial x_j} = \overline{D}_{ij} + \overline{W}_{ij}, \tag{8.82}$$

$$\overline{D}_{ij} = \frac{1}{2}\left(\frac{\partial \overline{u}_i}{\partial x_j} + \frac{\partial \overline{u}_j}{\partial x_i}\right), \quad \overline{W}_{ij} = \frac{1}{2}\left(\frac{\partial \overline{u}_i}{\partial x_j} - \frac{\partial \overline{u}_j}{\partial x_i}\right). \tag{8.83}$$

We define the magnitudes of $\overline{D}_{ij}$ and $\overline{W}_{ij}$ as

$$|\overline{D}| = \sqrt{2\overline{D}_{ij}\overline{D}_{ij}}, \quad |\overline{W}| = \sqrt{2\overline{W}_{ij}\overline{W}_{ij}} \tag{8.84}$$

and introduce the following variables

$$\overline{Q} = \frac{|\overline{W}|^2 - |\overline{D}|^2}{4} = -\frac{1}{2}\frac{\partial \overline{u}_i}{\partial x_j}\frac{\partial \overline{u}_j}{\partial x_i}, \tag{8.85}$$

$$\overline{E} = \frac{|\overline{W}|^2 + |\overline{D}|^2}{4} = \frac{1}{2}\frac{\partial \overline{u}_i}{\partial x_j}\frac{\partial \overline{u}_i}{\partial x_j}, \tag{8.86}$$

where $\overline{Q}$ is the second invariant of the grid-scale velocity gradient tensor (Q-criterion) and $\overline{E}$ is the squared magnitude of the grid-scale velocity gradient tensor. It should also be recalled from Sect. 6.3.5 that $\overline{E}$ is proportional to the rate of kinetic energy dissipation. Based on these variables, we define the *coherent structure function*

$$F_{CS} = \frac{\overline{Q}}{\overline{E}}, \tag{8.87}$$

which can be viewed as a normalized second invariant with a range of $|F_{CS}| < 1$. This function approaches 1 when the flow is under strong rotation. On the other hand, the function approaches -1 when the flow is shear-dominated compared to rotation. In the coherent structure model, the subgrid-scale stress

$$\tau_{ij}^a = -2C\Delta^2|\overline{D}|\overline{D}_{ij} \tag{8.88}$$

is provided with

$$C = \frac{1}{20}|F_{CS}|^{\frac{3}{2}}. \tag{8.89}$$

The coherent structure function approaches zero with an asymptotic profile of $F_{CS} \propto y^2$ near a flat wall, with y representing the wall-normal direction. The exponent $3/2$ in Eq. (8.89) ensures that such asymptotic behavior is achieved. This coherent structure model can be applied to turbulent flows around bodies with complex geometry.

As mentioned in Sect. 6.3.6, the rate-of-rotation tensor $\overline{W}_{ij}$ is not an objective quantity. When the coordinate system is under rotation with angular velocity $\dot{\theta}$, we can use $\overline{W}'_{ij} = \overline{W}^*_{ij} + \epsilon_{ijk}\theta^*_k$ to evaluate Q and $\overline{E}$. Here, the asterisk denotes the components in the rotating frame of reference. Kobayashi suggests the use of

$$C = \frac{1}{22}|F_{CS}|^{\frac{3}{2}} F_\Omega, \quad F_\Omega = 1 - F_{CS} \tag{8.90}$$

to extend the structure function model to turbulent flows under the influence of rotation [14].

Comparisons of SGS Eddy-Viscosity Models

For the eddy-viscosity coefficient in Eq. (8.33), the Smagorinsky model from Eq. (8.37) gives

$$\nu_e = (C_s \Delta)^2 |\overline{\boldsymbol{D}}| = 0.03 \Delta^2 |\overline{\boldsymbol{D}}| \tag{8.91}$$

based on $\alpha = 1.5$ and $C_s = 0.173$ (see Eq. (8.41)).

In comparison, the structure function model yields

$$\nu_e \approx 0.777 (C_s \Delta)^2 \sqrt{|\overline{\boldsymbol{D}}|^2 + |\overline{\boldsymbol{W}}|^2} \tag{8.92}$$

for a six-point formulation with Eq. (8.80) [17]. This model provides reduced eddy viscosity when $|\overline{\boldsymbol{W}}| \ll |\overline{\boldsymbol{D}}|$ and increased eddy viscosity when $|\overline{\boldsymbol{W}}| \gtrsim |\overline{\boldsymbol{D}}|$ in comparison to the Smagorinsky model. This means that the structure function model increases the eddy-viscosity coefficient according to the strength of the eddies in the field. However, one should be aware that ν_e does not approach zero at the solid wall based on the six-point formulation.

The SGS eddy-viscosity coefficient for the coherent structure model in Eq. (8.89) can be expressed as

$$\nu_e = 0.05 \Delta^2 \left(\frac{|\overline{\boldsymbol{W}}|^2 - |\overline{\boldsymbol{D}}|^2}{|\overline{\boldsymbol{W}}|^2 + |\overline{\boldsymbol{D}}|^2} \right)^{\frac{3}{2}} |\overline{\boldsymbol{D}}|. \tag{8.93}$$

Setting the assumptions used in deriving the above models aside, we can say that coherent structure model is a general formulation because it does not require any corrections to reduce or cancel the value of the eddy-viscosity coefficient at the no-slip wall.

In the near-wall region, nonuniform grids are often employed. When the grid width is chosen for the filter width, the SGS eddy-viscosity coefficient is strongly influenced by the grid distribution. Hence, deriving a SGS model that universally captures the asymptotic behavior of the flow is difficult. If the near-wall flow physics needs to be correctly predicted, LES needs to predict the flow through the grid-scale variables and not be strongly influence by the SGS components.

8.6.3   One-Equation SGS Model

Similar to the efforts placed on developing RANS models with turbulent kinetic energy transport equations and stress equations, there have also been attempts to incorporate transport equations for SGS turbulent kinetic energy and SGS turbulent

stress into LES models from the early stage of research. In LES, the large-scale
structures that characterize the dominant features of the flow are directly computed.
Hence the influence of the SGS fluctuations on the turbulent stress is smaller than
that from the fluctuations for Reynolds-averaged flow. For this reason, increasing
the complexity of SGS turbulence models has not received much attention in engineering applications. On the other hand, for problems in which the simulated flows
have sizable portion of their energy spectra in the subgrid-scale wave numbers (e.g.,
simulations of earth-scale fluid flows and multiphase turbulent flows with particles,
bubbles, or droplets), it might become meaningful to solve the transport equation
that can take SGS physics into account.

In the RANS models, two equations are commonly used to determine the eddy
viscosity. For LES, we can select the filter width Δ to be the representative length
scale in SGS models. Hence, it appears sufficient to utilize only the SGS turbulent
energy equation, Eq. (8.30), and set the SGS eddy-viscosity coefficient as

$$\nu_e = C_\nu \Delta_\nu \sqrt{k_{SGS}} \tag{8.94}$$

for the momentum equation, Eq. (8.34). This approach is the one-equation model. The
k_{SGS} transport equation that serves as the basis here has been derived theoretically
[32], which can be corrected for near-wall turbulence as [22]

$$\frac{\overline{D}k_{SGS}}{\overline{D}t} = -\tau_{ij}\overline{D}_{ij} - C_\varepsilon \frac{k_{SGS}^{3/2}}{\Delta}$$
$$- \varepsilon_w + \frac{\partial}{\partial x_j}\left[\left(C_d \Delta_\nu \sqrt{k_{SGS}} + \nu\right)\frac{\partial k_{SGS}}{\partial x_j}\right]. \tag{8.95}$$

If k_{SGS} is set to 0 at the wall as a boundary condition, Eq. (8.94) does not require
any corrections with a damping function. However, Eq. (8.95) utilizes

$$\Delta_\nu = \frac{\Delta}{1 + C_k \overline{\Delta}^2 |\overline{D}|^2 / k_{SGS}}, \quad \varepsilon_w = 2\nu \frac{\partial \sqrt{k_{SGS}}}{\partial x_j}\frac{\partial \sqrt{k_{SGS}}}{\partial x_j} \tag{8.96}$$

by taking the solution behavior near the wall into account. For the above model, the
following set of parameters

$$C_\nu = 0.05, \quad C_\varepsilon = 0.835, \quad C_d = 0.10, \quad C_k = 0.08 \tag{8.97}$$

should be used [22].

We can consider utilizing the k_{SGS} equation as a basis for the dynamic
eddy-viscosity model [12]. Recall that the dynamic Smagorinsky model can lead
to unstable computations when its eddy viscosity is naively implemented in the
GS momentum equation. This is due to the large fluctuation or the negative value
encountered by the eddy viscosity. The dynamic model has been originally proposed to capture the energy transfer between the GS and SGS components that are

represented by the first right-hand side terms in Eqs. (8.29) and (8.30). The dynamic Smagorinsky model has some shortcomings in evaluating the diffusive coefficient for the SGS fluctuations. Hence, we can alternatively consider the use of Eq. (8.94) to evaluate the eddy viscosity since k_{SGS} is always positive. Here, the eddy-viscosity approach can be adopted for the representing τ_{ij} so that the production term becomes

$$- \tau_{ij}\overline{D}_{ij} = 2C\overline{\Delta}^2|\boldsymbol{D}|^3, \tag{8.98}$$

instead of using Eq. (8.94). In the above equation, we can determine C through the procedure taken by the dynamic Smagorinsky model [12]. This model is referred to as the one-equation dynamic SGS model.

Even if Eq. (8.98) yields a locally negative value for C_S, it will only lead to the decrease of k_{SGS} and ν_e given by Eq. (8.94) and they do not become negative. It is expected that k_{SGS} and ν_e become zero when the production term becomes zero in laminar regions of the flow. This approach thus does not require any special corrections to suppress unphysical oscillations of eddy viscosity nor introduce any damping in laminar flow regions.

8.7 Numerical Methods for Large-Eddy Simulation

The LES is essentially the direct simulation of the large-scale vortices. The numerical methods for LES can be based on the discussion offered for the unsteady Navier–Stokes equations in Chaps. 3 and 4. Here, we discuss some of the numerical treatments unique to LES.

8.7.1 Computation of SGS Eddy Viscosity

For the Smagorinsky and the dynamic Smagorinsky models, we need to compute the characteristic filter length Δ and the magnitude of the rate-of-strain tensor $|\boldsymbol{D}|$ for the eddy-viscosity coefficient in Eq. (8.37). As it was explained in Sect. 3.6 for the viscous term, the eddy-viscosity term does not conserve energy either. For that reason, excessively improving the order of accuracy may not be as critical as for the other terms. However, it is important to consider compatibility, as described in Sect. 3.6, to compare the production and dissipation of energy and Reynolds stress or to evaluate the contributions from the subgrid-scale eddy-viscosity term.

The characteristic filter length for a three-dimensional Cartesian grid can be set to

$$\Delta = \sqrt[3]{\Delta_x \Delta_y \Delta_z}, \tag{8.99}$$

where Δ_x, Δ_y, and Δ_z are the filter widths in the respective three directions. The right-hand side of above expression is the cubic root of the filtered volume. We can extend this to a generalized coordinate system by using the coordinate transform Jacobian

$$\Delta = \sqrt[3]{J}, \quad J = \left| \frac{\partial x_i}{\partial \xi^j} \right|, \tag{8.100}$$

where we have let the filter width and the mesh size to be equal to each other. We can alternatively use $\Delta = \sqrt{\Delta_x^2 + \Delta_y^2 + \Delta_z^2}$ instead of Eq. (8.99), but this choice becomes difficult to extend to generalized coordinates.

The magnitude of the strain rate tensor is

$$|\boldsymbol{D}|^2 = 2 D_{ij} D_{ij} = 2 \left(D_{11}^2 + D_{22}^2 + D_{33}^2 \right) + 4 \left(D_{12}^2 + D_{23}^2 + D_{31}^2 \right). \tag{8.101}$$

For a Cartesian staggered grid, each component of D_{ij} can be evaluated with

$$D_{ij} = \frac{1}{2} \left(\delta_{x_j} u_i + \delta_{x_i} u_j \right). \tag{8.102}$$

For $i = j$, D_{ij} is located at the cell center and, for $i \neq j$, D_{ij} is positioned at the cell edge as illustrated in Fig. 3.12. Thus, we can use

$$|\boldsymbol{D}|^2 = 2 \left(D_{xx}^2 + D_{yy}^2 + D_{zz}^2 \right) + 4 \left(\left[\overline{D}_{xy}^{xy} \right]^2 + \left[\overline{D}_{yz}^{yz} \right]^2 + \left[\overline{D}_{zx}^{zx} \right]^2 \right) \tag{8.103}$$

to compute $|\boldsymbol{D}|$ at the cell centers and determine ν_e. If we need the value of ν_e at the cell edge, we can utilize interpolation based on the cell center values. For a collocated grid on a generalized coordinate, all components of D_{ij} are computed at the cell face to determine the (molecular) viscous diffusion flux. To evaluate the eddy viscosity ν_e, we can compute $|\boldsymbol{D}|$ at each of the cell interface. For the eddy viscosity ν_e at the cell center, we can calculate $|\boldsymbol{D}|$ at the cell face or at the cell center with

$$D_{ij} = \frac{1}{2} \left(\frac{\partial \xi^k}{\partial x_j} \delta'_{\xi_k} u_i + \frac{\partial \xi^k}{\partial x_i} \delta'_{\xi_k} u_j \right). \tag{8.104}$$

The subgrid-scale eddy-viscosity coefficient ν_e can be positioned at the cell center for a staggered grid as depicted in Fig. 3.12. Because the shear component of the subgrid-scale stress is computed at the cell edge, two-dimensional interpolation becomes necessary. For a collocated grid, we can use a one-dimensional interpolation from the cell center to the cell face. As an example, let us demonstrate an interpolation scheme for the x-y plane as shown in Fig. 8.9.

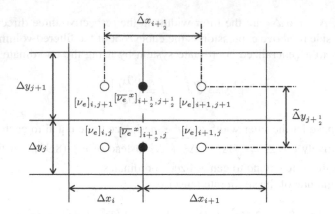

Fig. 8.9 Interpolation of kinematic eddy-viscosity coefficient

The eddy-viscosity coefficient $[\overline{\nu_e}^x]_{i+\frac{1}{2},j}$ at the cell face is determined by solving the following relationship[5]

$$\frac{1}{[\overline{\nu_e}^x]_{i+\frac{1}{2},j}} = \frac{1}{\Delta x_i + \Delta x_{i+1}}\left(\frac{\Delta x_i}{\nu_{ei,j}} + \frac{\Delta x_{i+1}}{\nu_{ei+1,j}}\right). \tag{8.105}$$

Along the cell edge, Eq. (8.105) can be extended to two dimensions for $[\overline{\nu_e}^{xy}]_{i+\frac{1}{2},j+\frac{1}{2}}$

$$\frac{1}{[\overline{\nu_e}^{xy}]_{i+\frac{1}{2},j+\frac{1}{2}}} = \frac{1}{(\Delta x_i + \Delta x_{i+1})(\Delta y_j + \Delta y_{j+1})}$$
$$\times \left(\frac{\Delta x_i \Delta y_j}{\nu_{ei,j}} + \frac{\Delta x_{i+1}\Delta y_j}{\nu_{ei+1,j}} + \frac{\Delta x_i \Delta y_{j+1}}{\nu_{ei,j+1}} + \frac{\Delta x_{i+1}\Delta y_{j+1}}{\nu_{ei+1,j+1}}\right). \tag{8.106}$$

---

[5] Let us derive the interpolation equation for the kinematic (eddy) viscosity coefficient based on how the heat transfer rate is determined for a multispecies interface [23]. Figure 8.10 is provided for reference. Denoting the coefficient at the interface as $\overline{\nu}$ and evaluating the flux with diffusive gradient, we have

$$\tau = -\overline{\nu}\frac{-u_M + u_P}{\Delta_m + \Delta_p}.$$

On the other hand, by assuming that the stress is constant over the segment M to P and the viscosity ν is constant inside the cell, we can take the difference across the cell interface to express τ as

$$\tau = -\nu_M \frac{-u_M + \overline{u}}{\Delta_m} = -\nu_P \frac{-\overline{u} + u_P}{\Delta_p}.$$

Eliminating $\overline{u}$ in the above equation, we can derive

$$\frac{1}{\overline{\nu}} = \frac{1}{\Delta_m + \Delta_p}\left(\frac{\Delta_m}{\nu_M} + \frac{\Delta_p}{\nu_P}\right),$$

which becomes the harmonic mean when the grid is uniform ($\Delta_m = \Delta_p$).

Fig. 8.10 Interpolation of diffusion coefficient (viscosity)

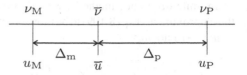

The computed eddy viscosity can then be incorporated into Eq. (8.16) to yield

$$\frac{\partial \overline{u}_i}{\partial t} = \frac{\partial}{\partial x_i} \left[-\overline{u}_i \overline{u}_j - \frac{1}{\rho} \delta_{ij} \overline{P} + 2 \left(\nu_e + \nu \right) \overline{D}_{ij} \right], \qquad (8.107)$$

which appears as the momentum equation with an extra viscous coefficient. The numerical treatment of this equation follows what has already been discussed in Chaps. 3 and 4. It should be reminded that the grid-scale pressure $\overline{p}$ is not directly determined. What is solved for in the coupled system of equations comprised of Eqs. (8.107) and (8.13) is

$$\overline{P} = \overline{p} + \frac{1}{3} \rho \tau_{ii} \qquad (8.108)$$

the modified pressure.

Because the eddy-viscosity coefficient ν_e varies over time and space, it is practical to treat only the molecular viscous term (ν) implicitly. With adequate grid resolution near the wall, there is significant decay of the eddy-viscosity value in that region. In other words, physical viscosity dominates over the eddy viscosity in the near-wall region, which lets the implicit treatment of only the molecular viscous term be sufficiently effective.

8.7.2 Implementation of Filtering

For the scale-similarity model or the dynamic model, it is necessary to filter the grid-scale flow field. Let us consider a one-dimensional filter

$$\overline{u}(x) = \int_{-\infty}^{\infty} G(r) u(x - r) \, dr. \qquad (8.109)$$

In practice, it is inefficient to compute this convolution. As a special case, Fourier transform can be utilized when there is a periodic direction. The filter function can be Fourier transformed a priori and be multiplied to the Fourier transformed field variable, described by Eq. (8.11). The inverse transform of the product yields the filtered field variable in physical space. However, this method is not common.

In finite-difference methods, we can implement a numerical filtering technique that approximates Eq. (8.109) to be more practical. Consider a Taylor series expansion of $u(x - r)$ about x

$$u(x - r) = u(x) - ru'(x) + \frac{r^2}{2}u''(x) - \frac{r^3}{6}u'''(x) + \frac{r^4}{24}u^{(4)} - \cdots \quad (8.110)$$

and substitute this into Eq. (8.109). For an even filter function G with its integral being unity, we have the series expansion

$$\bar{u}(x) = u(x) + \gamma_2 u''(x) + \gamma_4 u^{(4)}(x) + \gamma_6 u^{(6)}(x) + \cdots, \quad (8.111)$$

where

$$\gamma_m = \frac{1}{m!} \int_{-\infty}^{\infty} r^m G(r) \, dr \quad (8.112)$$

are coefficients dependent on the filter function.

For the box filter represented by Eq. (8.3), the coefficients are

$$\gamma_m = \frac{2}{m!} \int_0^{\Delta/2} r^m \, dr = \frac{\Delta^m}{2^m (m + 1)!} \quad (8.113)$$

$$\gamma_2 = \frac{\Delta^2}{24}, \quad \gamma_4 = \frac{\Delta^4}{1920}, \quad \gamma_6 = \frac{\Delta^6}{322560}, \quad \cdots \quad (8.114)$$

and for the Gaussian filter represented by Eq. (8.7), the coefficients are

$$\gamma_m = \frac{(m - 1)!!}{12^{m/2} m!} \Delta^m \quad (8.115)$$

$$\gamma_2 = \frac{\Delta^2}{24}, \quad \gamma_4 = \frac{\Delta^4}{1152}, \quad \gamma_6 = \frac{\Delta^6}{82944}, \quad \cdots, \quad (8.116)$$

where $(2n - 1)!! = (2n - 1)(2n - 3) \cdots 3 \cdot 1$. Approximating the derivatives in Eq. (8.111) with finite differences and incorporating the coefficients above, we can approximate the effect of filtering.

For a three-point central-difference scheme for u'', Eq. (2.20), we can compute up to the second term in Eq. (8.111). For this case, the box filter and the Gaussian filter provide the identical finite-difference scheme of

$$\bar{u}_j = u_j + \frac{\Delta^2}{24} \frac{u_{j-1} - 2u_j + u_{j+1}}{\Delta x^2}. \quad (8.117)$$

If the filter width and the grid size are equal ($\Delta = \Delta x$), we obtain

$$\bar{u}_j = \frac{u_{j-1} + 22u_j + u_{j+1}}{24} \quad (8.118)$$

and if the filter width is twice the grid size ($\Delta = 2\Delta x$), we have

$$\bar{u}_j = \frac{u_{j-1} + 4u_j + u_{j+1}}{6}. \tag{8.119}$$

If we are to use a five-point central-difference scheme for u'' and $u^{(4)}$ with Eqs. (2.22) and (2.24), respectively, we can evaluate Eq. (8.111) up to the third term. The box filter yields an approximation of

$$\bar{u}_j = u_j + \frac{\Delta^2}{288} \frac{-u_{j-2} + 16u_{j-1} - 30u_j + 16u_{j+1} - u_{j+2}}{\Delta x^2}$$
$$+ \frac{\Delta^4}{1920} \frac{u_{j-2} - 4u_{j-1} + 6u_j - 4u_{j+1} + u_{j+2}}{\Delta x^4}. \tag{8.120}$$

When the filter width is the same as the grid size ($\Delta = \Delta x$), we have

$$\bar{u}_j = \frac{-17u_{j-2} + 308u_{j-1} + 5178u_j + 308u_{j+1} - 17u_{j+2}}{5760} \tag{8.121}$$

and when the filter width is twice the grid size ($\Delta = 2\Delta x$), we find

$$\bar{u}_j = \frac{-2u_{j-2} + 68u_{j-1} + 228u_j + 68u_{j+1} - 2u_{j+2}}{360}. \tag{8.122}$$

The Gaussian filter on the other hand provides

$$\bar{u}_j = u_j + \frac{\Delta^2}{288} \frac{-u_{j-2} + 16u_{j-1} - 30u_j + 16u_{j+1} - u_{j+2}}{\Delta x^2}$$
$$+ \frac{\Delta^4}{1152} \frac{u_{j-2} - 4u_{j-1} + 6u_j - 4u_{j+1} + u_{j+2}}{\Delta x^4}, \tag{8.123}$$

which becomes

$$\bar{u}_j = \frac{-3u_{j-2} + 60u_{j-1} + 1038u_j + 60u_{j+1} - 3u_{j+2}}{1152} \tag{8.124}$$

when the filter width is equal to the grid size ($\Delta = \Delta x$) and

$$\bar{u}_j = \frac{u_{j-1} + 4u_j + u_{j+1}}{6} \tag{8.125}$$

when the filter width is twice the grid size ($\Delta = 2\Delta x$). The last expression is identical to the results using the three-point differencing in Eq. (8.119).

Taylor series expansion is valid for a smooth function in a neighborhood of the expansion point. This may not be a problem in a practical sense for a filter G that effectively decays over Δ. However, there is a question of how many terms we should retain in Eq. (8.111). If we keep high-order terms to approximate the filtering operation to high accuracy, we observe the appearance of negative coefficients, as seen in Eqs. (8.121), (8.122), and (8.124), which are not present in the original filter function.

8.7.3 Boundary and Initial Conditions

Problems with LES in practice are often associated with the numerical setup instead of the turbulence model. Let us consider an example of flow over an object on a flat plate, as shown in Fig. 8.11. Here, the boundary conditions for velocity and pressure at the inlet, outlet, far field (sides and top) and walls must be specified.

Prescribing the turbulent inlet condition is especially difficult. Since we are interested in performing unsteady turbulent flow simulations, the velocity and pressure boundary conditions including the fluctuations should be provided at the inlet. This is somewhat self-contradicting since we would need a theoretical solution that captures turbulent fluctuations. However, such solution has not been found to this date, which is precisely why LES and DNS remain powerful tools. In practice, we can only provide an inlet condition in approximate form and require a run-up region that is sufficiently long for the incoming boundary layer to develop into an accurate turbulent flow profile.

For the averaged velocity profile, we can adopt experimental measurements with similar flow conditions or prediction based on the k-ε model. In principle the added fluctuation should satisfy the continuity equation. However, small level of random noise can be used to model the fluctuation. The component that does not satisfy continuity (non-solenoidal) can be removed as the flow enters the computational domain, if the projection is implemented correctly in the incompressible flow solver. We show in Fig. 8.12 a typical evolution of the velocity profile that started with an

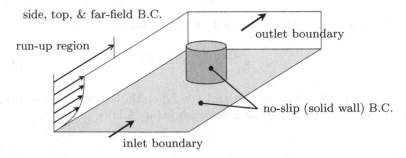

Fig. 8.11 Computational domain for a flow over a object on a wall

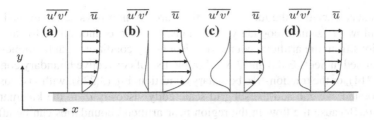

Fig. 8.12 The evolution process of the velocity and Reynolds stress profiles over the run-up region

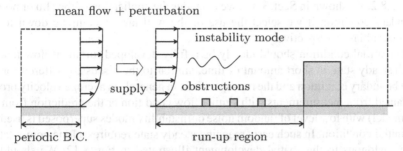

Fig. 8.13 An example of prescribing a non-periodic unsteady turbulent inlet boundary condition

inlet velocity condition with small level of random noise added to a mean turbulent boundary layer velocity profile. In physical flows, the average velocity profile is influenced by the Reynolds and viscous stresses. Adding random perturbation with no correlation ($\overline{u_i' u_j'}$) to the average velocity profile (Fig. 8.12a), the fluid is accelerated near the wall by the non-equilibrium nature of the flow (Fig. 8.12b), which then increases the fluctuation and Reynolds stress due to the steepened velocity gradient (Fig. 8.12c). The flow eventually reaches an equilibrium state (Fig. 8.12d). During such development of the flow, a long run-up region is required and in the process from (b) to (c), the flow exhibits an overshoot which can drive the numerical simulation to become unstable.

For a spatially developing flow field, we can consider the use of the setup illustrated in Fig. 8.13 for unsteady inlet conditions in DNS or LES. One choice is to perform a separate unsteady periodic simulation (upstream) and utilize a spanwise slice of the such flow field as the inlet boundary condition to the full non-periodic simulation. The use of such periodic inlet condition is of course an added computational burden. Another approach is not only to consider the addition of random fluctuation, but also to superpose instability modes (that satisfy the continuity equation) or to place obstructions or roughness (which need not be precise) in the run-up region to support the growth of the boundary layer, as often performed in experiments. These methods may, however, suffer from numerical instability. There are ongoing research on developing models to describe unsteady velocity profiles that approximate the developed turbulent boundary flow, especially in fields such as architectural engineering that consider large-scale flows.

It is necessary that the turbulent vortices are not non-physically strained due to the outflow or far-field boundary conditions. We also cannot allow for numerical reflections from the artificial boundaries. Boundary conditions can be implemented as discussed in Sects. 3.8.3 and 3.8.4. For the use of convective boundary condition, Eq. (3.214), or the traction-free boundary condition, Eq. (3.223), with viscous effect incorporated, we can add the subgrid-scale eddy viscosity ν_e to the kinematic viscosity ν. Because the flow in the region near artificial boundaries can be affected, data in such region should be excluded from the final analysis.

For the wall boundary condition, we can adopt the methodology discussed in Sect. 3.8.2. As shown in Sect. 3.8.4, we consider the application of the slip or no-slip boundary condition if we select the use of the wall law or compute down to the viscous sublayer, respectively.

The initial condition should ideally be a fully developed turbulent flow field to reach steady state in short amount of time, similar to the discussion offered for the inlet boundary condition and the run-up region. In general, an average velocity profile (obtained from measurements with similar flow condition or the prediction from the k-ε model) with low level of random noise or instability modes superposed is used for the initial condition. In such case, reaching steady state requires a long computation time analogous to the spatial development illustrated in Fig. 8.12. We should be careful not to let the intermediate process be prone to numerical instability. Perhaps the most effective setup is to initiate the computation from an instantaneous flow field obtained through a simulation performed under a similar condition (e.g., different Re or mesh). Such flow field may not be readily available because the change in flow condition often requires modification to the computational domain. Moreover, interpolation and coarsening of the data could possibly lead to error in the continuity equation. Nonetheless the use of an initial condition from a similar simulation can be very effective and beneficial for starting a DNS or LES computation.

8.7.4 Influence of Numerical Accuracy

As discussed above, LES provides the unsteady numerical solution to the Navier–Stokes equations with the added eddy-viscosity term. While DNS is to be set up with grids fine enough so that we can ignore the flow scales smaller than the mesh size, grid scales for LES lie within the range that carries turbulent energy. Thus, LES should in principle require higher order accuracy than DNS. In reality nonetheless, DNS may not necessarily be performed at the resolution corresponding to the finest scale of turbulence and often requires higher-order accurate methods. On the other hand, LES is often under the misconception that it requires only, for example, up to second-order spatial accuracy for the Smagorinsky model due to the ambiguity in the choice of Δ^2 in the eddy viscosity. However, as we can observe from Fig. 8.14, the results from LES based on the Smagorinsky model agree with the results from DNS much better by adopting the fourth-order central-difference scheme instead of the second-order central-difference scheme [13]. Furthermore, to take advantage

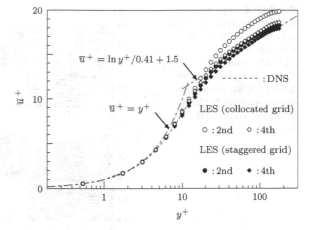

Fig. 8.14 Influence of variable placement and spatial order of accuracy (second and fourth) for LES of turbulent channel flow using the Smagorinsky model

of high-order turbulence models, high-order accuracy is required. It is not easy to determine the optimal order of accuracy. When a coarse grid is used, the spatial accuracy needs to be increased for an accurate solution. Models that extract the wave number components from the high-end, such as the dynamic model or the structure function model should benefit from high-order spatial accuracy.

8.8 Remarks

Numerous workshops are frequently held worldwide to evaluate turbulence models or computational methods. A comparison of results based on different computational methods and experiments for flow over a cube mounted on a flat surface is shown in Fig. 8.15 [25]. This type of flow is often encountered around buildings and inside industrial machines. From the figure it is clear that the LES performs quite well in predicting the flow. The k-ε model also captures the qualitative behavior of the flow. Depending on the level of fidelity required by users, it may also be possible to interpret that the results from LES and k-ε models do not exhibit a difference significant enough to be concerned of. There may be other applications in which predicting the flow to the level of LES results or higher may be necessary. The assessment of models should be based on the viewpoints of users and the objectives of the computational study. The purpose of the workshops or assessments should not focus on which model outperforms another, but to reveal the characteristics and computational cost associated with each model. The emphasis of the comparison should be placed on providing useful information so that users can select the appropriate model to satisfy their requirements.

One of the characteristic features of the flows similar to the one presented in Fig. 8.15 is the shedding of the large-scale vortices from the structure. Reynolds averaging may encounter difficulty predicting such flows. It would be necessary to predict not

Fig. 8.15 Comparison of
flow over a surface mounted
cube from experiment and
simulations [25]:
a Experimental
measurement, **b** k-ε model,
and **c** LES. Reprinted with
permission from the Japan
Society of Mechanical
Engineers

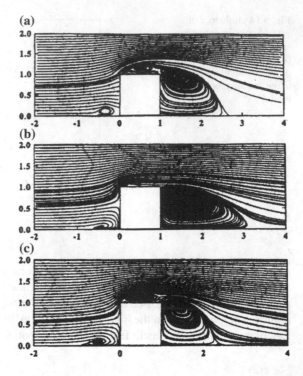

only the average base flow but also the fluctuating component of the flow to obtain
an understanding of the dynamics of the flow. In many instances, the magnitude of
the fluctuation can be more important than the average value (e.g., evaluating force
on a building, predicting the heat transfer rate of a heat exchanger, or characterizing
a separation bubble). For simulating these types of turbulent flows, LES is attractive
because of its ability to accurately predict the unsteady characteristics and the time-
averaged flow field. When the objective of the simulation is to evaluate the turbulent
statistics (average flow profile, RMS values, and higher-order moments), we should
keep in mind that the computation should be performed over a sufficiently long
advective time with enough samples collected.

As we have mentioned at the end of Chap. 6, one of the advantages of LES
compared to RANS is the relative clarity of how the length scales are defined for
unsteady motion of fluids. Thus, when we consider the ability to handle multiple
physical scales (e.g., different length scales in multiphase flow, such as the size of
particles, droplets, and bubbles), LES is more easily extendable to incorporate various spatial scales than in RANS. For all of the above reasons, LES would likely
remain at the center of turbulence simulation for the years to come, especially for
flows over a complex domain or for turbulence of complex fluids.

Let us end this chapter with brief discussions on the following three points. First,
the governing equations for LES should take into account the variation of filter width
over the computational domain, precisely speaking. Due to the resulting complexity

from considering the filter width variation, there has only been a limited number of studies that have considered addressing this point. However, it may be well necessary to revisit the assumptions made in the derivation of the governing equations, including Eq. (6.15), and reassess the conventional LES system of equations to further improve the accuracy of LES models.

Second, LES primarily models small-scale eddies responsible for dissipating kinetic energy. There is also an approach in LES called *implicit LES* [11] that numerically dissipate SGS fluctuations by utilizing upwinding of the advection term. We should however note that such dissipation mechanism is fundamentally different from physical dissipation. Implicit LES yields accurate turbulent flow solutions when the spatial grid resolution is close to that of DNS.

Third, highly refined grids are required near the wall for simulating high-Reynolds number flows over bodies. The resolution requirement in the near-wall regions for LES can be close to that for DNS. As a practical solution to engineering problems, it is possible to consider a hybrid approach where the near-wall flow is solved with RANS and the regions far from the walls is tackled with LES, with a gradual transition between the two regions. Use of RANS near the wall allows for the use of coarse grid. Because this approach can be viewed as solving for the unsteady eddies away from the wall, it is referred to as *detached-eddy simulation* [30]. It should be realized that this hybrid approach is based on two different concepts of averaging for different regions of the flow.

Simulation techniques such as the implicit LES and the detached-eddy simulation can be considered as convenient turbulent simulation tools. Users should however be aware of when and how these techniques can be employed to meet the desired level of results. To better predict the behavior of complex turbulent flows, it appears intuitive and appropriate that we pursue the improvement of physics-based eddy-viscosity models in LES.

In this chapter, we only provided an overview of LES and its implementation. For additional details, we ask readers to refer to review articles [16, 20, 26] and reference books [3, 5, 17, 27].

8.9 Exercises

8.1 Consider a filter based on the following equation

$$f = \overline{f} - \Delta^2 \frac{\partial^2 \overline{f}}{\partial x_k \partial x_k}$$

for $x \in \mathbb{R}^3$ to determine the filtered quantify $\overline{f}$. This type of filter is known as a *differential filter* (Helmholtz elliptic filter) [6, 7, 27].

1. Given the above equation, find $\overline{f}$.
2. Show that in this case this filter operation satisfies $\overline{\frac{\partial f}{\partial x_i}} = \frac{\partial \overline{f}}{\partial x_i}$.

8.2 Consider approximating the Gaussian filter operation on a uniform grid about point (x_i, y_j) when data from $(x_{i\pm 1}, y_{j\pm 1}) = (x_i \pm \Delta x, y_j \pm \Delta y)$ are available

1. Given $f_{i,j}$, $f_{i-1,j}$, $f_{i+1,j}$, $f_{i,j-1}$, and $f_{i,j+1}$, find the approximation to the filtered $\overline{f}$ with the highest order of accuracy.
2. Given $f_{i,j}$, $f_{i-1,j-1}$, $f_{i+1,j-1}$, $f_{i+1,j-1}$, and $f_{i+1,j+1}$, find the approximation to the filtered $\overline{f}$ with the highest order of accuracy.
3. Given $f_{i,j}$, $f_{i-1,j}$, $f_{i+1,j}$, $f_{i,j-1}$, and $f_{i,j+1}$ as well as $f_{i-1,j-1}$, $f_{i+1,j-1}$, $f_{i+1,j-1}$, and $f_{i+1,j+1}$, find the approximation to the filtered $\overline{f}$ with the highest order of accuracy.

8.3 Show that

1. The Leonard term L_{ij} is not Galilean invariant.
2. The modified Leonard stress L_{ij}^m is Galilean invariant.

8.4 Assuming local equilibrium, fully derive the Smagorinsky constant of $C_s = 0.173$ using the value of $\alpha = 1.5$ for the Kolmogorov constant.

References

1. Bardina, J., Ferziger, J.H., Reynolds, W.C.: Improved subgrid scale models for large eddy simulation. AIAA Paper 80–1357 (1980)
2. Bardina, J., Ferziger, J.H., Reynolds, W.C.: Improved turbulence models based on large eddy simulation of homogeneous, incompressible, turbulent flows. Technical Report TF-19, Thermosciences Division, Dept. Mechanical Engineering, Stanford University (1983)
3. Berselli, L., Iliescu, T., Layton, W.J.: Mathematics of Large Eddy Simulation of Turbulent Flows. Springer (2006)
4. Chapman, D.R.: Computational aerodynamics development and outlook. AIAA J. **17**(12), 1293–1313 (1979)
5. Garnier, E., Adams, N., Sagaut, P.: Large Eddy Simulation for Compressible Flows. Springer (2009)
6. Germano, M.: Differential filters for the large eddy numerical simulation of turbulent flows. Phys. Fluids **29**(6), 1755–1757 (1986)
7. Germano, M.: Differential filters of elliptic type. Phys. Fluids **29**(6), 1757–1758 (1986)
8. Germano, M.: A proposal for a redefinition of the turbulent stresses in the filtered Navier-Stokes equations. Phys. Fluids **29**(7), 2323–2324 (1986)
9. Germano, M., Piomelli, U., Moin, P., Cabot, W.H.: A dynamic subgrid-scale eddy viscosity model. Phys. Fluids A **3**(7), 1760–1765 (1991)
10. Ghosal, S., Lund, T.S., Moin, P., Akselvoll, K.: A dynamic localization model for large-eddy simulation of turbulent flows. J. Fluid Mech. **286**, 229–255 (1995)
11. Grinstein, F.F., Margolin, L.G., Rider, W.J.: Implicit Large Eddy Simulation. Cambridge Univ. Press (2007)
12. Kajishima, T., Nomachi, T.: One-equation subgrid scale model using dynamic procedure for the energy production. J. Appl. Mech. **73**(3), 368–373 (2006)
13. Kajishima, T., Ohta, T., Okazaki, K., Miyake, Y.: High-order finite-difference method for incompressible flows using collocated grid system. JSME Int. J. B **41**(4), 830–839 (1998)
14. Kobayashi, H.: The subgrid-scale models based on coherent structures for rotating homogeneous turbulence and turbulent channel flow. Phys. Fluids **17**, 045, 104 (2005)

15. Leonard, A.: Energy cascade in large-eddy simulations of turbulent fluid flows. Adv. Geophys. **18**, 237–248 (1974)
16. Lesieur, M., Métais, O.: New trends in large-eddy simulations of turbulence. Annu. Rev. Fluid Mech. **28**, 45–82 (1996)
17. Lesieur, M., Métais, O., Comte, P.: Large-Eddy Simulations of Turbulence. Cambridge Univ. Press (2005)
18. Lilly, D.K.: The representation of small scale turbulence in numerical simulation experiments. pp. 195–209. IBM Form No. 320–1951 (1967)
19. Lilly, D.K.: A proposed modification of the Germano subgrid scale closure method. Phys. Fluids A **4**(3), 633–635 (1992)
20. Meneveau, C., Katz, J.: Scale-invariance and turbulence models for large-eddy simulation. Annu. Rev. Fluid Mech. **32**, 1–32 (2000)
21. Métais, O., Lesieur, M.: Spectral large-eddy simulation of isotropic and stably stratified turbulence. J. Fluid Mech. **239**, 157–194 (1992)
22. Okamoto, M., Shima, N.: Investigation for the one-equation-type subgrid model with eddy-viscosity expression including the shear-damping effect. Trans. Jpn. Soc. Mech. Eng. B **42**(2), 154–161 (1999)
23. Patankar, S.: Numerical Heat Transfer and Fluid Flow. Hemisphere, Washington (1980)
24. Piomelli, U., Liu, J.: Large-eddy simulation of rotating channel flows using a localized dynamic model. Phys. Fluids **7**(4), 839–848 (1995)
25. Rodi, W.: Large-eddy simulations of the flow past bluff bodies: state-of-the-art. JSME Int. J. B **41**(2), 361–374 (1998)
26. Rogallo, R.S., Moin, P.: Numerical simulation of turbulent flows. Annu. Rev. Fluid Mech. **16**, 99–137 (1984)
27. Sagaut, P.: Large Eddy Simulation for Incompressible Flows, 3rd edn. Springer (2006)
28. Salvetti, M.V., Banerjee, S.: A priori tests of a new dynamic subgrid-scale model for finite-difference large-eddy simulations. Phys. Fluids **7**(11), 2831–2847 (1995)
29. Smagorinsky, J.: General circulation experiments with the primitive equations: I. the basic experiment. Mon. Weather Rev. **91**(3), 99–164 (1963)
30. Spalart, P.R.: Detached-eddy simulation. Annu. Rev. Fluid Mech. **41**, 181–202 (2009)
31. Vreman, B., Geurts, B., Kuerten, H.: On the formulation of the dynamic mixed subgrid-scale model. Phys. Fluids **6**(12), 4057–4059 (1994)
32. Yoshizawa, A., Horiuti, K.: A statistically-derived subgrid-scale kinetic energy model for the large-eddy simuation of turbulent flows. J. Phys. Soc. Jpn. **54**(8), 2834–2839 (1985)
33. Zang, Y., Street, R.L., Koseff, J.R.: A dynamic mixed subgrid-scale model and its application to turbulent recirculating flows. Phys. Fluids A **5**(12), 3186–3196 (1993)

Appendix A
Generalized Coordinate System

In this appendix, vector and tensor derivatives are presented for a generalized coordinate system. Also derived in this appendix are the governing equations for fluid flows with contravariant components of the velocity vector for the generalized coordinate system.

A.1 Vector and Tensor Analysis

For the fundamentals of vector and tensor analysis, we refer readers to mathematics textbooks as necessary. To gain further knowledge on tensors for fluid and continuum mechanics, one can refer to textbooks by Aris [1] and Flügge [2].

A.1.1 Coordinate Transform

As a preparation to transform the governing equations from physical space to computational space, let us introduce the fundamental relationship between two arbitrary oblique coordinate systems.

Oblique Coordinate System and Basis

Consider three linearly independent vectors g_i $(i = 1, 2, 3)$ positioned at the origin O as shown in Fig. A.1. If we represent a position vector with x for an arbitrary point P with the linear superposition of g_i,

$$x = x^i g_i, \tag{A.1}$$

© Springer International Publishing AG 2017
T. Kajishima and K. Taira, *Computational Fluid Dynamics*,
DOI 10.1007/978-3-319-45304-0

Fig. A.1 Oblique coordinate
basis vectors g^i and
reciprocal basis vectors g_j

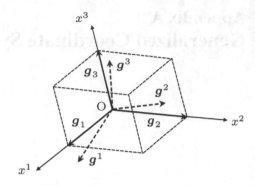

we refer to x^1, x^2, x^3 as the coordinates and g_1, g_2, g_3 as the basis vectors. In general, when the subscripts or superscripts appear in pairs, it implies summation.[1] For example, $a^i b_i = \sum_{i=1}^{3} a^i b_i$. In what follows, we imply summation when the same indices appear twice unless otherwise noted. The vectors g_i are not necessarily normalized (unit vectors) nor orthogonal but we assume that they follow the right-handed coordinate system.

Next, let us introduce reciprocal basis vectors (dual basis vectors) g^i ($i = 1, 2, 3$) such that $g_i \cdot g^j = \delta_i^j$. Because g^1 is orthogonal to g^2 and g^3, we can say $g^1 = \alpha(g^2 \times g^3)$, where $\alpha > 0$ for a right-handed system. Noting that $g_1 \cdot g^1 = 1$, we find that $\alpha = 1/g_1 \cdot (g_2 \times g_3)$. Generalizing this observation, we have

$$g^1 = \frac{g_2 \times g_3}{\sqrt{g}}, \quad g^2 = \frac{g_3 \times g_1}{\sqrt{g}}, \quad g^3 = \frac{g_1 \times g_2}{\sqrt{g}}, \tag{A.2}$$

where $\sqrt{g} = g_1 \cdot (g_2 \times g_3) = g_2 \cdot (g_3 \times g_1) = g_3 \cdot (g_1 \times g_2)$ is the triple product of vectors which corresponds to the volume of the box made by the three edges of $g_1, g_2,$ and g_3 as shown in Fig. A.1.

The position vector for point P can also be represented by the linear superposition of the reciprocal basis vectors

$$x = x_i g^i. \tag{A.3}$$

We refer to x^i as the *contravariant components* and x_i as the *covariant components*. Figure A.2a illustrates in two dimensions the relationship between the two components. The contravariant components can be rewritten as

$$x^j = x \cdot g^j \tag{A.4}$$

because $x^j = x^i \delta_i^j = x^i g_i \cdot g^j$. The covariant components can be expressed as

$$x_i = x \cdot g_i \tag{A.5}$$

---

[1]In Sect. 1.3.4, we did not distinguish between the superscripts and subscripts.

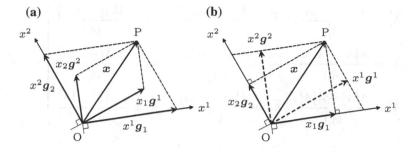

Fig. A.2 Two-dimensional representation of the relationship between contravariant and covariant components of vectors. **a** Contravariant and covariant components. **b** Geometrical relationship

because $x_i = x_j \delta_i^j = x_j g^j \cdot g_i$. The vectors shown by the dashed lines in Fig. A.2b represent the above inner product relations.

Transformation Matrix

Consider an oblique coordinate system with another set of basis vectors $\widehat{g}_i$ and a corresponding set of reciprocal basis vectors $\widehat{g}^i$. In this case, let us represent the position vector x with contravariant components ξ^i and covariant components ξ_i

$$x = \xi^i \widehat{g}_i = \xi_i \widehat{g}^i. \tag{A.6}$$

The coordinate transformation between the two coordinates can be expressed using the contravariant components as

$$d\xi^j = A_i^j dx^i, \quad A_i^j = \frac{\partial \xi^j}{\partial x^i}, \tag{A.7}$$

where $[A_i^j]$ is the *coordinate transformation matrix*. Since g_1, g_2, g_3 and g^1, g^2, g^3 are both linearly independent bases, the coordinate transformation matrix is regular (i.e., $|A_i^j| \neq 0$) with

$$A_i^j \overline{A}_j^k = \delta_i^k \tag{A.8}$$

whose inverse matrix is $[\overline{A}_j^i] = [A_i^j]^{-1}$. The inverse transform of the contravariant components is

$$dx^i = \overline{A}_j^i d\xi^j, \quad \overline{A}_j^i = \frac{\partial x^i}{\partial \xi^j}. \tag{A.9}$$

For completeness, we list the components of the inverse matrix·

$$\left.\begin{array}{lll}
\overline{A}_1^1 = \dfrac{A_2^2 A_3^3 - A_3^2 A_2^3}{|A_i^j|}, & \overline{A}_2^1 = \dfrac{A_2^3 A_3^1 - A_3^3 A_2^1}{|A_i^j|}, & \overline{A}_3^1 = \dfrac{A_2^1 A_3^2 - A_3^1 A_2^2}{|A_i^j|} \\[3mm]
\overline{A}_1^2 = \dfrac{A_3^2 A_1^3 - A_1^2 A_3^3}{|A_i^j|}, & \overline{A}_2^2 = \dfrac{A_3^3 A_1^1 - A_1^3 A_3^1}{|A_i^j|}, & \overline{A}_3^2 = \dfrac{A_3^1 A_1^2 - A_1^1 A_3^2}{|A_i^j|} \\[3mm]
\overline{A}_1^3 = \dfrac{A_1^2 A_2^3 - A_2^2 A_1^3}{|A_i^j|}, & \overline{A}_2^3 = \dfrac{A_1^3 A_2^1 - A_2^3 A_1^1}{|A_i^j|}, & \overline{A}_3^3 = \dfrac{A_1^1 A_2^2 - A_2^1 A_1^2}{|A_i^j|}
\end{array}\right\} \tag{A.10}$$

Here, the determinant $|A_i^j| = e^{ijk} A_i^1 A_j^2 A_k^3$, where

$$e^{ijk} = \begin{cases}
e^{123} = e^{231} = e^{312} = 1 & \text{(even permutation)} \\
e^{321} = e^{213} = e^{132} = -1 & \text{(odd permutation)} \\
0 & \text{(otherwise)}
\end{cases} \tag{A.11}$$

is called the *permutation symbol* or the *Levi–Civita symbol*.

Transformation of Vectors and Tensors

From Eqs. (A.1) and (A.9), we see that $d\mathbf{x} = dx^i \mathbf{g}_i = \overline{A}_j^i d\xi^j \mathbf{g}_i$, and likewise from Eqs. (A.6) and (A.7), $d\mathbf{x} = d\xi^j \widehat{\mathbf{g}}_j = A_i^j dx^i \widehat{\mathbf{g}}_j$. By comparing these two relations, we find the transformation equations for the basis vectors

$$\widehat{\mathbf{g}}_j = \overline{A}_j^i \mathbf{g}_i, \quad \mathbf{g}_i = A_i^j \widehat{\mathbf{g}}_j. \tag{A.12}$$

On the other hand, by denoting the transformation for the reciprocal basis vectors as $\widehat{\mathbf{g}}^j = B_k^j \mathbf{g}^k$, we have $\delta_i^j = \widehat{\mathbf{g}}^j \cdot \widehat{\mathbf{g}}_i = B_k^j \mathbf{g}^k \cdot \overline{A}_i^l \mathbf{g}_l = B_k^j \overline{A}_i^k$, which shows that $B_k^j = A_k^j$. The inverse transform can be derived in a similar manner, yielding

$$\widehat{\mathbf{g}}^j = A_i^j \mathbf{g}^i, \quad \mathbf{g}^i = \overline{A}_j^i \widehat{\mathbf{g}}^j. \tag{A.13}$$

Let us now consider an arbitrary vector

$$\mathbf{u} = u^i \mathbf{g}_i = u_i \mathbf{g}^i = U^i \widehat{\mathbf{g}}_i = U_i \widehat{\mathbf{g}}^i. \tag{A.14}$$

By substituting Eqs. (A.12) and (A.13) into the above equation, we can derive the coordinate transform between vector components. The transformation between covariant components have the same form as the basis vector transformation

$$U_j = \overline{A}_j^i u_i, \quad u_i = A_i^j U_j. \tag{A.15}$$

The transformation between contravariant components has the opposite form from the basis vector transformation

$$U^j = A_i^j u^i, \quad u^i = \overline{A}_j^i U^j. \tag{A.16}$$

The use of the term *covariant* or *contravariant* refers to whether the transformation follows the form of the basis vector transformation.

In general, for a tensor of contravariant order m and covariant order n

$$T = t_{j_1 \cdots j_n}^{i_1 \cdots i_m} \boldsymbol{g}_{i_1} \cdots \boldsymbol{g}_{i_m} \boldsymbol{g}^{j_1} \cdots \boldsymbol{g}^{j_n} = T_{l_1 \cdots l_n}^{k_1 \cdots k_m} \widehat{\boldsymbol{g}}_{k_1} \cdots \widehat{\boldsymbol{g}}_{k_m} \widehat{\boldsymbol{g}}^{l_1} \cdots \widehat{\boldsymbol{g}}^{l_n}, \tag{A.17}$$

the following coordinate transform

$$T_{l_1 \cdots l_n}^{k_1 \cdots k_m} = A_{i_1}^{k_1} \cdots A_{i_m}^{k_n} \overline{A}_{l_1}^{j_1} \cdots \overline{A}_{l_n}^{j_n} t_{j_1 \cdots j_n}^{i_1 \cdots i_m} \tag{A.18}$$

is satisfied.

Metric Ttensors and Physical Components

Denoting the magnitude of a vector $\boldsymbol{u}$ by u, we have

$$u^2 = (u^i \boldsymbol{g}_i) \cdot (u^j \boldsymbol{g}_j) = u^i u^j (\boldsymbol{g}_i \cdot \boldsymbol{g}_j). \tag{A.19}$$

Here, we can define the *covariant metric tensor* as

$$g_{ij} = \boldsymbol{g}_i \cdot \boldsymbol{g}_j, \tag{A.20}$$

which is a symmetric tensor that relates the contravariant components and the magnitude of the vector. We can also define an analogous tensor for the other coordinate system, i.e., $\widehat{g}_{ij} = \widehat{\boldsymbol{g}}_i \cdot \widehat{\boldsymbol{g}}_j$. Comparing the relations of $u^2 = U^k U^l \widehat{g}_{kl}$ and $u^2 = u^i u^j g_{ij} = \overline{A}_k^i \overline{A}_l^j U^k U^l g_{ij}$, we find the transformation equations for the covariant metric tensor are

$$\widehat{g}_{kl} = \overline{A}_k^i \overline{A}_l^j g_{ij}, \quad g_{ij} = A_i^k A_j^l \widehat{g}_{kl}. \tag{A.21}$$

We define g to be the determinant of the covariant metric tensor

$$g \equiv |g_{ij}| \tag{A.22}$$

and similarly define $\widehat{g} \equiv |\widehat{g}_{ij}|$. Accordingly, by taking the determinant of Eq. (A.21), we can derive that

$$g \equiv |A|^2 \widehat{g}. \tag{A.23}$$

By denoting the unit basis vector by $\boldsymbol{e}_i$ for the basis vector $\boldsymbol{g}_i$, we have $\boldsymbol{g}_i = a \boldsymbol{e}_i$. Since $\boldsymbol{g}_i \cdot \boldsymbol{g}_i = a^2 \boldsymbol{e}_i \cdot \boldsymbol{e}_i = a^2$, $a = \sqrt{g_{ii}}$. This tells us that

$$\boldsymbol{e}_i = \frac{\boldsymbol{g}_i}{\sqrt{g_{ii}}}. \quad \text{(no summation implied)} \tag{A.24}$$

Because a basis composed of unit vectors are related to physical scales, we refer to components represented by unit basis vectors as *physical components*. The physical contravariant component representation $u^{(i)}$ of vector $\boldsymbol{u}$ is

$$u = u^i g_i = u^{(i)} e_i, \tag{A.25}$$

where

$$u^{(i)} = u^i \sqrt{g_{ii}}. \quad \text{(no summation implied)} \tag{A.26}$$

The physical contravariant components for tensors are

$$t^{(ij)} = t^{ij} \sqrt{g_{ii}} \sqrt{g_{jj}}. \quad \text{(no summation implied)} \tag{A.27}$$

Transformation Between Covariant and Contravariant Components

For reciprocal vectors, let us define the *contravariant metric tensor*

$$g^{ij} = g^i \cdot g^j. \tag{A.28}$$

The tensor g^{ij} is symmetric and can be transformed to $\widehat{g}^{ij} = \widehat{g}^i \cdot \widehat{g}^j$ in the following manner:

$$\widehat{g}^{ij} = A_k^i A_l^j g^{kl}, \quad g^{ij} = \overline{A}_k^i \overline{A}_l^j \widehat{g}^{kl}. \tag{A.29}$$

By setting $g_i = a_{ij} g^j$, we have $g_{ik} = g_i \cdot g_k = a_{ij} g^j \cdot g_k = a_{ij} \delta_k^j = a_{ik}$. Similarly by setting $g^j = b^{ij} g_i$, we get $g^{jl} = g^j \cdot g^l = b^{ij} g_i \cdot g^l = b^{ij} \delta_i^l = b^{jl}$. Hence, we observe that

$$g_i = g_{ij} g^j, \quad g^j = g^{ij} g_i. \tag{A.30}$$

The basis vectors g_i and the reciprocal basis vectors g^j can be transformed from one to the other with the transformations g_{ij} and g^{ij}. Note that from $g_i \cdot g^j = \delta_i^j$, we see that $g_{ik} g^{kj} = \delta_i^j$ holds.

Based on Eq. (A.5), we have

$$u_i = g_{ij} u^j, \quad u^i = g^{ij} u_j \tag{A.31}$$

since $u_i = u \cdot g_i = (u^j g_j) \cdot g_i = u^j g_{ij}$. The covariant and contravariant components, u_i and u^j, can be transformed from one to the other using g_{ij} and g^{ij}.

Differentiation of Scalars

Observing that the transformation for the scalar derivative $\partial \phi / \partial x^j$ with contravariant components x^j is

$$\frac{\partial \phi}{\partial \xi^i} = \frac{\partial x^j}{\partial \xi^i} \frac{\partial \phi}{\partial x^j} = \overline{A}_i^j \frac{\partial \phi}{\partial x^j}, \tag{A.32}$$

we find that $\partial \phi / \partial x^j$ is a covariant component. Note that $\partial / \partial \xi^i$ represents the change in the direction of the reciprocal basis g^i and not in the direction of the basis. The gradient operator is represented as

$$\nabla = \boldsymbol{g}^i \frac{\partial}{\partial x^i} = \boldsymbol{g}_j g^{ij} \frac{\partial}{\partial x^i} \tag{A.33}$$

with summation implied over i. Here, $\partial/\partial x^i$ are the covariant components of the gradient operator. Accordingly, the gradient of a scalar is expressed as

$$\nabla \phi = \boldsymbol{g}^i \frac{\partial \phi}{\partial x^i} = \boldsymbol{g}_i g^{ik} \frac{\partial \phi}{\partial x^k}. \tag{A.34}$$

The Laplacian of a scalar becomes

$$\nabla^2 \phi = \nabla \cdot (\nabla \phi) = \frac{1}{\sqrt{g}} \frac{\partial}{\partial x^i} (\sqrt{g} g^{ik} \frac{\partial \phi}{\partial x^k}), \tag{A.35}$$

which can be derived by taking the divergence of Eq. (A.51) presented later and substituting the contravariant components $g^{ik} \partial \phi / \partial x^k$ of $\nabla \phi$ into u^i.

A.1.2 Differentiation of Vectors and Tensors

Next, let us consider spatial derivatives of vectors and tensors for a generalized coordinate system as shown in Fig. A.3a. For derivatives, we must not only take non-orthogonality of the basis vectors into account, as seen Sect. A.1.1, but also consider the variation in direction and magnitude of the basis vectors over space as illustrated in Fig. A.3b. The spatial derivative $\partial \boldsymbol{u}/\partial x^i$ of $\boldsymbol{u}$ becomes

$$\frac{\partial \boldsymbol{u}}{\partial x^i} = \frac{\partial}{\partial x^j} (u^i \boldsymbol{g}_i) = \frac{\partial u^i}{\partial x^j} \boldsymbol{g}_i + u^i \frac{\partial \boldsymbol{g}_i}{\partial x^j}, \tag{A.36}$$

where the second term results from the distortion of the coordinate system.

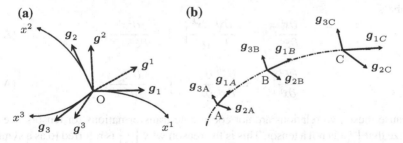

Fig. A.3 Spatial variations in a generalized coordinate system and basis vectors. **a** Curvilinear coordinate system **b** Spatial variation in basis vectors

Christoffel Symbol

A basis can change over space due to the motion of a point. That is,

$$g_j(x + dx) = g_j(x) + dg_j, \tag{A.37}$$

where

$$dg_j = \left\{ {}_{j\,k}^{\,i} \right\} dx^k g_i. \tag{A.38}$$

This expression tells us that dg_j is proportional to the differential dx^k and represent it by a linear superposition of basis vectors g_i with their coefficients denoted by $\left\{ {}_{j\,k}^{\,i} \right\}$. The symbol $\left\{ {}_{j\,k}^{\,i} \right\}$ is referred to as the *Christoffel symbol*. By using Eq. (A.38) for the derivative of g_{jl}, i.e., $dg_{jl} = dg_j \cdot g_l + g_j \cdot dg_l$, we can find the components of the Christoffel symbol to be

$$\left\{ {}_{j\,k}^{\,i} \right\} = \frac{g^{li}}{2} \left(\frac{\partial g_{jl}}{\partial x^k} + \frac{\partial g_{kl}}{\partial x^j} - \frac{\partial g_{kj}}{\partial x^l} \right). \tag{A.39}$$

Because $\left\{ {}_{j\,k}^{\,i} \right\} = \left\{ {}_{k\,j}^{\,i} \right\}$, there are 6 and 18 components in two and three dimensions, respectively. The Christoffel symbol for the divergence of a vector can be expressed as

$$\left\{ {}_{i\,k}^{\,i} \right\} = \frac{1}{\sqrt{g}} \frac{\partial \sqrt{g}}{\partial x^k}. \tag{A.40}$$

Next, let us represent the Christoffel symbol for a different coordinate system $x = \xi^i \widehat{g}_i = \xi_i \widehat{g}^i$ with

$$\left\{ {}_{j\,k}^{\,i} \right\}' = \frac{\widehat{g}^{li}}{2} \left(\frac{\partial \widehat{g}_{jl}}{\partial \xi^k} + \frac{\partial \widehat{g}_{kl}}{\partial \xi^j} - \frac{\partial \widehat{g}_{kj}}{\partial \xi^l} \right). \tag{A.41}$$

By substituting Eq. (A.21) into $\partial \widehat{g}_{jl}/\partial \xi^k$ to evaluate the terms within the parentheses in the above equation, we obtain the transformation equations for the Christoffel symbols

$$\frac{\partial x^i}{\partial \xi^a} \left\{ {}_{b\,c}^{\,a} \right\}' = \frac{\partial x^j}{\partial \xi^b} \frac{\partial x^k}{\partial \xi^c} \left\{ {}_{j\,k}^{\,i} \right\} + \frac{\partial^2 x^i}{\partial \xi^b \partial \xi^c}, \tag{A.42}$$

$$\frac{\partial \xi^a}{\partial x^i} \left\{ {}_{j\,k}^{\,i} \right\} = \frac{\partial \xi^b}{\partial x^j} \frac{\partial \xi^c}{\partial x^k} \left\{ {}_{b\,c}^{\,a} \right\}' + \frac{\partial^2 \xi^a}{\partial x^j \partial x^k}. \tag{A.43}$$

Because these two relations are not coordinate transformations for tensors, we can realize that $\left\{ {}_{j\,k}^{\,i} \right\}$ is not a tensor. This is the reason why $\left\{ {}_{j\,k}^{\,i} \right\}$ is referred to as a symbol.

Differentiation of Vectors

The derivative of a vector $\boldsymbol{u}$ with respect to contravariant components x^j is

$$\frac{\partial \boldsymbol{u}}{\partial x^j} = \frac{\partial}{\partial x^j}(u^i \boldsymbol{g}_i) = \left(\frac{\partial u^i}{\partial x^j} + \{{}^i_{kj}\}u^k\right)\boldsymbol{g}_i, \tag{A.44}$$

which is derived using Eqs. (A.36) and (A.38). For a coordinate transform of

$$\frac{\partial U^j}{\partial \xi^i} = \frac{\partial x^k}{\partial \xi^i}\frac{\partial}{\partial x^k}\left(u^l\frac{\partial \xi^j}{\partial x^l}\right) = \frac{\partial x^k}{\partial \xi^i}\left(\frac{\partial \xi^j}{\partial x^l}\frac{\partial u^l}{\partial x^k} + \frac{\partial^2 \xi^j}{\partial x^l \partial x^k}u^l\right), \tag{A.45}$$

we can utilize the transform equation, Eq. (A.43), for the Christoffel symbol to yield

$$\frac{\partial U^j}{\partial \xi^i} + \{{}^j_{ki}\}'U^k = \overline{A}^m_i A^j_n\left(\frac{\partial u^n}{\partial x^m} + \{{}^n_{km}\}u^k\right). \tag{A.46}$$

Since $\partial u^i/\partial x^j$ does not obey the transformation law, it is not a tensor. However, the variable defined by

$$u^i|_j = \frac{\partial u^i}{\partial x^j} + \{{}^i_{kj}\}u^k \tag{A.47}$$

is a contravariant order 1 and covariant order 1 tensor. The derivative that takes the spatial variation of the basis into consideration is called *covariant derivative* because the covariant order increases by 1. It can be said that $u^i|_j$ is the covariant derivative of the contravariant components.

For completeness, let us list the derivative operations that appear in the governing equations for fluid flow. The gradient of a vector for generalized coordinates is

$$\nabla \boldsymbol{u} = \left(\boldsymbol{g}^j\frac{\partial}{\partial x^j}\right)(u^i\boldsymbol{g}_i) = u^i|_j\boldsymbol{g}^j\boldsymbol{g}_i = g^{jk}u^i|_j\boldsymbol{g}_k\boldsymbol{g}_i, \tag{A.48}$$

where we can observe that the contravariant component of $\nabla \boldsymbol{u}$ is $g^{jk}u^i|_j$. The divergence of a vector is

$$\nabla \cdot \boldsymbol{u} = \left(\boldsymbol{g}^j\frac{\partial}{\partial x^j}\right) \cdot (u^i\boldsymbol{g}_i) = \boldsymbol{g}^j \cdot (u^i|_j\boldsymbol{g}_i) = u^i|_j\delta^j_i = u^i|_i, \tag{A.49}$$

which can be expressed as

$$u^i|_i = \frac{\partial u^i}{\partial x^i} + \{{}^i_{ki}\}u^k = \frac{\partial u^i}{\partial x^i} + \frac{1}{\sqrt{g}}\frac{\partial \sqrt{g}}{\partial x^l}u^i \tag{A 50}$$

with Eq. (A.40). We can alternatively write

$$\nabla \cdot \boldsymbol{u} = \frac{1}{\sqrt{g}} \frac{\partial (\sqrt{g} u^i)}{\partial x^i}. \tag{A.51}$$

for the divergence of a vector.

Differentiation of Tensors

The derivation for differentiation of tensors follows that for a vector, if we take into account the spatial variations with respect to the two bases in $\boldsymbol{T} = t^{ij} \boldsymbol{g}_i \boldsymbol{g}_j$. For the derivative of a tensor $\boldsymbol{T}$ with respect to the contravariant components x^k, we have

$$\frac{\partial \boldsymbol{T}}{\partial x^k} = \frac{\partial (t^{ij} \boldsymbol{g}_i \boldsymbol{g}_j)}{\partial x^k} = \left(\frac{\partial t^{ij}}{\partial x^k} + \{ {}_{k}^{i}{}_{l} \} t^{jl} + \{ {}_{k}^{j}{}_{l} \} t^{il} \right) \boldsymbol{g}_i \boldsymbol{g}_j. \tag{A.52}$$

If we perform a coordinate transform

$$\frac{\partial T^{ij}}{\partial \xi^k} = \frac{\partial x^n}{\partial \xi^k} \frac{\partial}{\partial x^n} \left(t^{lm} \frac{\partial \xi^i}{\partial x^l} \frac{\partial \xi^j}{\partial x^m} \right), \tag{A.53}$$

we find that

$$\frac{\partial T^{ij}}{\partial x^k} + \{ {}_{k}^{i}{}_{l} \}' T^{jl} + \{ {}_{k}^{j}{}_{l} \}' T^{il}$$
$$= \overline{A}_k^n A_l^i A_m^j \left(\frac{\partial t^{lm}}{\partial x^n} + \{ {}_{n}^{l}{}_{h} \} t^{mh} + \{ {}_{n}^{m}{}_{h} \} t^{lh} \right), \tag{A.54}$$

where we have used the Christoffel symbol for the transformation. Hence, we observe that the tensor $t^{ij}|_k$ defined by

$$t^{ij}|_k = \frac{\partial t^{ij}}{\partial x^k} + \{ {}_{k}^{i}{}_{l} \} t^{jl} + \{ {}_{k}^{j}{}_{l} \} t^{il} \tag{A.55}$$

is the covariant derivative of the contravariant components and is a tensor of contravariant order 2 and covariant order 1. It should be noted that $\partial t^{ij}/\partial x^k$ does not obey the transformation law and hence is not a tensor.

Let us also provide the expressions for gradient and divergence operations for tensors. The gradient of a tensor is

$$\nabla \boldsymbol{T} = t^{ij}|_k \boldsymbol{g}^k \boldsymbol{g}_i \boldsymbol{g}_j = g^{kl} t^{ij}|_k \boldsymbol{g}_l \boldsymbol{g}_i \boldsymbol{g}_j. \tag{A.56}$$

The divergence of a tensor is provided by

$$\nabla \cdot \boldsymbol{T} = t^{ij}|_j \boldsymbol{g}_i = \left(\frac{\partial t^{ij}}{\partial x^j} + \left\{ {}^{\ i}_{j\ k} \right\} t^{jk} + \left\{ {}^{\ j}_{j\ k} \right\} t^{ik} \right) \boldsymbol{g}_i \tag{A.57}$$

$$= \left(\frac{1}{\sqrt{g}} \frac{\partial(\sqrt{g} t^{ij})}{\partial x^j} + \left\{ {}^{\ i}_{j\ k} \right\} t^{jk} \right) \boldsymbol{g}_i, \tag{A.58}$$

which appears in the momentum equation.

A.2 Governing Equations with Contravariant Components of Velocity

In Sect. 4.3, we have introduced the strong conservation form of the governing equations. It is also possible to represent the governing equations using only the contravariant components of the velocity vectors. When the strong form of the momentum equation is numerically solved on collocated grids, the contravariant components are used to couple the pressure gradient term and the continuity equation. For this reason, we only derive the governing equations here in terms of the contravariant components of the velocity vectors.

A.2.1 Conservation of Mass

The conservation of mass is expressed as

$$\frac{\partial \rho}{\partial t} + \nabla \cdot (\rho \boldsymbol{u}) = 0. \tag{A.59}$$

With Eq. (A.51), the above equation becomes

$$\frac{\partial \rho}{\partial t} + \frac{1}{J} \frac{\partial(J \rho U^i)}{\partial \xi^i} = 0 \tag{A.60}$$

using the contravariant components. The Jacobian $J = \sqrt{\bar{g}/g}$ defined by Eq. (4.5) becomes $J = \sqrt{\bar{g}}$ for Cartesian coordinate systems since $\sqrt{g} = 1$.

For incompressible flow, we have $\nabla \cdot \boldsymbol{u} = 0$. Its representation with contravariant components

$$U^i|_i = 0 \tag{A.61}$$

becomes

$$\frac{1}{J}\frac{\partial(JU^i)}{\partial\xi^i} = 0 \tag{A.62}$$

with the use of Eq. (A.51).

A.2.2 Material Derivative and Forms of the Advection Term

The material derivative D/Dt defined by

$$\frac{D}{Dt} = \frac{\partial}{\partial t} + \boldsymbol{u}\cdot\nabla \tag{A.63}$$

for the generalized coordinate system is

$$\frac{D}{Dt} = \frac{\partial}{\partial t} + (U^i\widehat{\boldsymbol{g}}_i)\cdot\left(\widehat{\boldsymbol{g}}^j\frac{\partial}{\partial\xi^j}\right) = \frac{\partial}{\partial t} + U^j\frac{\partial}{\partial\xi^j}. \tag{A.64}$$

If we take a look at the acceleration term in the momentum equation

$$\rho\frac{D\boldsymbol{u}}{Dt} = \rho\left(\frac{\partial\boldsymbol{u}}{\partial t} + \boldsymbol{u}\cdot\nabla\boldsymbol{u}\right) \tag{A.65}$$

we observe that the advective term is a product of the advection velocity and the velocity gradient. For this reason, this form is called the *gradient form* (*advective form*). With contravariant components, we can write the above expression as

$$\rho\frac{D\boldsymbol{u}}{Dt} = \rho\left(\frac{\partial U^i}{\partial t} + U^j U^i|_j\right)\widehat{\boldsymbol{g}}_i. \tag{A.66}$$

We can further transform Eq. (A.65) using the continuity equation to yield

$$\rho\frac{D\boldsymbol{u}}{Dt} = \rho\left(\frac{\partial\boldsymbol{u}}{\partial t} + \boldsymbol{u}\cdot\nabla\boldsymbol{u}\right) + \boldsymbol{u}\left[\frac{\partial\rho}{\partial t} + \nabla\cdot(\rho\boldsymbol{u})\right]$$
$$= \frac{\partial(\rho\boldsymbol{u})}{\partial t} + \nabla\cdot(\rho\boldsymbol{u}\boldsymbol{u}). \tag{A.67}$$

The above form is referred to as the *divergence form* because the advective term is expressed in terms of a divergence. Using contravariant components, the above equation can be written as

$$\rho\frac{D\boldsymbol{u}}{Dt} = \left[\frac{\partial(\rho U^i)}{\partial t} + (\rho U^i U^j)|_j\right]\widehat{\boldsymbol{g}}_i. \tag{A.68}$$

A.2.3 Conservation of Momentum

The conversation of momentum can be expressed in the gradient form corresponding to Eq. (A.65)

$$\rho \frac{Du}{Dt} = \nabla \cdot T, \tag{A.69}$$

where T is the stress tensor and we assume that there is no external force. If we write the momentum conservation with the divergence form (Eq. (A.67)), we have

$$\frac{\partial(\rho u)}{\partial t} + \nabla \cdot (\rho uu - T) = 0. \tag{A.70}$$

For a Newtonian fluid, let us restate the constitutive relation, Eq. (1.12),

$$T = -pI + 2\mu \left(D - \frac{1}{3} I \nabla \cdot u \right), \tag{A.71}$$

where I is the fundamental tensor ($I = \widehat{g}^i \widehat{g}_i = \widehat{g}_i \widehat{g}^i = \widehat{g}^{ij} \widehat{g}_i \widehat{g}_j$), p is the static pressure, and μ is the dynamic viscosity. The velocity gradient tensor D from Eq. (1.13) is

$$D = \frac{1}{2}[(\nabla u)^T + \nabla u] = \frac{1}{2}(U^i|_j \widehat{g}_i \widehat{g}^j + U^j|_i \widehat{g}^i \widehat{g}_j) \tag{A.72}$$

can be written in the contravariant form as

$$D^{ij} = \frac{1}{2}(\widehat{g}^{kj} U^i|_k + \widehat{g}^{ki} U^j|_k). \tag{A.73}$$

Hence, the contravariant representation for $T = T^{ij} \widehat{g}_i \widehat{g}_j$ becomes

$$T^{ij} = -\widehat{g}^{ij}(p + \frac{2}{3}\mu U^k|_k) + \mu(\widehat{g}^{kj} U^i|_k + \widehat{g}^{ki} U^j|_k). \tag{A.74}$$

Momentum Equation for Compressible Flow

Representing Eq. (A.70) with contravariant components, we have

$$\frac{\partial(\rho U^i)}{\partial t} + F^{ij}|_j = 0 \tag{A.75}$$

or

$$\frac{\partial(\rho U^i)}{\partial t} + \frac{1}{J}\frac{\partial}{\partial \xi^j}(JF^{ij}) + \left\{ {}_{j}^{i}{}_{k} \right\} F^{ik} = 0. \tag{A.76}$$

Here, the term

$$F^{ij} = \rho U^i U^j - T^{ij}$$

$$= \rho U^i U^j + \widehat{g}^{ij} \left(p + \frac{2}{3} \mu \left. U^k \right|_k \right) - \mu \left(\widehat{g}^{kj} \left. U^i \right|_k + \widehat{g}^{ki} \left. U^j \right|_k \right) \tag{A.77}$$

is the contravariant component of the momentum flux tensor.

Momentum Equation for Incompressible Flow

For incompressible flows ($D\rho/Dt = 0$, $\nabla \cdot \boldsymbol{u} = 0$), the momentum equation using the contravariant components can be derived using the divergence form of the material derivative, Eq. (A.68) in Eq. (A.69)

$$\frac{\partial U^i}{\partial t} + (U^i U^j)|_j = \frac{1}{\rho} T^{ij}|_j, \tag{A.78}$$

where

$$T^{ij} = -\widehat{g}^{ij} p + \mu (\widehat{g}^{kj} U^i|_k + \widehat{g}^{ki} U^j|_k). \tag{A.79}$$

If we instead use the gradient form of the material derivative, Eq. (A.66), in Eq. (A.69), we obtain

$$\frac{\partial U^i}{\partial t} + U^j U^i|_j + \frac{\widehat{g}^{ij}}{\rho} \frac{\partial p}{\partial \xi^j} - \frac{1}{\rho} \left[\mu \left(\widehat{g}^{kj} U^i|_k + \widehat{g}^{ki} U^j|_k \right) \right] \Big|_j = 0. \tag{A.80}$$

The above equation reduces to

$$\frac{\partial U^i}{\partial t} + U^j U^i|_j + \frac{\widehat{g}^{ij}}{\rho} \frac{\partial p}{\partial \xi^j} - \nu \frac{1}{J} \frac{\partial}{\partial \xi^l} \left(J \widehat{g}^{kl} \frac{\partial U^i}{\partial \xi^k} \right) = 0 \tag{A.81}$$

for constant viscosity.

A.2.4 Physical Component Representation

The numerical values of the contravariant components depend on the amount of stretching and distortion of the coordinate system and are difficult to relate to the physical quantities in the flow field in a straightforward manner. We can, however, use the relations, $U^{(i)} = \sqrt{\widehat{g}_{ii}} U^i$ and $D^{(ij)} = \sqrt{\widehat{g}_{ii}} \sqrt{\widehat{g}_{jj}} D^{ij}$ (no summation over i or j implied), and derive the governing equations using the physical component representation (or for variables that can posses dimensions). However, the governing equations with the physical component representation becomes cumbersome and are not widely used with the generalized coordinate systems except for the cylindrical and spherical coordinate systems.

Using contravariant physical components $U^{(i)}$, the mass conservation can be stated as

$$\frac{\partial \rho}{\partial t} + \frac{1}{\sqrt{\widehat{g}}} \frac{\partial}{\partial \xi^i} \frac{\sqrt{\widehat{g}} \rho U^{(i)}}{\sqrt{\widehat{g}_{ii}}} = 0, \tag{A.82}$$

which reduces to

$$\frac{1}{\sqrt{\widehat{g}}} \frac{\partial}{\partial \xi^i} \frac{\sqrt{\widehat{g}} U^{(i)}}{\sqrt{\widehat{g}_{ii}}} = 0 \tag{A.83}$$

for incompressible flow. The velocity gradient tensor can be expressed as

$$D^{(ij)} = \frac{\sqrt{\widehat{g}_{ii}} \sqrt{\widehat{g}_{jj}}}{2} \left(\widehat{g}^{kj} \frac{U^{(i)}}{\sqrt{\widehat{g}_{ii}}} \bigg|_k + \widehat{g}^{ki} \frac{U^{(j)}}{\sqrt{\widehat{g}_{jj}}} \bigg|_k \right). \tag{A.84}$$

for the momentum equation with physical component representation.

References

1. Aris, R.: Vectors, Tensors and the Basic Equations of Fluid Mechanics. Dover (1989)
2. Flügge, W.: Tensor Analysis and Continuum Mechanics. Springer (1972)

Appendix B
Fourier Analysis of Flow Fields

In this book, we use *Fourier transform* to

1. analyze the accuracy and stability of finite-difference methods (see Chap. 2)
2. accelerate the solver for elliptic equations (see Sect. 3.4.2)
3. study the behavior of turbulent flow in wave space to aid the development of turbulence models (see Chap. 8)

This appendix provides the background on Fourier analysis necessary to perform simulations of fluid flows and present the governing equations in wave space. Readers should refer to [2, 12, 3, 4] for further details on Fourier analysis and its use for fluid flow simulations.

B.1 Fourier Analysis

Fourier Series

Fourier series represents a periodic function $f(x)$ as a superposition of cosine and sine functions. If the function of interest has a period of $2L$, the function satisfies $f(x) = f(x + 2L)$ for all x. The Fourier series representation of function $f(x)$ can be expressed as

$$f(x) = \frac{a_0}{2} + \sum_{n=1}^{\infty} a_n \cos \frac{n\pi x}{L} + \sum_{n=1}^{\infty} b_n \sin \frac{n\pi x}{L}. \tag{B.1}$$

Using Euler's formula ($e^{ix} = \cos x + i \sin x$), we can have an alternative complex variable representation of

$$f(x) = \sum_{n=-\infty}^{\infty} c_n \exp \frac{in\pi x}{L}, \tag{B.2}$$

© Springer International Publishing AG 2017
T. Kajishima and K. Taira, *Computational Fluid Dynamics*,
DOI 10.1007/978-3-319-45304-0

where $i = \sqrt{-1}$. The Fourier coefficients in Eqs. (B.1) and (B.2) are related by

$$c_0 = \frac{a_0}{2}, \qquad c_n = \frac{a_n - ib_n}{2}, \qquad c_{-n} = \frac{a_n + ib_n}{2}. \tag{B.3}$$

Next, let us find the Fourier coefficients. The orthogonality of basis functions

$$\frac{1}{2L} \int_{-L}^{L} \exp \frac{in\pi x}{L} \exp \frac{-im\pi x}{L} dx = \delta_{nm} \tag{B.4}$$

can be taken advantage of in determining the coefficients. We can multiply both sides of Eq. (B.2) by $\exp(-im\pi x/L)$ and integrate the equation from $-L$ to L to obtain

$$c_m = \frac{1}{2L} \int_{-L}^{L} f(x) \exp \frac{-im\pi x}{L} dx, \qquad m = 0, \pm 1, \cdots \tag{B.5}$$

If $f(x)$ is a real-valued function, we have $c_{-m} = c_m^*$, where c_m^* denotes the complex conjugate of c_m. The coefficients a_m and b_m can be found through

$$a_m = \frac{1}{L} \int_{-L}^{L} f(x) \cos \frac{m\pi x}{L} dx, \qquad b_m = \frac{1}{L} \int_{-L}^{L} f(x) \sin \frac{m\pi x}{L} dx. \tag{B.6}$$

Note that when $f(x)$ is even, $b_m = 0$ and when $f(x)$ is odd, $a_m = 0$.

Fourier Integral

Let us introduce the Fourier integral by taking a limit of the Fourier series. For a periodic function $f_L(x)$ with finite periodicity $2L$, the Fourier series expansion of this function is

$$f_L(x) = \sum_{n=-\infty}^{\infty} \frac{1}{2L} g_L(y) \exp \frac{in\pi x}{L}, \tag{B.7}$$

where c_n in Eq. (B.2) is denoted by $g_L(y)/2L$ with $y = n/2L$. Next, we consider the increase in y to be continuous with $dy = 1/2L$ as $L \to \infty$. Thus, we can rewrite Eq. (B.7) as

$$f(x) = \int_{-\infty}^{\infty} g(y) e^{2\pi ixy} dy, \tag{B.8}$$

where we have replaced f_L and g_L with f and g. Since $g(y) = \lim_{L\to\infty} 2Lc_n$, we have

$$g(y) = \int_{-\infty}^{\infty} f(x) e^{-2\pi ixy} dx \tag{B.9}$$

from Eq. (B.5). The transforms $f(x) \Rightarrow g(y)$ and $g(y) \Rightarrow f(x)$ are referred to as the *forward and inverse Fourier transforms*, respectively. The definitions of the transforms can vary slightly in the literature depending on where to insert the factor 2π in the transform definition.

Rewriting the above with the replacement of $2\pi x$ with x and y with k, we arrive at the standard forward and inverse Fourier transform expressions of

$$F(k) = \frac{1}{2\pi} \int_{-\infty}^{\infty} f(x)e^{-ikx}dx, \tag{B.10}$$

$$f(x) = \int_{-\infty}^{\infty} F(k)e^{ikx}dk, \tag{B.11}$$

where k is the wave number. The above Eqs. (B.10) and (B.11) appear often in turbulent flow analysis.

Properties of Fourier Transform

For convenience, let us represent the Fourier transform of a function $f(x)$ and $g(x)$ by $\mathcal{F}[f(x)](k)$ and $\mathcal{F}[g(x)](k)$, respectively. One of the important and useful properties of Fourier transform is linearity:

$$\mathcal{F}[f(x) + g(x)](k) = \mathcal{F}[f(x)](k) + \mathcal{F}[g(x)](k), \tag{B.12}$$

$$\mathcal{F}[cf(x)](k) = c\mathcal{F}[f(x)](k), \tag{B.13}$$

where c is a constant. Another important property is

$$\mathcal{F}[e^{iax}f(x)](k) = \mathcal{F}[f(x)](k - a), \tag{B.14}$$

which states that the multiplication of a function f and a sinusoid with wave number a shifts the modes in the wave space by a. This relationship becomes important when we examine the influence of the nonlinear terms. We can also note that

$$\mathcal{F}[f(bx)](k) = \frac{1}{|b|}\mathcal{F}[f(x)](k/b), \tag{B.15}$$

which tells us that stretching x by a factor b corresponds to having modes at $1/b$ of the original frequency with a reduced amplitude of $1/|b|$.

Differentiation with Fourier Transform

Here, let us assume that the function $f(x)$ and its derivatives $f^{(k)}(x)$, $k = 1, 2, \ldots$, are continuous. Observing that a derivative of a periodic function $f(x + 2L) = f(x)$ is also periodic such that $f'(x + 2L) = f'(x)$ with period $2L$, we can express the derivative with a Fourier series as

$$f'(x) = \sum_{n=-\infty}^{\infty} c'_n \exp \frac{in\pi x}{L}. \tag{B.16}$$

Comparing this equation with Eq. (B.2), we observe that

$$c_n' = \frac{in\pi}{L}c_n. \tag{B.17}$$

Thus, we can express the derivative by

$$f'(x) = \sum_{n=-\infty}^{\infty} \frac{in\pi}{L}c_n \exp\frac{in\pi x}{L} \tag{B.18}$$

using the Fourier coefficients c_n for $f(x)$. Furthermore, we can also note that $a_n' = (n\pi/L)b_n$ and $b_n' = -(n\pi/L)a_n$. For higher-order derivatives, we find that the Fourier coefficients for $f^{(k)}$ are

$$c_n^{(k)} = \left(\frac{in\pi}{L}\right)^k c_n. \tag{B.19}$$

Fourier coefficients for a continuously differentiable function should decay rapidly over higher wave numbers because $c_n = \mathcal{O}(n^{-k})$.

In summary, we note that differentiation in Fourier space (wave space) becomes a simple multiplication of the Fourier coefficients with the wave numbers. Since this operation does not induce any discretization errors that appear in finite differencing, we can perform highly accurate calculations when Fourier series has desirable convergence properties (e.g., infinitely differentiable). A numerical method that takes advantage of these properties and accuracy is called the spectral method [3, 4, 12, 7].

Convolution

Convolution is important when we consider the treatment of filtering, nonlinear terms, and energy spectrum. Let us briefly discuss convolution in the context of Fourier transform. The convolution integral is an operation defined by

$$f * g(x) = \int_{-\infty}^{\infty} f(x-y)g(y)dy. \tag{B.20}$$

If we replace y by $x - y$, it can be realized that the convolution operation is commutative (i.e., $f * g = g * f$). The Fourier transform of the convolution integral is

$$\mathcal{F}[f * g(x)](k) = 2\pi\mathcal{F}[f(x)](k) \cdot \mathcal{F}[g(x)](k), \tag{B.21}$$

which is a product of the individually transformed functions. This relation is often used for filtering operations.

Analogously, the Fourier transform of a product of two functions yields

$$\mathcal{F}[f(x)g(x)](k) = F * G(k), \tag{B.22}$$

which is the convolution of the functions in Fourier space. This relation provides us with some insights when we analyze the nonlinear terms in Fourier space. Here, the convolution in wave space is

$$F * G(k) = \int_{-\infty}^{\infty} F(k - h)G(h)dh \tag{B.23}$$

with $F(k) = \mathcal{F}[f(x)](k)$ and $G(k) = \mathcal{F}[g(x)](k)$. The relation, Eq. (B.22), represents the Fourier transform relation of a nonlinear term and tells us that the Fourier coefficient of the nonlinear term for a wave number k is influenced by all coefficients $F(k_1)$ and $G(k_2)$, where k_1 and k_2 satisfies $k_1 + k_2 = k$.

If we replace g by its complex conjugate g^* in the above discussion and use Eq. (B.22), we find

$$\int_{-\infty}^{\infty} f(x)g^*(x)dx = \mathcal{F}[f(x)g^*(x)](0) = \int_{-\infty}^{\infty} F(k)G^*(k)dk. \tag{B.24}$$

Setting $f = g$ and noticing that $|f|^2 = ff^*$, we derive

$$\int_{-\infty}^{\infty} |f(x)|^2 dx = \int_{-\infty}^{\infty} |F(k)|^2 dk, \tag{B.25}$$

which is known as Parseval's formula (Plancherel's formula).

B.2 FFT for Discrete Fourier Transform

Fast Fourier transform (FFT) is literally an algorithm that accelerates the computation of the discrete Fourier transform. In the computations, the following operations take place

$$\tilde{u}_k = \sum_{j=0}^{N-1} u_j \exp \frac{\pm 2\pi i j k}{N}, \qquad k = 0, 1, \cdots, N-1, \tag{B.26}$$

where $\{u_j\}_{j=0}^{N-1}$ is taken to be a complex array. In the above operations, directly computing the summation requires $8N^2$ operations. This computational cost can be reduced to $5N \log_2 N$ (when $N = 2^n$) using the method of Cooley and Tukey [5]. This accelerated algorithm is referred to as FFT. We do not discuss the details of FFT in this appendix. We ask readers to refer to [5, 10, 2, 6].

The forward transform to be performed in the computation is

$$\widetilde{u}_k = \frac{1}{N} \sum_{j=0}^{N-1} u_j \exp \frac{-2\pi ijk}{N}, \qquad k = -\frac{N}{2}, -\frac{N}{2}+1, \cdots, \frac{N}{2}-1. \qquad \text{(B.27)}$$

Since $\widetilde{u}_{k+pN} = \widetilde{u}_k$ (where p is an integer) holds, the execution of Eq. (B.26) provides the following complex-valued array:

$$N\left[\widetilde{u}_0 \ \ \widetilde{u}_1 \ \ \cdots \ \ \widetilde{u}_{\frac{N}{2}-1} \ \ \widetilde{u}_{-\frac{N}{2}} \ \ \widetilde{u}_{-\frac{N}{2}+1} \ \ \cdots \ \ \widetilde{u}_{-1} \right]. \qquad \text{(B.28)}$$

For the computation of the inverse transform

$$u_j = \sum_{k=-N/2}^{N/2-1} \widetilde{u}_k \exp \frac{+2\pi ijk}{N}, \quad j = 0, 1, \cdots, N-1, \qquad \text{(B.29)}$$

we can divide the above array (B.28) by N and perform the computation described in Eq. (B.26) to recover the complex array $[u_0, u_1, \cdots, u_{N-1}]$.

When all u_j are real, the following holds with respect to Eq. (B.27)

$$\widetilde{u}_{-k} = \widetilde{u}_k^*. \qquad \text{(B.30)}$$

The Fourier coefficient $\widetilde{u}_0$ corresponds to the average value of u_j

$$\widetilde{u}_0 = \frac{1}{N} \sum_{j=0}^{N-1} u_j \qquad \text{(B.31)}$$

and $\widetilde{u}_{\pm N/2}$ becomes

$$\widetilde{u}_{\pm N/2} = \frac{1}{N} \sum_{j=0}^{N-1} (-1)^j u_j. \qquad \text{(B.32)}$$

For real-valued u_j, both $\widetilde{u}_0$ and $\widetilde{u}_{\pm N/2}$ are also real. Hence, Eq. (B.29) can be written as

$$u_j = \widetilde{u}_0 + (-1)^j \widetilde{u}_{-\frac{N}{2}}$$
$$+ 2 \sum_{k=1}^{N/2-1} \Re \widetilde{u}_k \cos \frac{2\pi jk}{N} + 2 \sum_{k=1}^{N/2-1} \Im \widetilde{u}_k \cos \frac{2\pi jk}{N}. \qquad \text{(B.33)}$$

Accordingly, for N values of u_j, we can independently determine N of the real-valued coefficients:

$$\widetilde{u}_0, \ \widetilde{u}_{-\frac{N}{2}}, \ \{\Re \widetilde{u}_k\}_{k=1}^{N/2-1}, \ \{\Im \widetilde{u}_k\}_{k=1}^{N/2-1}.$$

Exploiting the relations described by Eqs. (B.30), (B.31), and (B.32), we obtain the following array of complex numbers:

$$
N \begin{bmatrix} \tilde{u}_0 & \Re\tilde{u}_1 & \cdots & \Re\tilde{u}_{\frac{N}{2}-1} & \tilde{u}_{-\frac{N}{2}} & \Re\tilde{u}_{\frac{N}{2}-1} & \cdots & \Re\tilde{u}_1 \\ +iN \begin{bmatrix} 0 & \Im\tilde{u}_1 & \cdots & \Im\tilde{u}_{\frac{N}{2}-1} & 0 & -\Im\tilde{u}_{\frac{N}{2}-1} & \cdots & -\Im\tilde{u}_1 \end{bmatrix} \end{bmatrix}
\tag{B.34}
$$

For executing Eq. (B.26), we can save the results in terms of real-valued numbers

$$
N \begin{bmatrix} \tilde{u}_0 & \Re\tilde{u}_1 & \cdots & \Re\tilde{u}_{\frac{N}{2}-1} & \tilde{u}_{-\frac{N}{2}} & \Im\tilde{u}_1 & \cdots & \Im\tilde{u}_{\frac{N}{2}-1} \end{bmatrix}
\tag{B.35}
$$

for example. For real-valued Fourier transforms, procedures similar to what was described above are taken.

B.3   Wave Space Representation of Flow Fields

To provide the fundamental knowledge necessary for turbulence modeling, we discuss the energy spectra in this section. Here, we do not discuss the theory of turbulence in detail and only highlight the necessary background. For further details on this topic, readers should consult with references [8, 1, 11, 9].

B.3.1   Fourier Analysis of Flow Fields

Let us derive the governing equations for isotropic incompressible flow in wave space for which Fourier transform can be applied. Here, the flow is assumed to be periodic with no discontinuity. In particular, we consider an isotropic turbulent flow with no mean velocity component. That is, we only consider the fluctuating component of the velocity field.

Let us apply Fourier transform to the velocity field $u(x, t)$

$$
u(x, t) = \int v(k, t) \exp(ik \cdot x) dk,
\tag{B.36}
$$

where the Fourier components are

$$
v(k, t) = \frac{1}{(2\pi)^3} \int u(x, t) \exp(-ik \cdot x) dx.
\tag{B.37}
$$

Here, $k = (k_x, k_y, k_z)$ is the wave number vector with each element representing the wave number in each direction. The space spanned by this vector k is called the wave space. The shorthand notation used for physical and wave space integrations can be expanded in the following manner:

$$\int dx = \int_{-\infty}^{\infty} \int_{-\infty}^{\infty} \int_{-\infty}^{\infty} dx_1 dx_2 dx_3,$$

$$\int dk = \int_{-\infty}^{\infty} \int_{-\infty}^{\infty} \int_{-\infty}^{\infty} dk_1 dk_2 dk_3. \tag{B.38}$$

Because the velocity field u is real-valued, the complex-valued v and its complex conjugate v^* satisfy $v^*(k, t) = v(-k, t)$.

For incompressible flow, the continuity equation is

$$\nabla \cdot u = 0 \tag{B.39}$$

and the momentum equation (for constant viscosity) is

$$\frac{\partial u}{\partial t} + u \cdot \nabla u = -\frac{1}{\rho}\nabla p + \nu\nabla^2 u. \tag{B.40}$$

We can take the three-dimensional Fourier transform of the above two equations. The continuity equation becomes

$$\frac{\partial u_j}{\partial x_j} = 0 \quad \Longleftrightarrow \quad ik_j v_j = 0 \tag{B.41}$$

in indicial notation and

$$ik \cdot v(k, t) = 0. \tag{B.42}$$

in vector notation. Equation (B.42) states that the velocity vector and the wave number vector are orthogonal to each other in Fourier space. The transformed momentum equation becomes

$$\frac{\partial v(k, t)}{\partial t} + i\int [k' \cdot v(k - k', t)]v(k', t)dk' = -\frac{ik}{\rho}q(k, t) - \nu k^2 v(k, t), \tag{B.43}$$

where q is the Fourier transformed pressure variable in wave space

$$p(x, t) = \int q(k, t)\exp(ik \cdot x)dk \tag{B.44}$$

and $k = |k|$. The second term on the left-hand side of Eq. (B.43) is the convolution which corresponds to the product of $v(k - k', t) \cdot (ik')$ (that represents $u \cdot \nabla$) and $v(k', t)$.

Next, let us eliminate the pressure variable q from the momentum equation. Taking an inner product[2] of Eq. (B.43) with k and using Eq. (B.42), the pressure Poisson equation in wave space becomes

---

[2]Taking an inner product of a vector with ik in wave space is equivalent to computing the divergence ($\nabla \cdot$) of a vector in physical space.

$$\int [k' \cdot v(k - k', t)][k \cdot v(k', t)]\mathrm{d}k' = -\frac{1}{\rho}k^2 q(k, t). \tag{B.45}$$

Hence the momentum equation in wave space yields an equation for $v(k, t)$

$$\left(\frac{\partial}{\partial t} + \nu k^2\right)v(k, t)$$

$$= -i \int [k \cdot v(k - k', t)] \left\{ v(k', t) - \frac{k}{k^2}[k \cdot v(k', t)] \right\} \mathrm{d}k', \tag{B.46}$$

where we have utilized the orthogonal relation $k \cdot v(k - k', t) = k' \cdot v(k - k', t)$.

The viscous term on the left-hand side of Eq. (B.46) represents the linear damping effect. The factor of k^2 in the viscous term suggests that components with high k or smaller spatial structures experience viscous effects more prominently. The two terms on the right-hand side resulting from the inertial and pressure terms show the nonlinear contribution and the interaction between components with two different wave numbers.

B.3.2   Correlation and Spectrum

Let $u_A = u_{iA}e_i$ and $u_B = u_{iB}e_i$ represent velocity fluctuations at spatial points A and B, respectively. The vector e_i represents the unit basis vector for Cartesian coordinates. The two-point correlation is written as $Q_{ij}(x_A, x_B) = \overline{u_{iA}u_{jB}}$ which reduces to only a function of the relative position between the two points $r_{AB} = x_B - x_A$ for isotropic flows:

$$Q_{ij}(r_{AB}) = \overline{u_{iA}u_{jB}}. \tag{B.47}$$

The evolution equation for $Q_{ij}(r_{AB})$ can be derived by performing a Reynolds average on $[u_{jB} \times \text{Eq. (6.1)}_{iA}] + [u_{iA} \times \text{Eq. (6.1)}_{jB}]$

$$\left(\frac{\partial}{\partial t} - 2\nu\frac{\partial^2}{\partial r_k \partial r_k}\right)Q_{ij} = \frac{\partial}{\partial r_k}\left(\overline{u_{iA}u_{kA}u_{jB}} - \overline{u_{iA}u_{jB}u_{kB}}\right)$$

$$- \frac{1}{\rho}\left(\frac{\partial}{\partial r_j}\overline{p_B u_{iA}} - \frac{\partial}{\partial r_i}\overline{p_A u_{jB}}\right). \tag{B.48}$$

Using notations of

$$L_i(r_{AB}) = \frac{1}{\rho}\overline{p(x_A)u_i(x_B)} \tag{B.49}$$

and

$$T_{ijk}(r_{AB}, r_{AC}) = \overline{u_i(x_A)u_j(x_B)u_k(x_C)}, \tag{B.50}$$

we can rewrite Eq. (B.48) as

$$
\left(\frac{\partial}{\partial t} - 2\nu \frac{\partial^2}{\partial r_k \partial r_k}\right) Q_{ij}(r) = \frac{\partial}{\partial r_k} \left[T_{ijk}(r, 0) - T_{jik}(-r, 0)\right]
$$
$$
+ \frac{\partial}{\partial r_i} L_j(r) - \frac{\partial}{\partial r_j} L_i(-r). \tag{B.51}
$$

Next, let us introduce the following Fourier transforms of the correlations defined by

$$
\Phi_{ij}(k) = \frac{1}{(2\pi)^3} \int Q_{ij}(r) \exp(-ik \cdot r) dr, \tag{B.52}
$$

$$
\Lambda_i(k) = \frac{1}{(2\pi)^3} \int L_i(r) \exp(-ik \cdot r) dr, \tag{B.53}
$$

$$
\Psi_{ijk}(k, k') = \frac{1}{(2\pi)^6} \iint T_{ijk}(r, r') \exp[-i(k \cdot r + k' \cdot r')] dr dr'. \tag{B.54}
$$

The inverse Fourier transform of Eq. (B.52) is

$$
Q_{ij}(r) = \int \Phi_{ij}(k) \exp(ik \cdot r) dk, \tag{B.55}
$$

which reduces to

$$
\overline{u_i u_j} = \int \Phi_{ij}(k) dk \tag{B.56}
$$

when $r = 0$, which means that $\Phi_{ij}(k)$ represents the distribution of $\overline{u_i u_j}$ in wave space. For this reason, $\Phi_{ij}(k)$ is referred to as the energy spectra.

Fourier transform of Eq. (B.51) yields

$$
\left(\frac{\partial}{\partial t} + 2\nu k^2\right) \Phi_{ij}(k) = ik_l \int \left[\Psi_{ijl}(k, k') - \Psi_{jil}(-k, k')\right] dk'
$$
$$
+ i \left[k_i \Lambda_j(k) - k_j \Lambda_i(-k)\right]. \tag{B.57}
$$

The first integral term on the right-hand side with Ψ results from the advective term in the momentum equation. Integration of this term over the entire k equals 0. In other words, there is no contribution from the inertial effect to the overall energy spectra, $(\partial/\partial t) \int \Phi_{ij}(k) dk$. The inertial term is responsible for the redistribution of energy over the wave space for the tensor component $\Phi_{ij}(k)$.

The term that involves Λ on the right-hand side of Eq. (B.57) originates from the pressure gradient term. If we set $i = j$ in Eq. (B.57), it can be observed that $i[k_j \Lambda_j(k) - k_j \Lambda_j(-k)] = 0$ from the continuity equation. This pressure term takes an energy tensor component $\Phi_{ij}(k)$ at a wave number k from one direction and redirects it to another direction. The pressure term performs energy redistribution between

tensor components at the same wave number but does not change the overall energy
at that wave number.

If we take a contraction of Eq. (B.57) and integrate the equation over the wave
space, we have

$$\frac{\partial}{\partial t} \int \Phi_{ii}(\boldsymbol{k}) \mathrm{d}\boldsymbol{k} = -2\nu \int k^2 \Phi_{ii}(\boldsymbol{k}) \mathrm{d}\boldsymbol{k}. \tag{B.58}$$

We therefore observe that the only term that can contribute to the time rate of change
of the total kinetic energy is the viscous diffusion term in homogeneous isotropic
turbulence. The right-hand side of the above equation is always negative indicating
that viscous effect always dissipates energy. It should however be noted that energy
dissipation is not completely free from the influence of the inertial and pressure terms.
This is because these terms can modify the distribution of $\Phi_{ii}(\boldsymbol{k})$ by redistributing
the wave number contents in terms of the magnitude and direction. This in turn
alters the energy dissipation rate on the right-hand side of Eq. (B.58). In general,
viscous dissipation rate is increased when the flow becomes isotropic as the energy
is transferred from large to small-scale structures.

B.3.3   Generation of Homogeneous Isotropic Turbulence

It is not easy to provide an incompressible turbulent flow field as an initial condition
for DNS and LES. The use of random noise in physical space to model turbulence
has issues satisfying the continuity equation and, moreover, does not yield a physical
energy spectra. In wave space, however, it is possible to generate a fluctuating velocity
field that satisfies the continuity equation, Eq. (B.41), and provides the desired spectra
when the flow is periodic in all directions.

A vector $\boldsymbol{v}$ that lies on a plane orthogonal to the wave number vector $\boldsymbol{k}$, as shown
in Fig. B.1, satisfies the continuity equation. We can specify the length of the velocity
vector in accordance to the energy spectra with random angle θ for all wave numbers
present in the computational domain. Performing inverse transform of the $\boldsymbol{v}(\boldsymbol{k})$ profile
yields an appropriate velocity field in physical space.

Fig. B.1 Velocity vector $\boldsymbol{v}$
being orthogonal to the wave
number vector $\boldsymbol{k}$

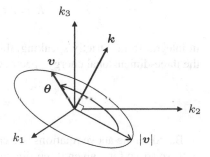

Below, we provide an outline of the procedure discussed above. Let us consider flow inside a cubic domain V_L with length L for which periodic boundary conditions are prescribed in all three directions. For this setup, the Fourier transform of the velocity component in the x_j direction becomes

$$v_j(k) = \frac{1}{(2\pi)^3} \int_{V_L} u_j(x) \exp(-ik \cdot x) dx. \tag{B.59}$$

The velocity fluctuation correlation in wave space is then

$$\overline{v_j(k)v_l(h)} = \frac{L^3}{(2\pi)^3} \Phi_{jl}(h)\delta_{k+h}, \tag{B.60}$$

where

$$\delta_s = \frac{1}{L^3} \int_{V_L} \exp(-is \cdot x) dx = \begin{cases} 1 & s = 0 \\ 0 & s \neq 0 \end{cases} \tag{B.61}$$

which is nonzero only when $s = 0$. Setting $h = -k$ in Eq. (B.60), we can show that the average magnitude of the velocity vector for wave number k is

$$\overline{|v|^2(k)} = \frac{L^3}{(2\pi)^3} \Phi_{jj}(k) \tag{B.62}$$

for which we used the fact that the velocity field is real-valued and took the contraction of the one-point velocity correlation, which is a symmetric tensor.

We also have the turbulent energy to be

$$K = \frac{1}{2}\overline{u_j u_j} = \frac{1}{2}Q_{jj}(0) = \frac{(2\pi)^3}{2L^3} \sum_k \Phi_{jj}(k). \tag{B.63}$$

For isotropic turbulence, we can treat Φ_{jj} as a function only of the magnitude of the wave number $k = |k|$ (i.e., $\Phi_{jj} = \Phi_{jj}(k)$). This means that the function is constant on the surface of a sphere of radius k. In that case, Eq. (B.63) can be approximated as

$$K = 2\pi \int_0^\infty k^2 \Phi_{jj}(k) dk \tag{B.64}$$

in integral form. Strictly speaking, this holds in the limit of $L \to \infty$. Furthermore, the three-dimensional energy spectra can be related to K by

$$K = \int_0^\infty E(k) dk. \tag{B.65}$$

Based on the above relations, we can assign the magnitude of the velocity vector in wave space for a given three-dimensional energy spectra $E(k)$ with

$$|v|(k) = \sqrt{\frac{L^3}{(2\pi)^3} \frac{E(k)}{2\pi k^2}}. \tag{B.66}$$

For $E(k)$, we can use an available experimental measurement. There are also studies that utilize a functional form such as $E(k) \propto k^4 \exp(-k^2)$. In Eq. (B.66), $E(k)$ is based on a statistical average. Without adding some fluctuation to $E(k)$, the magnitude $|v|$ for a sphere of constant k would not display much variation. We should also note that error in the continuity equation can appear when the inverse Fourier transformed velocity field is discretized onto the grid and finite differenced.

References

1. Batchelor, G.K.: The Theory of Homogeneous Turbulence. Cambridge Univ. Press (1982)
2. Bracewell, R.: The Fourier Transform and its Applications, 3rd edn. McGraw-Hill (1999)
3. Canuto, C., Hussaini, M.Y., Quarteroni, A., Zang, T.A.: Spectral Methods in Fluid Dynamics. Springer, New York (1988)
4. Canuto, C., Hussaini, M.Y., Quarteroni, A., Zang, T.A.: Spectral Methods: Fundamentals in Single Domains. Springer (2006)
5. Cooley, J.W., Tukey, J.W.: An algorithm for the machine calculation of complex Fourier series. Math. Comput. **19**, 297–301 (1965)
6. Frigo, M., Johnson, S.G.: The design and implementation of FFTW3. Proc. IEEE **93**(2), 216–231 (2005)
7. Gottlieb, D., Orszag, S.A.: Numerical Analysis of Spectral Methods: Theory and Applications. SIAM, Philadelphia (1993)
8. Hinze, J.O.: Turbulence. McGraw-Hill, New York (1975)
9. Pope, S.B.: Turbulent Flows. Cambridge Univ. Press (2000)
10. Press, W.H., Teukolsky, S.A., Vetterling, W.T., Flannerry, B.P.: Numerical Recipes, 3rd edn. Cambridge Univ. Press (2007)
11. Tennekes, H., Lumley, J.L.: A First Course in Turbulence. MIT Press, Cambridge (1972)
12. Trefethen, L.N.: Spectral Methods in Matlab. SIAM (2001)

Appendix C
Modal Decomposition Methods

The simulation methodologies discussed in this book can provide us with rich fluid flow data. Turbulent flow simulations, in particular, capture complex nonlinear interactions over a wide range of scales, both in time and space, yielding large data sets. Analyzing the vast amount of such data in a rigorous manner can be a challenge. In this appendix, we present data-based modal decomposition methods that can extract spatial structures from a set of large flow field data. The modes found from these methods can reveal the dominant structures in the fluid flows, which can also be utilized to model the behavior of the flows in a computationally inexpensive manner. Below, we briefly describe two of the decomposition methods [16]; namely, the *proper orthogonal decomposition* (POD) [4, 12] and the *dynamic mode decomposition* (DMD) [25, 24, 20]. A short presentation of the modal decomposition techniques and an application of POD analysis to a turbulent airfoil wake was given in Sect. 6.3.7. The techniques described here are not only restricted to CFD data but also are applicable to those from experimental measurements, such as particle image velocimetry, as well as data from any dynamical systems in general.

C.1 Preliminaries

Let us first format the flow field data to be in a matrix form X such that it is convenient to perform the modal decomposition analysis. For simplicity, the example considered below assumes that the grid is structured with uniform spacing and the temporal step size between discrete flow fields are constant. Here, we take a two-dimensional incompressible velocity field data $u = (u, v)$ and construct X. In general, the data to be examined can be raw from unstructured or nonuniform grid, if scaled appropriately, or processed to reside on a uniform grid. The formulation discussed here can be easily extended to scalar variables, three-dimensional vector fields, or compressible flows.

© Springer International Publishing AG 2017

T. Kajishima and K. Taira, *Computational Fluid Dynamics*,
DOI 10.1007/978-3-319-45304-0

Consider the velocity field u made available from simulation

$$u_{pq} = (u_{pq}, v_{pq}) = (u(x_p, y_q), v(x_p, y_q)),$$ (C.1)

where the coordinates (x_p, y_q) with $p = 1, \ldots, n_x$ and $q = 1, \ldots, n_y$ are for a collocated setup, which is representative of the format used for fluid flow visualization. To construct the data matrix X, we need to organize the data in the form of a column vector at each time instance. An example of stacking the data to form a column vector $x(t)$ at time $t = t_j$ can be written as

$$
\begin{bmatrix}
u_{11} & \cdots & u_{n_x 1} \\
\vdots & \ddots & \vdots \\
u_{1n_y} & \cdots & u_{n_x n_y} \\
v_{11} & \cdots & v_{n_x 1} \\
\vdots & \ddots & \vdots \\
v_{1n_y} & \cdots & v_{n_x n_y}
\end{bmatrix}_{t=t_j}
\begin{array}{c} \xrightarrow{\text{stack}} \\ \xleftarrow{\text{unstack}} \end{array}
x(t_j) \equiv
\begin{bmatrix}
\begin{pmatrix} u_{11} \\ \vdots \\ u_{1n_y} \end{pmatrix} \\
\vdots \\
\begin{pmatrix} u_{n_x 1} \\ \vdots \\ u_{n_x n_y} \end{pmatrix} \\
\begin{pmatrix} v_{11} \\ \vdots \\ v_{1n_y} \end{pmatrix} \\
\vdots \\
\begin{pmatrix} v_{n_x 1} \\ \vdots \\ v_{n_x n_y} \end{pmatrix}
\end{bmatrix}_{t=t_j} \in \mathbb{R}^n
$$ (C.2)

Note that the above stacking and unstacking operations are for a two-dimensional vector, which yield a column vector size $n = 2n_x n_y$. The unstacking operation is useful for visualizing the decomposition modes once decomposition is performed on the flow field data.

The collection of the column vectors $x(t_j)$ from $j = 1$ to m produces the matrix X

$$X = [x(t_1) \, x(t_2) \, \ldots \, x(t_m)] \in \mathbb{R}^{n \times m}.$$ (C.3)

If there are spatial or temporal variations in the grid or step sizes, appropriate weights can be introduced in the above matrix [22]. This formulation prepares us for a vector-based modal decomposition. If the decomposition is to be performed on scalar data or only on certain components of the vector, then only that particular component needs to be stacked. In what follows, we use the term *dimension* to mainly describe the size of the state vector x (instead of the dimension of the spatial domain). We also use the term *order* interchangeably with dimension. The full dimensions of the state vectors in CFD are high as mentioned above. The objective of the decomposition techniques here is to find a low-dimensional representation of the flow field by projecting the high-dimensional flow field data onto the modal basis vectors.

C.2   Proper Orthogonal Decomposition

The *proper orthogonal decomposition* (POD) [14, 6, 4, 12] is a technique to extract low-dimensional components from high-dimensional data in an optimal manner. The POD method is not only used widely in fluid mechanics but is also extensively utilized under the names of *principal component analysis* (*PCA*) and *Karhunen-Loève expansion* in the fields of economics, statistics, and image processing. Lumley introduced POD to the field of fluid mechanics for extracting coherent structures from turbulent flows [17]. Since then, POD has been widely used to analyze a large number of fluid flow problems [4, 12, 2, 9, 34].

Classical POD

With POD, we determine the basis functions that can optimally represent the given data of interest. Consider a time series of finite-dimensional data vector

$$x(t) \in \mathbb{R}^n, \quad t = t_1, t_2, \ldots, t_m \tag{C.4}$$

with its time average value removed. Here, we discuss the concept of POD in the discrete time setting.[3]

We seek the optimal basis vectors by performing POD of the given data. That is, when the vector $x(t)$ is projected on an r-dimensional basis and reprojected back onto the original n-dimensional space, the basis must be chosen such that the residual between the original $x(t)$ and the projected $\tilde{\Pi}x(t)$ is minimized. Out of the possible set of basis vectors $\{\tilde{\phi}_k\}_{k=1}^r$, the optimal set of basis vectors $\{\phi_k\}_{k=1}^r$ can be determined by solving the following optimization problem [12]

$$\{\phi_k\}_{k=1}^r = \arg\min_{\{\tilde{\phi}_k\}_{k=1}^r} \sum_{i=1}^m \left\| x(t_i) - \tilde{\Pi}x(t_i) \right\|^2$$
$$= \arg\max_{\{\tilde{\phi}_k\}_{k=1}^r} \sum_{i=1}^m \left\| \tilde{\Pi}x(t_i) \right\|^2, \tag{C.5}$$

where the projection $\tilde{\Pi}$ is given by

$$\tilde{\Pi} = \sum_{k=1}^r \tilde{\phi}_k \tilde{\phi}_k^T. \tag{C.6}$$

---

[3]See Holmes et al. [12] for the continuous time formulation.

The solution to this problem [10] can be determined by finding the eigenvectors ϕ_k (POD modes) and the eigenvalues λ_k from

$$R\phi_k = \lambda_k\phi_k, \quad \lambda_1 \geq \ldots \lambda_n \geq 0, \tag{C.7}$$

where R is the covariance matrix of vector $x(t)$

$$R = \sum_{i=1}^{m} x(t_i)x^T(t_m) = XX^T \in \mathbb{R}^{n \times n}. \tag{C.8}$$

The size of the covariance matrix n is based on the degrees of freedom of the data. For CFD data, n is large and is about the number of grid points times the number of variables to be considered in the data, as illustrated in Eq. (C.2).

Noting that the covariance matrix is symmetric, it can be realized all eigenvalues are positive. Moreover the corresponding eigenvectors are orthonormal, i.e.,

$$\langle \phi_i, \phi_j \rangle \equiv \int_V \phi_i \cdot \phi_j dV = \delta_{ij}, \quad i, j = 1, \ldots, n. \tag{C.9}$$

The eigenvalues convey how well each of the eigenvectors captures the original data in the L_2 sense, as it can be seen from Eq. (C.5). When the velocity vector is used for $x(t)$, the eigenvalues correspond to the kinetic energy captured by the respective POD modes. If the eigenvalues are arranged from the largest to the smallest in decreasing order, the expression below tells us how well the first r modes capture the energy of the data

$$\sum_{i=1}^{m} \|\Pi x(t_i)\|^2 = \sum_{k=1}^{r} \lambda_k. \tag{C.10}$$

This relation can be used in quantifying the number of modes necessary to capture the kinetic energy of the flow field. Generally, we choose r number of modes for the basis such that

$$\sum_{k=1}^{r} \lambda_k / \sum_{k=1}^{n} \lambda_k \approx 1. \tag{C.11}$$

The collection of these r eigenvectors (with corresponding eigenvalues in descending orders) are called the *POD basis*.

Method of Snapshots

The aforementioned approach to determine the POD modes leads to an $n \times n$ eigenvalue problem, which is very large for typical fluid flow simulations. With the recent enhancement in computational capability for performing CFD simulations, n can be as large as $\mathcal{O}(10^7)$ or even larger. Handling a large-scale eigenvalue problem of such matrix size ($n \times n$) is computationally expensive. To address this issue, the *method of snapshots* [28] can be used to determine the POD modes in a computationally efficient manner. The method of snapshots takes a collection of snapshots $x(t_j)$ at discrete time levels $t_j, j = 1, 2, \ldots, m$, with $m \ll n$, and solves an eigenvalue problem of a smaller size ($m \times m$) to find the POD modes. The number of snapshots m should be chosen such that important fluctuations in the flow field are well resolved in time.

Instead of solving the large eigenvalue problem, Eq. (C.7), we can solve an eigenvalue problem of a much smaller size

$$X^T X \psi_k = \lambda_k \psi_k, \quad \psi_k \in \mathbb{R}^m, \ m \ll n, \tag{C.12}$$

where $X^T X \in \mathbb{R}^{m \times m}$. This smaller eigenvalue problem is examined instead, since the same nonzero eigenvalues are shared by $X^T X$ and XX^T and the eigenvectors are related by

$$\phi_k = X \psi_k / \sqrt{\lambda_k}. \tag{C.13}$$

Collecting the eigenvectors in matrix form

$$\Phi = [\phi_1 \ldots \phi_m] \in \mathbb{R}^{n \times m} \quad \text{and} \quad \Psi = [\psi_1 \ldots \psi_m] \in \mathbb{R}^{m \times m}, \tag{C.14}$$

we can also express the transform relation, Eq. (C.13), in terms of matrices

$$\Phi = X \Psi \Lambda^{-1/2}. \tag{C.15}$$

The method of snapshots enables us to solve for the eigenvalues and eigenvectors through the reduced-size eigenvalue problem. Once the eigenvectors u_k are determined, the POD modes can be retrieved by Eqs. (C.13) and (C.15). Due to the ease of computation, the method of snapshots has been widely used to determine POD modes.

Singular Value Decomposition

POD is closely related to SVD [11, 32, 16], which can be applied to a rectangular matrix to provide left and right singular vectors. In matrix form, the data matrix X can be decomposed with SVD as

$$X = \Phi \Sigma \Psi^T, \tag{C.16}$$

where $\Phi \in \mathbb{R}^{n \times n}$, $\Psi \in \mathbb{R}^{m \times m}$, and $\Sigma \in \mathbb{R}^{n \times m}$. The matrices Φ and Ψ contain the left and right singular vectors[4] of X and matrix Σ holds the associated singular values $(\sigma_1, \ldots, \sigma_m)$ along its diagonal. In fact, these singular vectors Φ and Ψ are identical to the eigenvectors of XX^T and $X^T X$, respectively, from Eq. (C.14). Moreover, the singular values and the eigenvalues are related by $\sigma_i^2 = \lambda_i$. This means that SVD can be applied to directly X to determine the POD modes Φ. It should be noted that Φ

---

[4]The matrices Φ and Ψ are orthonormal, i.e., $\Phi^T \Phi = \Phi \Phi^T = I$ and $\Psi^T \Psi = \Psi \Psi^T = I$.

obtained from Eq. (C.16) has a full size of $n \times n$ compared to Φ from the snapshot approach which is $n \times m$, holding only the leading m modes.

The terms POD and SVD are often used interchangeably in the literature. However, SVD is a decomposition technique for rectangular matrices in general and POD can be seen as a decomposition formalism for which SVD can be one of the approaches to determine its solution. While the method of snapshot is preferred for handling large data sets, the SVD-based technique to determine the POD modes is known to be more robust against roundoff errors (although SVD requires much higher computationally expense) [16].

Example of POD-Based Analysis

As an example, let us consider the application of POD on two-dimensional incompressible flow over a flat plate at an angle of attack of $30°$ and a chord-based Reynolds number of 100. Simulations were performed by the immersed boundary projection method [30, 8] from Sect. 5.2.3. Flow over the flat plate at a high angle of attack can generate a wake with periodic shedding of leading and trailing-edge vortices. The dominant POD modes ϕ_k computed from the snapshot method are presented in Fig. C.1. Modes 1 and 2 possess the same amount of kinetic energy (i.e., $\lambda_1 = \lambda_2$) and together capture 96.3 % of the kinetic energy. It should be noted that POD modes appear in pairs for periodic flow, as shown by the eigenvalues in Fig. C.2. These pairings are representative of periodic dynamics and are analogous to the relationship between sine and cosine functions. The mode pairs have similar spatial structures but with phase difference in the streamwise direction, as in the case with modes 1 and 2. In general, for higher order modes, the spatial structures become finer as evident from modes 3 and 5.

C.3 Dynamic Mode Decomposition

While POD modes capture energetic structures in the flow field well using spatial and temporal correlations of the given data, POD analysis is not devised to capture the dynamical behavior of the flow, such as growth rates and frequencies. A modal decomposition method that can extract the dynamical characteristics of the flow and the associated spatial structures is the *dynamic mode decomposition* [25, 24].

This decomposition method takes advantage of the realization that the temporal evolution of the flow field can be approximated by considering a linear transform A between snapshots. That is, if we have a collection of snapshots $X_{1 \to m}$ for $t = t_1, t_2, t_3, \ldots, t_m$ (with equal time step Δt), we can say

$$X_{1 \to m} \equiv [\boldsymbol{x}_1, \boldsymbol{x}_2, \boldsymbol{x}_3, \ldots, \boldsymbol{x}_m] \in \mathbb{R}^{n \times m}$$
$$\approx [\boldsymbol{x}_1, A\boldsymbol{x}_1, A^2\boldsymbol{x}_1 \ldots, A^{m-1}\boldsymbol{x}_1].$$

$$(C.17)$$

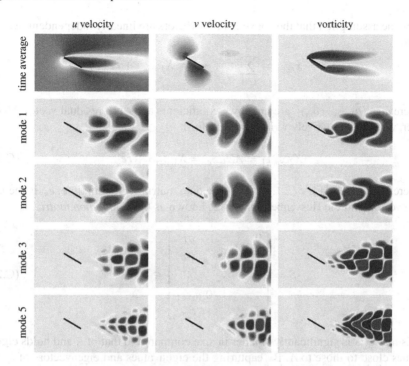

Fig. C.1 POD modes for the two-dimensional wake behind a flat plate wing at $\alpha = 30°$. Shown are the time average flow and modes 1, 2, 3, and 5 for the velocity and vorticity fields [29]. Reprinted with permission from the Japan Society of Fluid Mechanics

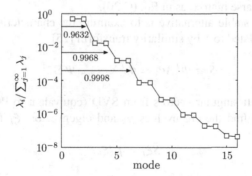

Fig. C.2 The normalized eigenvalues $\lambda_i / \sum_{j=1}^{m} \lambda_j$ shows the fraction of kinetic energy captured by each POD mode for the results shown in Fig. C.1. The values indicated below the arrows show the fraction of total energy captured up to the indicated modes. The POD modes can be seen appearing in pairs for this periodic flow [29]. Reprinted with permission from the Japan Society of Fluid Mechanics

With the assumption that the above column vectors are linearly independent, we can write

$$x_{m+1} = \sum_{k=1}^{m} a_k x_k + r = X_{1 \to m} a + r,$$ (C.18)

where $a = (a_1, \ldots, a_m)^T$ are unknown coefficients and r is a residual vector. Moreover, we can alternatively write in matrix form

$$X_{2 \to m+1} = A X_{1 \to m} = X_{1 \to m} S + r \hat{e}_m^T,$$ (C.19)

where $X_{2 \to m+1} = [x_2, x_3, \ldots x_{m+1}]$ is the data matrix shifted in time, $\hat{e}_m$ is the mth unit vector, and the Hessenberg matrix S, known as the *companion matrix*, is

$$S = \begin{bmatrix} 0 & & & a_1 \\ 1 & 0 & & a_2 \\ & \ddots & \ddots & \vdots \\ & & 1 & 0 & a_{m-1} \\ & & & 1 & a_m \end{bmatrix} \in \mathbb{R}^{m \times m}.$$ (C.20)

This matrix S is significantly smaller in size compared to that of A and holds eigenvalues close to those to A. By capturing the eigenvalues and eigenvectors of S, we can retrieve those of A. While we can attempt to find S, the found matrix can be affected by numerical errors in practice [25]. Since the unknown residual vector in Eq. (C.19) is not accounted for in Eq. (C.21), the approximate companion matrix is moreover not a sparse matrix, as in Eq. (C.20).

A numerically stable alternative is to examine the characteristics of a full-size matrix $\tilde{S}$ that is related to S by similarity transform [25]

$$\tilde{S} \equiv \Phi^T A \Phi = \Phi^T X_{2 \to m+1} \Psi \Sigma^{-1},$$ (C.21)

where Φ is the left singular vectors from SVD (equivalent to POD modes). With $\tilde{S}$ known, we can find the eigenvalues μ_k and eigenvectors ξ_k for the eigenvalue problem of

$$\tilde{S} \xi_k = \mu_k \xi_k.$$ (C.22)

The eigenvalues μ_k and the eigenvectors ξ_k of the reduced matrix $\tilde{S}$ are also referred to as the *Ritz values* and *Ritz vectors*, respectively. These Ritz values are approximations to the eigenvalues of the operator A, which are the focus of the present analysis. The Ritz vectors ξ_k can be transformed to *dynamic modes* (DMD modes)

$$\zeta_k = \Phi \xi_k.$$ (C.23)

Note that DMD modes are not orthonormal as in the case of POD modes. We can take the Ritz values μ_k and determine the frequencies and growth rates of the modes with [26, 31]

$$\sigma_k = \text{Re}(\sigma_k) + i\text{Im}(\sigma_k) = \frac{1}{\Delta t}\ln(\mu_k). \tag{C.24}$$

If the flow is in the linear regime, the dynamic modes reduce to the linear global modes from hydrodynamic stability analysis [13, 25, 31].

Since the introduction of DMD by Schmid [25], there has been numerous studies on the decomposition technique [24, 7, 33, 3] and applications of DMD on a variety of fluid flows [27]. We note that the computation associated with DMD requires performing SVD on the large matrix $X_{1\to m} \in \mathbb{R}^{n\times m}$. This can be computationally challenging but there are now some examples in which large data sets have been examined with DMD [15]. There has also been studies on examining the connections between the DMD based approach and the *Koopman analysis* used to describe infinite-dimensional nonlinear dynamics. We refer readers to [19, 24, 7, 20, 33] for further details.

Example of DMD-Based Analysis

Let us reconsider the two-dimensional incompressible flow over a flat plate that was analyzed with POD in Sect. C.2. We take a collection of vorticity field snapshots and perform DMD analysis on them using the SVD-based approach, Eqs. (C.21)–(C.24). The DMD analysis returns the time average vorticity field and the dominant modes, as shown in Fig. C.3 (left). Here, the real and imaginary components of the first three dominant DMD modes, ζ_k, are shown, which have strong resemblance to the POD

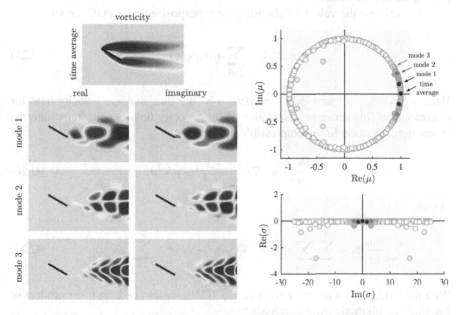

Fig. C.3 DMD analysis of the two-dimensional vorticity field behind a flat plate wing. The time average flow and modes 1, 2, and 3 (*left*) with the corresponding Ritz values μ and growth rates σ, shaded by the amplitudes of the modes, are shown (*right*)

modes in Fig. C.1. This is because the modal structures capturing fluctuations (POD) in the flow and the dynamically important features (DMD) are similar. In general, the POD and DMD modes can be different, especially for transient flows.

The Ritz values μ and the growth rates σ for the DMD modes are also shown in Fig. C.3. The spectra of μ align along the circumference of the unit circle suggesting that the flow is in a sustained oscillatory state. The spectra of μ can be transformed to frequencies and growth rates σ of the modes. The shading of the symbols represents the amplitudes of the DMD modes [15], which identify the dominant DMD modes for the given flow field. Note that $\text{Im}(\sigma_1)$ for mode 1 identifies the shedding frequency and $\text{Im}(\sigma_2)$ and $\text{Im}(\sigma_3)$ correspond to the first and second harmonics, respectively, of the dominant shedding frequency.

C.4 Reduced-Order Modeling

With the modal basis being available, we can model fluid flows by projecting the governing equations onto the modes and construct a *reduced-order model* (ROM) [2, 12]. Using this approach, the high-dimensional discrete Navier–Stokes equations with n dimensions can be modeled by a low r ($\ll n$) dimensional model to capture the dominant features of the flow field. Here, we present the derivation of the reduced-order model by projecting the incompressible Navier–Stokes equations onto the basis comprised of POD modes.

Let us express the velocity field using the superposition of r POD modes

$$u(x, t) = \sum_{j=0}^{r} a_j(t)\phi_j(x). \tag{C.25}$$

Here, we set $a_0 = 1$ and $\phi_0(x) = \bar{u}(x)$ to be the time average velocity field for convenience. This series representation of the velocity field is substituted into the momentum equation for incompressible flow

$$\frac{\partial u}{\partial t} + u \cdot \nabla u = -\nabla P + \nu \nabla^2 u \tag{C.26}$$

to yield

$$\sum_{j=0}^{r} \frac{da_j}{dt}\phi_j + \sum_{j=0}^{r}\sum_{k=0}^{r} a_j a_k \phi_j \cdot \nabla\phi_k = -\nabla P + \nu \sum_{j=0}^{r} a_j \nabla^2 \phi_j. \tag{C.27}$$

With the definition of an inner product, Eq. (C.9), we can take an inner product of Eq. (C.27) with mode ϕ_i, which leads to

$$\sum_{j=0}^{r} \frac{da_j}{dt} \langle \phi_i, \phi_j \rangle + \sum_{j=0}^{r} \sum_{k=0}^{r} a_j a_k \langle \phi_i, \phi_j \cdot \nabla \phi_k \rangle$$

$$= -\langle \phi_i, \nabla P \rangle + \nu \sum_{j=0}^{r} a_j \langle \phi_i, \nabla^2 \phi_j \rangle \tag{C.28}$$

for $i = 1, 2, \ldots, r$. The inner product of the pressure gradient term and mode ϕ_i in the above equation can be evaluated as

$$\langle \phi_i, \nabla P \rangle = \int_V \left[\nabla \cdot (P \phi_i) - P \nabla \cdot \phi_i \right] dV = \int_S P \phi_i \cdot \hat{n} dS = 0, \tag{C.29}$$

where we used the property that each mode follows incompressibility $\nabla \cdot \phi_i = 0$ and satisfies velocity boundary conditions along the computational boundary. With Eqs. (C.29), (C.28) can be expressed only in terms of the velocity-based modes without a pressure term. Recalling that POD modes are orthonormal, i.e., $\langle \phi_i, \phi_j \rangle = \delta_{ij}$, we can write the evolution equations for the time-varying POD coefficients a_i

$$\frac{da_i}{dt} = \sum_{j=0}^{r} \sum_{k=0}^{r} F_{ijk} a_j a_k + \sum_{j=0}^{r} G_{ij} a_j, \tag{C.30}$$

where the coefficients are

$$F_{ijk} = -\langle \phi_i, \phi_j \cdot \nabla \phi_k \rangle \quad \text{and} \quad G_{ij} = \nu \langle \phi_i, \nabla^2 \phi_j \rangle \tag{C.31}$$

for $i = 1, \ldots, r$ and $j, k = 0, \ldots, r$. The terms F_{ijk} and G_{ij} are dependent only on the spatial basis modes and can be precomputed. To complete the evolution equation for the coefficients, the initial conditions for the coefficients can be determined by projecting the initial velocity field onto the POD modes

$$a_i(t_0) = \langle u(x, t_0) - \bar{u}(x), \phi_i(x) \rangle. \tag{C.32}$$

We refer to Eq. (C.30) as the *reduced-order model*, which was derived from projecting the incompressible Navier–Stokes equations onto the POD modes [2, 21]. In this model, only r coefficients need to be advanced in time using a set of r ordinary differential equations. The process of projection used above is known as the *Galerkin projection*. Having the coefficients a_i, we can substitute them back into the series representation Eq. (C.25) to reproduce the fluid flow behavior using only an r-dimensional basis, instead of the n-dimensional representation needed by the original numerical simulation. If non-orthogonal modes (such as the DMD modes) are used, the reduced-order model requires a coefficient matrix $\langle \phi_i, \phi_j \rangle$ on the left-hand side of Eq. (C.30).

Reduced-order models are particularly useful in uncovering the essential dynamical properties of the fluid flow and in predicting the state of the flow field in a fast and accurate manner, which is important for applications such as closed-loop flow control [1, 5]. While the use of POD-based reduced-order models is attractive in many ways, there is room for improvement. POD-based reduced-order models can perform well for short time periods but may show divergence from the full solution over a longer course of time or in transient conditions. This can be due to the POD modes being chosen to capture the kinetic energy of the flow field but not the dynamics. There are also variants of the above model that can better capture the dynamical properties of the flow field. For details on the performance of the models and extensions, we refer the readers to [12, 22, 18, 23].

References

1. Ahuja, S., Rowley, C.W.: Feedback control of unstable steady states of flow past a flat plate using reduced-order estimators. J. Fluid Mech. **645**, 447–478 (2010)
2. Aubry, N., Holmes, P., Lumley, J.L., Stone, E.: The dynamics of coherent structures in the wall region of a turbulent boundary layer. J. Fluid Mech. **192**, 115–173 (1988)
3. Bagheri, S.: Effects of weak noise on oscillating flows: Linking quality factor, Floquet modes, and Koopman spectrum. Phys. Fluids **26**, 094, 104 (2014)
4. Berkooz, G., Holmes, P., Lumley, J.L.: The proper orthogonal decomposition in the analysis of turbulent flows. Annu. Rev. Fluid Mech. **25**, 539–575 (1993)
5. Brunton, S.L., Noack, B.R.: Closed-loop turbulence control: progress and challenges. App. Mech. Rev. **67**(5), 050, 801 (2015)
6. Chatterjee, A.: An introduction to the proper orthogonal decomposition. Current Sci. **78**(7), 808–817 (2000)
7. Chen, K.K., Tu, J.H., Rowley, C.W.: Variants of dynamic mode decomposition: boundary condition, Koopman, and Fourier analyses. J. Nonlin. Sci. **22**(6), 887–915 (2012)
8. Colonius, T., Taira, K.: A fast immersed boundary method using a nullspace approach and multi-domain far-field boundary conditions. Comput. Methods Appl. Mech. Engrg. **197**, 2131–2146 (2008)
9. Delville, J., Ukeiley, L., Cordier, L., Bonnet, J.P., Glauser, M.: Examination of large-scale structures in a turbulent plane mixing layer. Part 1. Proper orthogonal decomposition. J. Fluid Mech. **391**, 91–122 (1999)
10. Eckart, C., Young, G.: The approximation of one matrix by another of lower rank. Psychometrika **1**(3), 211–218 (1936)
11. Golub, G.H., Loan, C.F.V.: Matrix Computations, 3rd edn. Johns Hopkins Univ. Press (1996)
12. Holmes, P., Lumley, J.L., Berkooz, G., Rowley, C.W.: Turbulence, Coherent Structures, Dynamical Systems and Symmetry, 2nd edn. Cambridge Univ. Press (2012)
13. Huerre, P., Monkewitz, P.A.: Local and global instabilities in spatially developing flows. Annu. Rev. Fluid Mech. **22**, 473–537 (1990)
14. Jolliffe, I.T.: Principal Component Analysis. Springer (2002)
15. Jovanović, M.R., Schmid, P.J., Nichols, J.W.: Sparsity-promoting dynamic mode decomposition. Phys. Fluids **26**, 024, 103 (2014)
16. Kutz, J.N.: Data-Driven Modeling and Scientific Computation. Oxford Univ. Press (2013)
17. Lumley, J.L.: The structure of inhomogeneous turbulent flows. In: Yaglom, A.M., Tatarsky, V.I. (eds.) Atmosphetic Turbulence and Radio Wave Propagation, pp. 166–178. Nauka, Moscow (1967)
18. Ma, Z., Ahuja, S., Rowley, C.W.: Reduced-order models for control of fluids using the eigensystem realization algorithm. Theo. Comp. Fluid Dyn. **25**, 233–247 (2011)
19. Mezić, I.: Spectral properties of dynamical systems, model reduction and decompositions. Nonlin. Dyn. **41**, 309–325 (2005)

20. Mezić, I.: Analysis of fluid flows via spectral properties of the Koopman operator. Annu. Rev. Fluid Mech. **45**, 357–378 (2013)
21. Noack, B.R., Afanasiev, K., Morzyński, M., Tadmor, G., Thiele, F.: A hierarchy of low-dimensional models for the transient and post-transient cylinder wake. J. Fluid Mech. **497**, 335–363 (2003)
22. Rowley, C.W.: Model reduction for fluids, using balanced proper orthogonal decomposition. Int. J. Bif. Chaos **15**(3), 997–1013 (2005)
23. Rowley, C.W., Colonius, T., Murray, R.M.: Model reduction for compressible flows using POD and Galerkin projection. Physica D **189**, 115–129 (2004)
24. Rowley, C.W., Mezić, I., Bagheri, S., Henningson, D.S.: Spectral analysis of nonlinear flows. J. Fluid Mech. **641**, 115–127 (2009)
25. Schmid, P.J.: Dynamic mode decomposition of numerical and experimental data. J. Fluid Mech. **656**, 5–28 (2010)
26. Schmid, P.J., Henningson, D.S.: Stability and Transition in Shear Flows. Springer (2001)
27. Schmid, P.J., Li, L., Juniper, M.P., Pust, O.: Applications of the dynamic mode decomposition. Theor. Comput. Fluid Dyn. **25**, 249–259 (2011)
28. Sirovich, L.: Turbulence and the dynamics of coherent structures, Parts I-III. Q. Appl. Math. **XLV**, 561–590 (1987)
29. Taira, K.: Proper orthogonal decomposition in fluid flow analysis: 1. Introduction. J. Japan Soc. Fluid Mech. (Nagare) **30**, 115–123 (2011)
30. Taira, K., Colonius, T.: The immersed boundary method: a projection approach. J. Comput. Phys. **225**, 2118–2137 (2007)
31. Theofilis, V.: Global linear instability. Annu. Rev. Fluid Mech. **43**, 319–352 (2011)
32. Trefethen, L.N., Bau, D.: Numerical Linear Algebra. SIAM (1997)
33. Tu, J.H., Rowley, C.W., Luchtenburg, D.M., Brunton, S.L., Kutz, J.N.: On dynamic mode decomposition: Theory and applications. J. Comput. Dyn. **1**(2), 391–421 (2014)
34. Ukeiley, L., Cordier, L., Manceau, R., Delville, J., Glauser, M., Bonnet, J.P.: Examination of large-scale structures in a turbulent plane mixing layer. Part 2. Dynamical systems model. J. Fluid Mech. **441**, 67–108 (2001)

Index

A

Adams–Bashforth method, 50, 59
  first-order, 50
  second-order, 50, 59, 80, 83, 157, 158
  third-order, 50, 80
Adaptive mesh refinement, 16
Advection, 24
Advection-diffusion equation, 24, 59
Advection equation, 24, 55
Advective form, 8, 101, 103, 153, 320
Advective interpolation, 105, 166
Advective outflow condition, 129
Amplitude, 55, 64
Approximate relations, 18
Artificial compressibility method, 203
Artificial viscosity, 110, 112, 113, 169, 232

B

Bardina model, 282, 283
Basic variables, 148, 153
Basis, 309, 342
Basis vector, 310, 313
Bi-CGSTAB method, *see* biconjugate gradient stabilized method
Biconjugate gradient stabilized method, 98
Big $\mathcal{O}$, 26
Blasius equation, 144
Blasius profile, 128, 143
Body force, 6
Boundary condition, 120, 139, 170
Boundary-fitted grid, 147
Boundary force, 180, 181
Boussinesq approximation, 241
Box filter, 270
Burgers' equation, 24, 69

C

Cartesian components, 154
Cartesian grid, 14, 147, 180
Cell Reynolds number, 60
CFD, *see* computational fluid dynamics
CFL condition, *see* Courant–Friedrichs–Lewy condition
CFL number, *see* Courant–Friedrichs–Lewy number
Channel flow, 213
Checkerboard instability, 92, 155
Christoffel symbol, 155, 316
Closure, 6
Coherent structure function, 225, 291
Coherent structure model, 290
Collocated grid, 155–157
Column-major, 96
Compact finite-difference, 65
Companion matrix, 346
Compatibility, 23, 36, 101, 105, 137, 166, 175
Computational fluid dynamics, 1
Conductivity, 7
Conservation, 101, 137, 166
  conservation laws, 4
  conservation of energy, 5
  conservation of kinetic energy, 105
  conservation of mass, 4, 319
  conservation of momentum, 5, 105, 321
Conservative form, 101
Consistency, 67
Constitutive equation, 6, 181
Continuity equation, 7, 12, 159, 332
Continuous forcing approach, 180
Contravariant components, 148, 154, 310
Contravariant metric tensor, 314
Control volume, 4

© Springer International Publishing AG 2017
T. Kajishima and K. Taira, *Computational Fluid Dynamics*,
DOI 10.1007/978-3-319-45304-0

Printed in the United States
By Bookmasters